Plasticity

Ronaldo I. Borja

Plasticity

Modeling & Computation

Springer

Ronaldo I. Borja
Department of Civil and Environmental Engineering
Stanford University
Stanford, California
USA

ISBN 978-3-642-38546-9 ISBN 978-3-642-38547-6 (eBook)
DOI 10.1007/978-3-642-38547-6
Springer Heidelberg New York Dordrecht London

Library of Congress Control Number: 2013938490

Printed on acid-free paper

Springer is part of Springer Science+Business Media (www.springer.com)

Preface

One of the first courses I taught at Stanford was Advanced Soil Mechanics. The course title was accurate in the beginning, but the contents gradually became more and more about computational plasticity than soil mechanics, and when a few of my students commented to me that the contents did not seem to match the title of the course, I figured that it was time to change the title to what is now called Plasticity Modeling and Computation. In teaching this class, I wanted to focus on three elements, namely, theory, modeling, and computation. Unfortunately, the textbooks or references available in the market had only two of the three desired elements, and typically the third element was either missing or not covered very well.

About ten years ago, I began typesetting my lecture notes for this course, thinking that sooner rather than later I would be able to complete a book. However, due to the short duration of the quarter system and because I could only teach this class every other spring, there was not much progress with my writing, and I was stuck with the first three or four sections of each chapter for quite some time. So, I decided to sign a contract with a publisher and announce at my talks that a book is forthcoming. The pressure to get it completed was on, and finally, I finished the book.

The first drafts of the book were developed from my doctoral research on critical state soil mechanics as well as from the courses on plasticity and the finite element method I took from Tom Hughes. At first, I thought about focusing only on geomaterials, but I have seen strong interest and impetus in other disciplines to use the theory for modeling the behavior of different materials relevant to their area, whether they are natural or engineered materials, so I figured that writing a general textbook on this subject would be more beneficial to a larger audience and would generate broader impacts.

Chapters 1–3 deal with the theoretical foundations of plasticity in the context of one-dimensional problem and the J_2 theory. I felt that the numerics should be introduced as early as possible, so the concept of return mapping algorithm was also introduced and discussed in these first three chapters. Chapter 4 deals with isotropic functions and provides a critical transition chapter for more advanced topics beyond the formalism of the J_2 theory, such as those covered in Chapters 5 and 6. Chapter 7 covers the more contemporary topics of discontinuities, contact, friction, and various finite element enhancement techniques. The readers will note that Chapter 8, which deals with crystal plasticity, actually picks up from the previous chapter even though at first glance the two chapters may seem to have very little in common. Chapter 9 covers the subjects of material instability, loss of uniqueness, and deformation bands. The readers will also note that discontinuous inelastic deformation is the aspect common to the last three chapters.

I am very grateful to my former and current students who have offered help, advice, and corrections, and even volunteered to typeset the solutions to the exercises and tirelessly proofread the drafts of the chapters over and over again. I am not sure how far back I should go in acknowledging all of these wonderful individuals, so I will limit the list to those who worked with me over the last ten years. They include former students Chao Li, Medji Sama, José Andrade, Craig Foster, Pablo Sanz, Joshua White, and Fushen Liu; and current students Helia Rahmani, Martin Tjioe, Kane Bennett, Xiaoyu Song, and Jinhyun Choo.

Ronaldo I. Borja
May 27, 2013
Stanford, California

Contents

1 Motivations and Scope

1.1 Introduction

Natural and manufactured materials generally exhibit an irreversible deformation behavior such that when an applied load is removed only a fraction of deformation is recovered. The extent of reversible deformation is called the elastic range, whereas the extent of irreversible or inelastic deformation, or yield, is the plastic range. The elastic range depends on the properties of a given material: rubber, for example, can experience very large deformation and still stay within the elastic range, whereas steel yields at a much smaller strain. The mechanisms responsible for irreversible deformation also varies from one material to another. In metals, irreversible deformation is due largely to crystal dislocation, whereas in sandstone, it is produced from relative movement of the grains. In some cases, deformation in the plastic range can be so large that the deformed specimen may have very little semblance to the original specimen from which it was formed, see Fig. 1.1.

(a) (b)

Fig. 1.1 U.S. penny (97.5% zinc, 2.5% copper plating): (a) before flattening, and (b) after flattening

Significant understanding can be gained from modeling and simulating the irreversible deformation behavior of natural and manufactured materials. For example, irreversible deformation impacts the performance of a building, and quantifying its magnitude is of general interest to a structural engineer. Under normal working conditions, a nuclear power plant may be designed to perform within the elastic range, but it could also experience inelastic deformation under extreme loading conditions, such as during a strong earthquake. In mechanical engineering, plastic deformation is key to cross-die forming and metal extrusion processes. In geology and petroleum engineering, simulating the process of folding of rocks could shed light onto the sequence of tectonic events occurring over millions of years, as well as help in locating valued hydrocarbon deposits.

The objectives of this book are threefold: (a) to present the mathematical foundations of the theory of plasticity, (b) to develop plasticity models appropriate for different materials, and (c) to introduce a computational platform for implementing plasticity models into nonlinear finite element codes. Theory of plasticity generally offers a complete mathematical framework for modeling inelastic deformations. The seminal book by Hill (1950) spurred the maturation of the theory, but it was not until after the development of robust computational algorithms and advanced experimental techniques that the theory of plasticity began gaining grounds in research and practice. Apart from continuum deformation analysis, we now also use theory of plasticity for local stability analysis. In this book, we focus specifically on shear band instability that is commonplace in ductile and quasi-brittle material behavior. To this end, we extend the mathematical formulation to accommodate irreversible slip on a surface of discontinuity that could randomly nucleate and propagate through an elastoplastic solid.

The book strives to provide an equal treatment of the mathematical foundations of plasticity theory, the underpinnings of various plasticity models, and the implementation of the models into nonlinear finite element codes. Readers with an ample finite element background will find the work self-contained; otherwise, students without sufficient finite element background can gain from studying the underpinnings of the theory and the aspects of constitutive modeling. Chapter 2 lays down the foundations for one-dimensional problems and introduces the notion of return map, the underlying computational platform adopted throughout this work. The theme is repeated for three-dimensional problems in Chapters 3 and 4. Chapter 5 generalizes the formulation to the finite deformation range, providing more latitude for the remaining chapters to switch discussions from small-strain to large-strain problems, and vice versa. Chapter 6 covers a class of cap models, including the familiar Cam-Clay family of models, and highlights different aspects of strain hardening. Chapter 7 covers a class of interface problems characterized by a discontinuous displacement field, as well as unifies the formulations for classic nonlinear contact mechanics, the extended finite element method, and the assumed enhanced strain method. Chapter 8 discusses some recent developments in rate-independent crystal plasticity for infinitesimal and finite deformation problems. Finally, Chapter 9 focuses on bifurcation, material instability, and the loss of uniqueness of the solution.

1.2 Problem Statement

Let $\mathcal{B}$ denote a simple body delineated by boundary $\partial\mathcal{B}$ as shown in Fig. 1.2. We assume that $\partial\mathcal{B}$ can be decomposed into two non-overlapping parts: the

Dirichlet boundary $\partial\mathcal{B}_{\boldsymbol{u}}$ where a time-varying displacement $\boldsymbol{g}(\boldsymbol{x} \in \partial\mathcal{B}_{\boldsymbol{u}})$ is prescribed, and the Neumann boundary $\partial\mathcal{B}_{\boldsymbol{t}}$ where a time-varying surface traction $\boldsymbol{t}(\boldsymbol{x} \in \partial\mathcal{B}_{\boldsymbol{t}})$ is specified. The notion of non-overlapping boundaries can be expressed more succinctly using the set relations

$$\overline{\partial\mathcal{B}_{\boldsymbol{u}} \cup \partial\mathcal{B}_{\boldsymbol{t}}} = \partial\mathcal{B}\,, \qquad \partial\mathcal{B}_{\boldsymbol{u}} \cap \partial\mathcal{B}_{\boldsymbol{t}} = \varnothing\,, \tag{1.1}$$

where the superposed line denotes a closure.

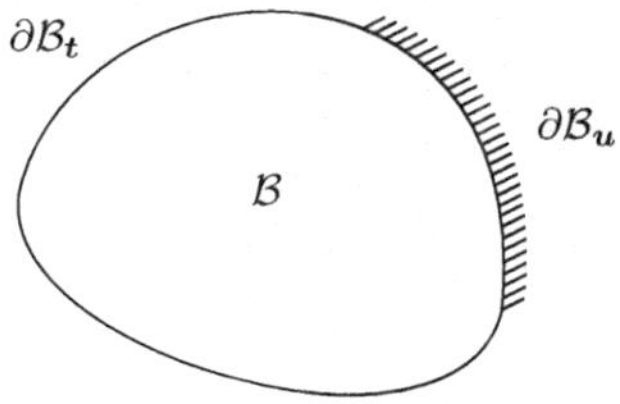

Fig. 1.2 Definition of problem domain $\mathcal{B}$, and boundaries $\partial\mathcal{B}_{\boldsymbol{u}}$ and $\partial\mathcal{B}_{\boldsymbol{t}}$

The strong form of the initial boundary-value problem under quasi-static condition may be stated as follows. Given $\boldsymbol{g}$, $\boldsymbol{t}$, and body force density $\boldsymbol{f}$, find $\boldsymbol{u}$ for all $\boldsymbol{x} \in \mathcal{B}$ such that

$$\boldsymbol{\nabla} \cdot \boldsymbol{\sigma} + \boldsymbol{f} = \boldsymbol{0} \quad \text{in} \quad \mathcal{B} \tag{1.2}$$

and

$$\boldsymbol{u} = \boldsymbol{g} \quad \text{on} \quad \partial\mathcal{B}_{\boldsymbol{u}} \tag{1.3}$$

$$\boldsymbol{\sigma} \cdot \boldsymbol{n} = \boldsymbol{t} \quad \text{on} \quad \partial\mathcal{B}_{\boldsymbol{t}}\,, \tag{1.4}$$

where $\boldsymbol{\sigma}$ is the Cauchy stress tensor, $\boldsymbol{n}$ is the outward unit normal vector to $\partial\mathcal{B}_{\boldsymbol{t}}$, and $\boldsymbol{\nabla}$ is the gradient operator. Equation (1.2) is the balance of linear momentum, while equations (1.3) and (1.4) are the Dirichlet and Neumann boundary conditions, respectively. In addition, we assume the following initial condition

$$\boldsymbol{u}(\boldsymbol{x}, 0) = \boldsymbol{u}_0(\boldsymbol{x})\,, \tag{1.5}$$

where $\boldsymbol{u}_0$ is a given displacement field that satisfies (1.2), (1.3) and (1.4) at $t = 0$.

To close the formulation of the initial boundary-value problem, we require that $\boldsymbol{\sigma}$ satisfy the rate constitution equation

$$\dot{\boldsymbol{\sigma}} = \boldsymbol{c} : \dot{\boldsymbol{\epsilon}}\,, \qquad \dot{\boldsymbol{\epsilon}} = \frac{1}{2}(\boldsymbol{\nabla}\dot{\boldsymbol{u}} + \dot{\boldsymbol{u}}\boldsymbol{\nabla})\,, \tag{1.6}$$

where $\dot{\boldsymbol{\epsilon}}$ is the infinitesimal strain rate tensor, and $\boldsymbol{c}$ is the tangent constitutive tensor. The rate symbols in the constitutive equation are used artificially to denote an incremental constitutive response, and should not be interpreted to mean that the response may contain rate-dependent effects. To obtain the evolution of the total Cauchy stress tensor, we integrate the rate equation (1.6), i.e.,

$$\boldsymbol{\sigma}_t = \boldsymbol{\sigma}(\boldsymbol{\epsilon}_t)\,. \tag{1.7}$$

Thus, for a given evolution of the strain tensor $\boldsymbol{\epsilon}_t$, we can determine the evolution of the Cauchy stress tensor $\boldsymbol{\sigma}_t$. Equation (1.7) is the classic strain-driven format of the constitutive relation. In contrast, the stress-driven format is of the form $\boldsymbol{\epsilon}_t = \boldsymbol{\epsilon}(\boldsymbol{\sigma}_t)$. Throughout this book, however, we shall adopt a strain-driven format, inasmuch as the stress-driven format could create a problem of non-unique map in the presence of a softening response (i.e., for a given $\boldsymbol{\sigma}_t$ there could be more than one $\boldsymbol{\epsilon}_t$).

Let $\boldsymbol{\sigma}_n$ denote the Cauchy stress tensor when the strain tensor is $\boldsymbol{\epsilon}_n$ at a given point in time t_n. The evolution of the stress tensor can be obtained from the integral

$$\boldsymbol{\sigma}_{n+1} = \boldsymbol{\sigma}_n + \int_{\boldsymbol{\epsilon}_n}^{\boldsymbol{\epsilon}_{n+1}} \boldsymbol{c} : \mathrm{d}\boldsymbol{\epsilon}\,. \tag{1.8}$$

For the integration to be possible, we must have a way of describing the evolution of the tangent constitutive tensor $\boldsymbol{c}$. In general, the tensor $\boldsymbol{c}$ may depend not only on the state of stress but also on the direction of loading. For example, if we write

$$\boldsymbol{c} = \boldsymbol{c}(\boldsymbol{\sigma}, \dot{\boldsymbol{\sigma}})\,, \tag{1.9}$$

then we see that the tangent constitutive tensor may depend not only on the state of $\boldsymbol{\sigma}$ but also on the loading direction $\dot{\boldsymbol{\sigma}}$. We then say that the material response is incrementally nonlinear because $\boldsymbol{c}$ has more than one possible value for a given state of $\boldsymbol{\sigma}$. Theory of plasticity results in an incrementally nonlinear response because the tangent constitutive tensor can have two possible values: an elastoplastic constitutive tensor for plastic loading and an elastic constitutive tensor for elastic unloading. On the other hand, if $\boldsymbol{c} = \boldsymbol{c}(\boldsymbol{\sigma})$, then we say that the response is incrementally linear because there is only one possible tangent constitutive tensor for a given state of $\boldsymbol{\sigma}$. Nonlinear elasticity results in an incrementally linear response.

1.3 Finite Element Equations

The finite element equations may be developed from the weak form of the initial value problem (Hughes 1987). To this end, we need the trial solution space $\boldsymbol{\mathcal{S}}$, such that $\boldsymbol{u} \in \boldsymbol{\mathcal{S}}$ satisfies the boundary condition $\boldsymbol{u} = \boldsymbol{g}$ on $\partial\mathcal{B}_{\boldsymbol{u}}$, and the variation space $\boldsymbol{\mathcal{V}}$, such that $\boldsymbol{w} \in \boldsymbol{\mathcal{V}}$ satisfies the homogeneous boundary condition $\boldsymbol{w} = \boldsymbol{0}$ on $\partial\mathcal{B}_{\boldsymbol{u}}$. The variational equation takes the form

$$\int_{\mathcal{B}} \boldsymbol{\nabla}^s \boldsymbol{w} : \boldsymbol{\sigma}\,\mathrm{d}V = \int_{\mathcal{B}} \boldsymbol{w} \cdot \boldsymbol{f}\,\mathrm{d}V + \int_{\partial\mathcal{B}_t} \boldsymbol{w} \cdot \boldsymbol{t}\,\mathrm{d}A\,, \tag{1.10}$$

where $\boldsymbol{\nabla}^s$ is the symmetric component of the gradient operator.

In the Galerkin approximation, the trial solution $\boldsymbol{u}$ and variation $\boldsymbol{w}$ are approximated using the same collection of shape functions. Let $\boldsymbol{N}$ represent the finite element shape function matrix for the nodal interpolation of trial function and variation; and $\boldsymbol{B} = \boldsymbol{\nabla}^s \boldsymbol{N}$ denote the symmetric gradient matrix. The finite element matrix equation may be written in the form

$$\boldsymbol{F}_{INT} = \boldsymbol{F}_{EXT}\,, \tag{1.11}$$

in which

$$\boldsymbol{F}_{INT} = \int_{\mathcal{B}} \boldsymbol{B}^{\mathsf{T}} \{\boldsymbol{\sigma}\}\, \mathrm{d}V\,, \tag{1.12}$$

$$\boldsymbol{F}_{EXT} = \int_{\mathcal{B}} \boldsymbol{N}^{\mathsf{T}} \boldsymbol{f}\, \mathrm{d}V + \int_{\partial \mathcal{B}_h} \boldsymbol{N}^{\mathsf{T}} \boldsymbol{t}\, \mathrm{d}A\,, \tag{1.13}$$

where $\{\boldsymbol{\sigma}\}$ are the components of the Cauchy stress tensor arranged in vector form, and T denotes a matrix transposition.

The external nodal force vector $\boldsymbol{F}_{EXT}$ is a function of the body force vector $\boldsymbol{f}$ (usually constant) and the traction vector $\boldsymbol{t}$, and can be evaluated explicitly. The variation of the internal nodal force vector $\boldsymbol{F}_{INT}$ is unknown and must be calculated from the evolution of the stresses, which in turn are functions of the unknown nodal displacements. In other words, $\boldsymbol{F}_{INT} = \boldsymbol{F}_{INT}(\boldsymbol{d})$, where $\boldsymbol{d}$ is the unknown nodal displacement vector. The task is to calculate (usually by iteration) the nodal displacement vectors $\boldsymbol{d}_n$, $\boldsymbol{d}_{n+1}$, etc., for given external loads $(\boldsymbol{F}_{EXT})_n$, $(\boldsymbol{F}_{EXT})_{n+1}$, etc.

Before proceeding with the strategies for iterative nonlinear equation solving, we elaborate the procedure for constructing the internal and external nodal force vectors. In practice, these vectors are evaluated from assembling the contributions of the individual finite elements, which are sampled at the quadrature points. For the internal nodal force vector, we have

$$\boldsymbol{F}_{INT} = \mathop{\mathsf{A}}_{e=1}^{n_{\mathrm{el}}} \boldsymbol{f}^e_{INT}\,, \qquad \boldsymbol{f}^e_{INT} = \sum_{\ell=1}^{n_{\mathrm{int}}} \boldsymbol{B}_\ell^{e\mathsf{T}} \{\boldsymbol{\sigma}_\ell\} W_\ell\,, \tag{1.14}$$

where 'A' denotes an assembly operator, ℓ is a counter representing the Gaussian quadrature points, W_ℓ is the weight of each integration point, and n_{int} is the number of integration points. Similarly, the external nodal force vector is evaluated as

$$\boldsymbol{F}_{EXT} = \mathop{\mathsf{A}}_{e=1}^{n_{\mathrm{el}}} \boldsymbol{f}^e_{EXT}\,, \qquad \boldsymbol{f}^e_{EXT} = \sum_{\ell=1}^{n_{\mathrm{int}}} \boldsymbol{N}_\ell^{e\mathsf{T}} \boldsymbol{f}_\ell W_\ell + \sum_{\ell'=1}^{n'_{\mathrm{int}}} \boldsymbol{N}_{\ell'}^{e\mathsf{T}} \boldsymbol{t}_{\ell'} W_{\ell'}\,, \tag{1.15}$$

where ℓ and ℓ' denote the volume and surface integration points, respectively. This means that the constitutive responses are stored and tracked at the Gauss integration points.

1.4 Newton's Method

To solve the finite element equation (1.11), we write it in residual form as

$$\boldsymbol{r}(\boldsymbol{d}) = \boldsymbol{F}_{EXT} - \boldsymbol{F}_{INT}(\boldsymbol{d}) \,. \tag{1.16}$$

The goal is to find the solutions at discrete time instants t_n, t_{n+1}, etc. To this end, assume that $\boldsymbol{d}_n$ satisfies the condition $\boldsymbol{r}(\boldsymbol{d}_n) = \boldsymbol{0}$ at the current time instant t_n. The problem is to find the next solution $\boldsymbol{d}_{n+1}$ that satisfies $\boldsymbol{r}(\boldsymbol{d}_{n+1}) = \boldsymbol{0}$ for a given external forcing function $(\boldsymbol{F}_{EXT})_{n+1}$.

We are interested in a Newton-type iteration of the form

$$\boldsymbol{d}_{n+1}^{k+1} = \boldsymbol{d}_{n+1}^{k} + \alpha_k \Delta \boldsymbol{d}^k \,, \qquad \boldsymbol{A}_k \Delta \boldsymbol{d}^k = \boldsymbol{r}(\boldsymbol{d}_{n+1}^k) \,, \tag{1.17}$$

where $\Delta \boldsymbol{d}^k$ is the search direction, $\alpha_k \in (0, 1]$ is a scalar line search parameter, $\boldsymbol{A}_k$ is a nonsingular tangent operator, and k is an iteration counter. We say that the iteration has converged when

$$\|\boldsymbol{r}(\boldsymbol{d}_{n+1}^k)\| \leq \varepsilon \|\boldsymbol{r}(\boldsymbol{d}_{n+1}^0)\| \,, \tag{1.18}$$

where $\|\boldsymbol{r}\| = (r_1^2 + \ldots + r_n^2)^{1/2}$ is the Euclidean length, or L_2-norm, of the residual force vector $\boldsymbol{r}$. The error tolerance ε depends on the precision of the calculations (typically chosen to be on the order 10^{-10} for double precision calculations). Because of the incremental nature of the algorithm, the previously converged solution is typically taken as the initial estimate of $\boldsymbol{d}_{n+1}$, i.e., $\boldsymbol{d}_{n+1}^0 = \boldsymbol{d}_n$. Box 1.1 summarizes the main steps of the algorithm.

Step 1. Initialize $k = 0$, $\boldsymbol{d}_{n+1}^0 = \boldsymbol{d}_n$, $\boldsymbol{r}^0 = \boldsymbol{r}(\boldsymbol{d}_n)$
Step 2. Compute $\boldsymbol{A}_k = -\boldsymbol{r}'(\boldsymbol{d}_{n+1}^k)$
Step 3. If $k > 0$, line search for α_k; otherwise, $\alpha_k = 1.0$
Step 4. Update $\boldsymbol{d}_{n+1}^{k+1} = \boldsymbol{d}_{n+1}^k + \alpha_k \boldsymbol{A}_k^{-1} \boldsymbol{r}(\boldsymbol{d}_{n+1}^k)$ and $\boldsymbol{r}^{k+1} = \boldsymbol{r}(\boldsymbol{d}_{n+1}^{k+1})$
Step 5. $k \leftarrow k + 1$
Step 6. If $\|\boldsymbol{r}^k\| / \|\boldsymbol{r}^0\| \leq \varepsilon$, exit; else, go to Step 2.

Box 1.1 Newton's method with line search

Newton's method is recovered when $\alpha^k = 1$ and $\boldsymbol{A}^k = -\boldsymbol{r}'(\boldsymbol{d}_{n+1}^k)$. This is easily verified by expanding the residual equation about the current estimate $\boldsymbol{d}_{n+1}^k$ using the Taylor series expansion, and ignoring the higher order terms,

$$\boldsymbol{r}(\boldsymbol{d}_{n+1}) = \boldsymbol{r}(\boldsymbol{d}_{n+1}^k) + \boldsymbol{r}'(\boldsymbol{d}_{n+1}^k)(\boldsymbol{d}_{n+1} - \boldsymbol{d}_{n+1}^k) + \ldots = \boldsymbol{0} \,, \tag{1.19}$$

which yields the iterative scheme

$$d_{n+1} \approx d_{n+1}^{k+1} = d_{n+1}^{k} \underbrace{-r'(d_{n+1}^{k})^{-1} r(d_{n+1}^{k})}_{:=\Delta d^k} . \tag{1.20}$$

The higher-order terms ignored in Newton's method is second order in $\Delta \boldsymbol{d}$, so the convergence of the iteration is asymptotically quadratic. For the finite element problem the consistent tangent operator is of the form

$$A_k = -r'(d_{n+1}^{k}) = \int_{\mathcal{B}} B^{\mathsf{T}} \mathbf{C}^k B \, \mathrm{d}V , \tag{1.21}$$

where

$$\mathbf{C}^k = \frac{\partial\{\sigma_{n+1}^{k}\}}{\partial\{\epsilon_{n+1}^{k}\}} \tag{1.22}$$

is the consistent tangential stress-strain matrix.

The convergence of Newton's method may be asymptotically quadratic, but the iteration could still diverge if the initial estimates are quite far from the true solution. In fact, it is possible for the iteration to converge to a different solution in some cases. This is because Newton's method does not guarantee a norm-reducing solution in the sense $\|r(d^k + \Delta d)\| < \|r(d^k)\|$ for all k. Typically, wild oscillations may be expected during the initial phase of the iteration, which could cause the solution to diverge. The line search parameter α^k may be used to damp the iteration and avoid unexpected 'bumps' in the solution. The next section shows that, in principle, a step length parameter $\alpha^k > 0$ may be chosen so that the iteration is always norm-reducing.

1.5 The Line Search

The line search parameter α^k may be used in (1.17) to force Newton's method to produce a solution that monotonically reduces the norm of the residual force vector. The following developments demonstrate that there is a step length parameter that always reduces the norm of the residual force vector.

We recall that a real-valued function f of one variable is differentiable at a point x if there is a real number $f'(x)$ such that

$$\lim_{\alpha \to 0} \frac{1}{\alpha}[f(x+\alpha) - f(x) - f'(x)\alpha] = 0 .$$

This definition extends naturally to n dimensions. Let $f(d) = \|r(d)\|$ be the L_2-norm of the residual vector (note that any norm may be used for this purpose), and assume that it is *Gateaux-* (or *G-*) *differentiable* in the neighborhood of $d \in \mathbb{R}^n$, where n denotes the length of the vector space. Then,

$$\lim_{\alpha \to 0} \frac{1}{\alpha}[f(d + \alpha \Delta d) - f(d) - \alpha \langle f'(d), \Delta d \rangle] = 0 , \tag{1.23}$$

where the symbol $\langle \cdot, \cdot \rangle$ denotes an inner product. A necessary requirement for a norm-reducing solution would be that $\langle f'(\boldsymbol{d}), \Delta \boldsymbol{d} \rangle < 0$, for then there exists an $\alpha > 0$ such that $f(\boldsymbol{d}+\alpha\Delta\boldsymbol{d}) < f(\boldsymbol{d})$. Stated in another way, the iteration is norm-reducing if the slope of the scalar function $f(\boldsymbol{d}+\alpha\Delta d)$, plotted versus α for given values of $\boldsymbol{d}$ and $\Delta\boldsymbol{d}$, is *negative* in the neighborhood $\alpha = 0$. Clearly, this slope is $\langle f'(\boldsymbol{d}), \Delta\boldsymbol{d}\rangle$. Now, for Newton's method this slope is identically negative, as can be seen from

$$\langle f'(\boldsymbol{d}), \Delta\boldsymbol{d}\rangle = \langle \boldsymbol{r}(\boldsymbol{d}), \boldsymbol{r}'(\boldsymbol{d})\Delta\boldsymbol{d}\rangle / \|\boldsymbol{r}(\boldsymbol{d})\| = -\|\boldsymbol{r}(\boldsymbol{d})\| < 0\,. \tag{1.24}$$

Thus, one can choose an $\alpha > 0$ so that Newton's method reduces the norm. If $\alpha < 1$, the search direction $\Delta\boldsymbol{d}$ is reduced in length by the factor α, resulting in a 'damped' Newton iteration.

The bound on the step length α is important in actual numerical analysis and can be determined, at least in principle, from the following. We invoke the majorization principle (Ortega and Rheinboldt 1970), which states that if there exists a majorization function $\psi(\alpha) \in \mathbb{R}$ such that

$$f(\boldsymbol{d}+\alpha\Delta\boldsymbol{d}) \leq \psi(\alpha) < f(\boldsymbol{d}) \qquad \forall \alpha \in (0, \widetilde{\alpha})\,, \tag{1.25}$$

then $\boldsymbol{d} + \alpha\Delta\boldsymbol{d}$ decreases the error norm f. Using the mean-value theorem (Dieudonné 1969) and the Cauchy-Schwarz inequality (Nowinski 1981), we have

$$\begin{aligned}
f(\boldsymbol{d}+\alpha\Delta\boldsymbol{d}) &= f(\boldsymbol{d}) + \int_0^1 \alpha\langle f'(\boldsymbol{d}+\theta\alpha\Delta\boldsymbol{d}), \Delta\boldsymbol{d}\rangle \,\mathrm{d}\theta \\
&= f(\boldsymbol{d}) + \alpha\langle f'(\boldsymbol{d}), \Delta\boldsymbol{d}\rangle + \alpha \int_0^1 [\langle f'(\boldsymbol{d}+\theta\alpha\Delta\boldsymbol{d}) - f'(\boldsymbol{d}), \Delta\boldsymbol{d}\rangle] \,\mathrm{d}\theta \\
&\leq f(\boldsymbol{d}) + \alpha\langle f'(\boldsymbol{d}), \Delta\boldsymbol{d}\rangle + \alpha\|\Delta\boldsymbol{d}\| \int_0^1 \|f'(\boldsymbol{d}+\theta\alpha\Delta\boldsymbol{d}) - f'(\boldsymbol{d})\| \,\mathrm{d}\theta \\
&\leq f(\boldsymbol{d}) + \alpha\langle f'(\boldsymbol{d}), \Delta\boldsymbol{d}\rangle + \alpha^2\|\Delta\boldsymbol{d}\|^2 K \int_0^1 \theta \,\mathrm{d}\theta\,.
\end{aligned} \tag{1.26}$$

Thus, choosing

$$\psi(\alpha) = f(\boldsymbol{d}) + \alpha\langle f'(\boldsymbol{d}), \Delta\boldsymbol{d}\rangle + \frac{1}{2}\alpha^2 K\|\Delta\boldsymbol{d}\|^2 \tag{1.27}$$

guarantees that

$$f(\boldsymbol{d}+\alpha\Delta\boldsymbol{d}) \leq \psi(\alpha)\,. \tag{1.28}$$

To determine the majorization interval $[0, \widetilde{\alpha}]$ we impose the constraint $\psi(\alpha) < f(\boldsymbol{d})$, which yields

$$\widetilde{\alpha} = -2\langle f'(\boldsymbol{d}), \Delta\boldsymbol{d}\rangle / K\|\Delta\boldsymbol{d}\|^2 \equiv 2f/K\|\Delta\boldsymbol{d}\|^2\,. \tag{1.29}$$

Thus, choosing $0 < \alpha < \widetilde{\alpha}$ guarantees a norm-reducing Newton's method. Of course, one can choose, as the 'best estimate' for the step length,

$$\alpha = \widetilde{\alpha}/2 = f/K\|\Delta \boldsymbol{d}\|^2, \tag{1.30}$$

which gives the minimum of the majorization curve $\psi(\alpha)$.

Since the parameter K is unknown, a simple line search algorithm can be formulated as summarized in Box 1.2. First, assume a trial majorization curve $\widetilde{\psi}(\alpha)$ of the quadratic form (1.27). Then, enforce the following boundary conditions: (a) $\widetilde{\psi}(0) = f(\boldsymbol{d}) \equiv f_1$; (b) $\widetilde{\psi}'(0) = \langle f'(\boldsymbol{d}), \Delta \boldsymbol{d}\rangle = -f_1$; and (c) $\widetilde{\psi}(\alpha^* > 0) = \|\boldsymbol{r}(\boldsymbol{d} + \alpha^* \Delta \boldsymbol{d})\| \equiv f_2$. The following polynomial results:

$$\widetilde{\psi}(\alpha) = f_1 - f_1\alpha + ((f_2 + f_1\alpha^* - f_1)/\alpha^{*2})\alpha^2\,. \tag{1.31}$$

To implement the above equation, we assume an initial trial value of $\alpha^* = 1.0$ and compare f_2 with f_1. If the step length $\alpha^* = 1.0$ decreases f, then we accept the iterate (undamped Newton's iteration) and proceed with the next iteration; otherwise, we compute a new α^* which gives the minimum $\widetilde{\psi}(\alpha)$ from (1.31), i.e.,

$$\alpha^* \leftarrow \alpha^{*2} f_1/2(f_2 + f_1\alpha^* - f_1)\,. \tag{1.32}$$

If $\widetilde{\psi}(\alpha)$ is a majorization curve, then the next value of α^* will lead to a descent in f (see Fig. 1.3a); otherwise, we continue searching for α that decreases f (see Fig. 1.3b). The algorithm is inexpensive in that it only

Step 1. Initialize $f_1 = \|\boldsymbol{r}(\boldsymbol{d})\|,\ \ f_2 = \|\boldsymbol{r}(\boldsymbol{d} + \Delta \boldsymbol{d})\|,\ \ \alpha = 1.0$
Step 2. If $f_2 < f_1$, return; else, continue
Step 3. Update $\alpha \leftarrow \alpha^2 f_1/2(f_2 + f_1\alpha - f_1)$
Step 4. Compute $f_2 = \|\boldsymbol{r}(\boldsymbol{d} + \alpha\Delta \boldsymbol{d})\|$ and go to Step 2.

Box 1.2 A simple line search algorithm based on quadratic polynomial fit on the majorization curve

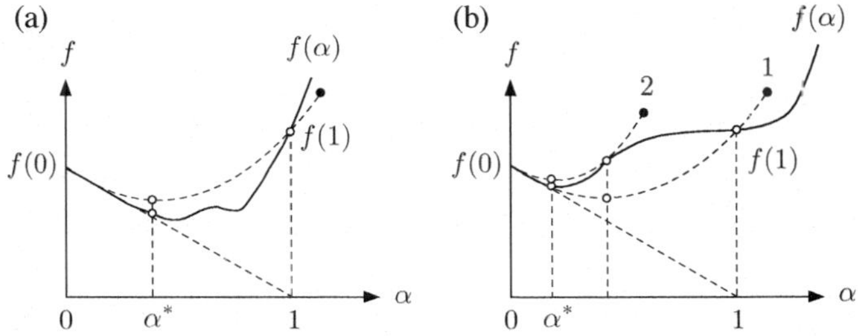

Fig. 1.3 Simple line search algorithm with Newton's method: trial majorization curve finds α^* after (a) one trial; (b) two trials. Solid curves are exact norm functions; dashed curves are trial majorization functions.

requires residual force calculations; however, the majorization interval can shrink rapidly, leading to an 'over-damped' Newton's method. More elaborate step length algorithms, such as that based on the principle of Goldstein (1966), are also available.

1.6 Preconditioned Conjugate Gradients

Newton's method requires the solution of the linear system $\boldsymbol{Ax} = \boldsymbol{b}$ for each iteration. In practice, the coefficient matrix $\boldsymbol{A}$ is not inverted. Instead, direct linear equation solving consists of a sequence of triangular factorization, forward reduction, and back substitution. Several factorization schemes are available, and for non-symmetric systems commonly encountered in computational plasticity, the Crout factorization scheme seems sufficiently robust. For large systems of equation, however, triangular factorization could be a very CPU-intensive process, and so one might seek an alternative to Newton's method.

One way of avoiding repeated triangular factorization is to factor the coefficient matrix only once, during the beginning of the iteration, and use this factored matrix all throughout the entire iteration process. This method is called the modified Newton iteration. The computing cost for each iteration is much lower compared to the full Newton's scheme since forward reduction and back substitution are not as time-consuming to perform as compared to triangular factorization. However, because the property of the tangent operator is not updated during the iteration process, the modified Newton iteration tends to converge very slowly, and has a propensity to diverge when dealing with highly nonlinear systems.

Alternatively, we can revert back to the full Newton's scheme, but instead of solving the linearized system directly, we can solve it iteratively. Effectively, we have two levels of iteration, where Newton's iteration is the outer loop and linear equation solving is the inner loop. The residual in the inner loop is $\boldsymbol{R}^m = \boldsymbol{b} - \boldsymbol{Ax}^m$, where $\boldsymbol{x}^m$ is the approximation of $\boldsymbol{x}$ (recall that $\boldsymbol{x} \equiv \Delta \boldsymbol{d}^k$ in the context of the outer loop). There is a broad class of iterative methods for this problem and a thorough treatment of this subject is beyond the scope of this book. We refer the readers to Saad (2003) for a thorough discussion of this subject.

To illustrate the process of iterative linear equation solving, we consider the conjugate gradients (CG) method in general and the preconditioned conjugate gradients (PCG) algorithm in particular. The CG method belongs in a broader class of Krylov-subspace methods and is appropriate for symmetric systems. The Krylov subspace $\boldsymbol{\mathcal{K}}$ can be defined as

$$\boldsymbol{\mathcal{K}}(\boldsymbol{A}, \boldsymbol{R}^0) = \text{span}\{\boldsymbol{R}^0, \boldsymbol{AR}^0, \boldsymbol{A}^2\boldsymbol{R}^0, \ldots, \boldsymbol{A}^{N-1}\boldsymbol{R}^0\}, \tag{1.33}$$

where N is the dimension of $\boldsymbol{A}$. The approximate solution $\boldsymbol{x}^m$ is determined from the affine subspace $\boldsymbol{R}^0 + \mathcal{K}$ such that $\boldsymbol{R}^m$ is orthogonal to some constraint space.

If $\boldsymbol{A}$ is symmetric and positive definite, then the solution of $\boldsymbol{A}\boldsymbol{x} = \boldsymbol{b}$ is the minimum of the quadratic functional

$$\mathcal{F}(\boldsymbol{x}) = c - \langle \boldsymbol{b}, \boldsymbol{x} \rangle + \frac{1}{2}\langle \boldsymbol{x}, \boldsymbol{A}\boldsymbol{x} \rangle \,. \tag{1.34}$$

A conjugate direction method has the form

$$\boldsymbol{x}^{m+1} = \boldsymbol{x}^m + \alpha^m \boldsymbol{p}^m \,, \tag{1.35}$$

where $\boldsymbol{p}^m$ is the descent direction and α^m is the step length chosen such that $\boldsymbol{A}\boldsymbol{x}^{m+1} = \boldsymbol{b}$, i.e.,

$$\alpha^m = \langle \boldsymbol{R}^m, \boldsymbol{p}^m \rangle / \langle \boldsymbol{p}^m, \boldsymbol{A}\boldsymbol{p}^m \rangle \,. \tag{1.36}$$

The residual then evolves according to the relation

$$\boldsymbol{R}^{m+1} = \boldsymbol{R}^m - \alpha^m \boldsymbol{A}\boldsymbol{p}^m \,. \tag{1.37}$$

To carry out a conjugate direction algorithm, it is necessary to obtain the conjugate directions $\boldsymbol{p}^m$. The CG procedure generates $\boldsymbol{p}^m$ simultaneously with the approximate solution $\boldsymbol{x}^m$ according to the following procedure. First, the approximate solution is evaluated from the equations

$$\begin{aligned} \boldsymbol{R}^m = \boldsymbol{b} - \boldsymbol{A}\boldsymbol{x}^m \,, \quad \alpha^m &= \langle \boldsymbol{R}^m, \boldsymbol{p}^m \rangle / \langle \boldsymbol{p}^m, \boldsymbol{A}\boldsymbol{p}^m \rangle \,, \\ \boldsymbol{x}^{m+1} &= \boldsymbol{x}^m + \alpha^m \boldsymbol{p}^m \,. \end{aligned} \tag{1.38}$$

Next, the descent direction is evaluated according to the relations

$$\begin{aligned} \boldsymbol{R}^{m+1} = \boldsymbol{b} - \boldsymbol{A}\boldsymbol{x}^{m+1} \,, \quad \beta^m &= -\langle \boldsymbol{R}^{m+1}, \boldsymbol{A}\boldsymbol{p}^m \rangle / \langle \boldsymbol{p}^m, \boldsymbol{A}\boldsymbol{p}^m \rangle \,, \\ \boldsymbol{p}^{m+1} &= \boldsymbol{R}^{m+1} + \beta^m \boldsymbol{p}^m \,. \end{aligned} \tag{1.39}$$

We note here that the step length β^m is chosen such $\langle \boldsymbol{A}\boldsymbol{p}^m, \boldsymbol{p}^{m+1} \rangle = 0$. The operations are repeated until one finds the desired solution $\boldsymbol{x}$. In principle, for a symmetric and positive-definite matrix $\boldsymbol{A} \in L(\mathbb{R}^N)$, $\boldsymbol{x}^m$ satisfies $\boldsymbol{A}\boldsymbol{x}^m = \boldsymbol{b}$ for some $m \leq N$ (Ortega and Rheinboldt 1970).

The efficiency of the CG method depends on the condition number of the matrix $\boldsymbol{A}$. Generally satisfactory performance is expected for condition numbers close to one (well conditioned matrix); otherwise, for large condition numbers (poorly conditioned matrix) some preconditioning may be necessary. The objective of preconditioning is to 'improve' the condition number of the coefficient matrix. To illustrate the idea of preconditioning, consider a matrix $\boldsymbol{B}$ having the same dimension as the coefficient matrix $\boldsymbol{A}$. The unknown vector $\boldsymbol{x}$ can be solved from the equivalent linear system $\boldsymbol{B}^{-1}\boldsymbol{A}\boldsymbol{x} = \boldsymbol{B}^{-1}\boldsymbol{b}$. There are two possible extreme values for the matrix $\boldsymbol{B}$: the identity matrix $\boldsymbol{I}$ that effectively renders no preconditioning, and the original coefficient marix

$\boldsymbol{A}$ that effectively results in solving the original system. Neither one is acceptable from our standpoint and we seek a compromise somewhere in between. An acceptable preconditioning matrix $\boldsymbol{B}$ is one that satisfies the following criteria: (a) it is easy to factor; and (b) the condition number of $\boldsymbol{B}^{-1}\boldsymbol{A}$ is close to one. The challenge lies in finding this preconditioning matrix.

With the preconditioned system the residual is effectively $\boldsymbol{B}^{-1}\boldsymbol{R}^m$, and thus the descent direction is modified to have the form

$$\boldsymbol{p}^{m+1} = \boldsymbol{B}^{-1}\boldsymbol{R}^{m+1} + \beta^m \boldsymbol{p}^m . \tag{1.40}$$

Setting $\langle \boldsymbol{A}\boldsymbol{p}^m, \boldsymbol{p}^{m+1}\rangle = 0$ once again gives the step length

$$\beta^m = -\langle \boldsymbol{B}^{-1}\boldsymbol{R}^{m+1}, \boldsymbol{A}\boldsymbol{p}^m\rangle / \langle \boldsymbol{p}^m, \boldsymbol{A}\boldsymbol{p}^m\rangle . \tag{1.41}$$

We also re-write

$$\begin{aligned}
\alpha^m &= \langle \boldsymbol{R}^m, \boldsymbol{p}^m\rangle / \langle \boldsymbol{p}^m, \boldsymbol{A}\boldsymbol{p}^m\rangle \\
&= \langle \boldsymbol{R}^m, \boldsymbol{B}^{-1}\boldsymbol{R}^m + \beta^{m-1}\boldsymbol{p}^{m-1}\rangle / \langle \boldsymbol{p}^m, \boldsymbol{A}\boldsymbol{p}^m\rangle \\
&= \langle \boldsymbol{R}^m, \boldsymbol{B}^{-1}\boldsymbol{R}^m\rangle / \langle \boldsymbol{p}^m, \boldsymbol{A}\boldsymbol{p}^m\rangle
\end{aligned} \tag{1.42}$$

upon noting that $\langle \boldsymbol{R}^m, \boldsymbol{p}^{m-1}\rangle = 0$. This form for α^m is used in Box 1.3, which summarizes the steps carried out in the PCG algorithm.

Step 1. Initialize $m = 0,\ \boldsymbol{x}^0 = \boldsymbol{0},\ \boldsymbol{r}^0 = \boldsymbol{b},\ \boldsymbol{p}^0 = \boldsymbol{z}^0 = \boldsymbol{B}^{-1}\boldsymbol{r}^0$
Step 2. Compute $\alpha^m = \langle \boldsymbol{R}^m, \boldsymbol{z}^m\rangle / \langle \boldsymbol{p}^m, \boldsymbol{A}\boldsymbol{p}^m\rangle$
$\boldsymbol{x}^{m+1} = \boldsymbol{x}^m + \alpha^m \boldsymbol{p}^m$
$\boldsymbol{R}^{m+1} = \boldsymbol{R}^m - \alpha^m \boldsymbol{A}\boldsymbol{p}^m$
Step 3. If $\|\boldsymbol{R}^{m+1}\| / \|\boldsymbol{R}^0\| \le \varepsilon$, return; else continue
Step 4. Compute $\boldsymbol{z}^{m+1} = \boldsymbol{B}^{-1}\boldsymbol{R}^{m+1}$
$\beta^m = -\langle \boldsymbol{z}^{m+1}, \boldsymbol{A}\boldsymbol{p}^m\rangle / \langle \boldsymbol{p}^m, \boldsymbol{A}\boldsymbol{p}^m\rangle$
$\boldsymbol{p}^{m+1} = \boldsymbol{z}^{m+1} + \beta^m \boldsymbol{p}^m$
Step 5. Update $m \leftarrow m + 1$ and go to Step 2.

Box 1.3 Preconditioned conjugate gradients algorithm

We should note that the linearized system generally encountered in computational plasticity results in a non-symmetric $\boldsymbol{A}$. There are more elaborate iterative solvers for non-symmetric systems, such as the generalized minimum residuals (GMRES) and stabilized biconjugate gradients (BICGSTAB) methods (Saad 2003). However, note in Box 1.3 that the PCG algorithm is quite simple to implement. Furthermore, with a suitable preconditioner it can be just as effective for non-symmetric and block systems (Borja 1991; White and Borja 2011).

1.7 Solved Problem

Consider the following system of nonlinear equations:

$$\boldsymbol{F}_{INT}(\boldsymbol{d}) = \left\{ \begin{array}{c} 2d_1 + d_1^4 + d_2 + 3d_1 d_2^2 - 9d_4 + d_4^4 \\ d_1 + 3d_1^2 d_2 + 10d_2 + 4d_2^2 + 2d_2 d_3 - 8d_3 + 7d_4 \\ -8d_2 + d_2^2 + 3d_3 + d_3^2 + 2d_4 \\ -9d_1 + 4d_1 d_4^3 + 7d_2 + 2d_3 + 5d_4 \end{array} \right\},$$

$$\boldsymbol{d} = \left\{ \begin{array}{c} d_1 \\ d_2 \\ d_3 \\ d_4 \end{array} \right\}, \qquad \boldsymbol{F}_{EXT} = \left\{ \begin{array}{c} 0.2 \\ -0.1 \\ 0.0 \\ 0.5 \end{array} \right\},$$

and $\boldsymbol{F}_{INT}(\boldsymbol{d}) = \boldsymbol{F}_{EXT}$. Assume initial conditions $\boldsymbol{F}_{EXT} = \boldsymbol{0}$ when $\boldsymbol{d} = \boldsymbol{0}$ and impose the final external load in one full step (thus, $\Delta\boldsymbol{d} = \boldsymbol{d}$). The tangent operator is symmetric and takes the form

$$\boldsymbol{A} = \begin{bmatrix} (2 + 4d_1^3 + 3d_2^2) & (1 + 6d_1 d_2) & 0 & (-9 + 4d_4^3) \\ (1 + 6d_1 d_2) & (10 + 3d_1^2 + 8d_2 + 2d_3) & (-8 + 2d_2) & 7 \\ 0 & (-8 + 2d_2) & (3 + 2d_3) & 2 \\ (-9 + 4d_4^3) & 7 & 2 & (5 + 12d_1 d_4^2) \end{bmatrix}.$$

Take the preconditioner as

$$\boldsymbol{B} = \boldsymbol{A}(\boldsymbol{d} = \boldsymbol{0}) = \begin{bmatrix} 2 & 1 & 0 & -9 \\ 1 & 10 & -8 & 7 \\ 0 & -8 & 3 & 2 \\ -9 & 7 & 2 & 5 \end{bmatrix}.$$

Using double-precision operations, the solution to this problem is given by the displacement vector

$$\boldsymbol{d} = -\left\{ \begin{array}{l} 0.19439\ldots \\ 0.07741\ldots \\ 0.16836\ldots \\ 0.07425\ldots \end{array} \right\}.$$

Box 1.4 shows an asymptotic quadratic convergence of Newton iteration, demonstrated by the negative exponent of the residual error norm, which doubles after each iteration. As for the inner iteration, both the CG and PCG methods converge to machine precision after four iterations, which is the size of the matrix system. However, the rate of convergence of the PCG iteration is superior to that of the standard CG iteration, and if the error tolerance was relaxed a little bit the PCG method would have taken fewer iterations to converge. The standard CG iteration converges to machine precision simply because it reaches the theoretical maximum limit of N iterations, but obviously, we do not want to have to reach this limit each time, especially when N is very large.

Iteration Number	Newton	PCG	CG
0	1.000e+00	1.000e+00	1.000e+00
1	1.784e−01	1.254e−02	4.881e+00
2	9.935e−03	4.002e−04	8.716e−01
3	3.770e−05	2.604e−06	4.592e−01
4	5.413e−10	3.038e−16	4.148e−14
5	1.241e−16	—	—

Box 1.4 Normalized norms of residual for Newton, PCG, and CG iterations. For Newton iteration, the normalized error norm is $\|\boldsymbol{r}^k\|/\|\boldsymbol{r}^0\|$; for the PCG and CG iterations, the normalized error norm is $\|\boldsymbol{R}^m\|/\|\boldsymbol{R}^0\|$ on the second Newton iteration $(k = 2)$.

Exercise 1.1. Consider the following system of nonlinear equations in five variables: $\boldsymbol{F}_{INT}(\boldsymbol{d}) = \boldsymbol{F}_{EXT}$, where

$$\boldsymbol{F}_{INT}(\boldsymbol{d}) = \begin{Bmatrix} 5d_1 + 3d_2^2 - 3d_3 - 8d_4 \\ d_1 + d_2 + d_3 + d_4 + d_5 \\ 2d_1 - d_2 + 6d_3 + d_4^2 - d_5 \\ d_3 + 9d_4 + 10d_5 \\ d_1^2 + d_2^3 + 2d_3 - d_4 + d_5 \end{Bmatrix}, \quad \boldsymbol{F}_{EXT} = \begin{Bmatrix} -24 \\ 15 \\ 29 \\ 89 \\ 16 \end{Bmatrix},$$

and $d_1, \ldots, d_5$ are elements of the vector $\boldsymbol{d}$. (a) Determine the consistent tangent operator $\boldsymbol{F}'_{INT}(\boldsymbol{d})$. (b) Assuming an initial guess of $\boldsymbol{d}_0 = \boldsymbol{0}$, determine the displacement vector $\boldsymbol{d}$ by Newton's method. How many iterations are required to satisfy a convergence criterion $\|\boldsymbol{r}(\boldsymbol{d}^k)\|/\|\boldsymbol{r}(\boldsymbol{d}_0)\| \leq 10^{-5}$? How many iterations are required to satisfy a convergence criterion $\|\boldsymbol{r}(\boldsymbol{d}^k)\|/\|\boldsymbol{r}(\boldsymbol{d}_0)\| \leq 10^{-15}$?(c) Repeat problem (b) assuming an initial guess of $\boldsymbol{d}_0 = \{5, 5, 5, 5, 5\}$. (d) Repeat problem (b) using the modified Newton's method.

2 One-Dimensional Problem

2.1 Introduction

This chapter presents the basic elements of plasticity theory in the context of a one-dimensional bar problem. The goals are to facilitate an understanding of the basic ideas of plasticity theory without the need for extensive tensor notations, and to provide the reader an opportunity to conduct simple manual calculations that solidify understanding of the important concepts. The chapter introduces the notions of yield function, flow rule, hardening/softening responses, convexity of a function, and the concept of plastic dissipation. To provide a balanced treatment of theory and computation, we introduce at the outset the idea of return mapping for integrating the rate-constitutive equations in one dimension. Readers not familiar with the algorithm may wish to refer to the numerical example found at the end of this chapter and follow the sequence of calculations manually. One of the highlights of the chapter is the discussion of isotropic and kinematic hardening and how they can be implemented numerically within the framework of return mapping.

2.2 Elastoplastic Problem in One Dimension

Consider an elastoplastic bar subjected to uniaxial tension shown in Fig. 2.1. Assume that the stress-strain curve is given by a bilinear relation, where E is Young's modulus and E_T is the tangential modulus beyond the stress point $\sigma_{YT} > 0$, the initial yield stress in tension. If the bar is compressed, the elastic range in compression is defined by the coordinate $-\sigma_{YC} < 0$, where σ_{YC} is the yield stress in compression. For simplicity we assume that $\sigma_{YT} = \sigma_{YC} = \sigma_Y$. We define the initial *elastic region* as

$$\mathsf{E}_\varnothing = \{\sigma \in \mathbb{R} \,|\, -\sigma_{Y0} < \sigma < \sigma_{Y0}\} \tag{2.1}$$

and the two initial *yield points* as

$$\partial\mathsf{E}_\varnothing = \{\sigma \in \mathbb{R} \,|\, \sigma = \pm\sigma_{Y0}\}\,. \tag{2.2}$$

The set

$$\overline{\mathsf{E}}_\varnothing = \mathsf{E}_\varnothing \cup \partial\mathsf{E}_\varnothing \tag{2.3}$$

denotes the closure of the region $\mathsf{E}_\varnothing$.

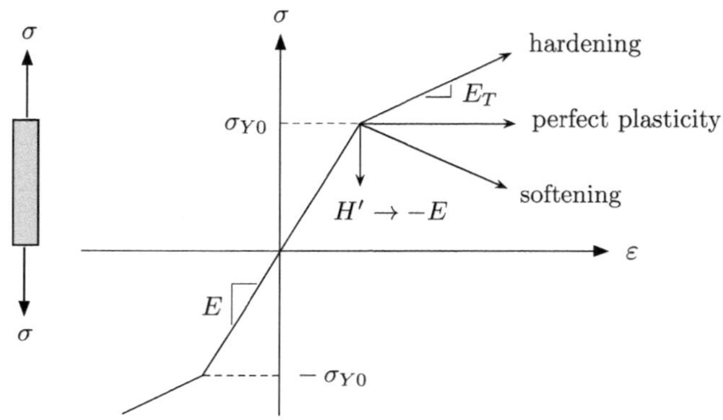

Fig. 2.1 Uniaxial loading of an elastoplastic bar: hardening when $H' > 0$, perfect plasticity when $H' = 0$, and softening when $H' < 0$

At any point beyond the initial yield stresses, the total axial strain ε in the bar can be decomposed additively into an elastic part ε^{e} and a plastic part ε^{p}, i.e.,

$$\varepsilon = \varepsilon^{\mathrm{e}} + \varepsilon^{\mathrm{p}} . \tag{2.4}$$

The constitutive equation for the bar is given by

$$\sigma = E\varepsilon^{\mathrm{e}} = E(\varepsilon - \varepsilon^{\mathrm{p}}) . \tag{2.5}$$

For any applied load that produces additional plastic strain, the time derivative of the axial stress is given by

$$\dot{\sigma} = E(\dot{\varepsilon} - \dot{\varepsilon}^{\mathrm{p}}) . \tag{2.6}$$

With reference to Fig. 2.1, the same time-derivative of the axial stress can be written as

$$\dot{\sigma} = E_T\, \dot{\varepsilon} , \tag{2.7}$$

where E_T is the tangential modulus.

2.3 Yield Function, Flow Rule, and Hardening/Softening

For the same one-dimensional bar problem, an initial *yield function* may be defined as

$$f(\sigma, \sigma_{Y0}) = |\, \sigma \,| - \sigma_{Y0} \le 0 . \tag{2.8}$$

Therefore, in terms of this yield function the initial elastic region may be defined as

$$\mathsf{E}_{\varnothing} = \{\sigma \in \mathbb{R} \,|\, f(\sigma, \sigma_{Y0}) < 0\}\,, \tag{2.9}$$

and its closure as

$$\overline{\mathsf{E}}_{\varnothing} = \{\sigma \in \mathbb{R} \,|\, f(\sigma, \sigma_{Y0}) \leq 0\}\,. \tag{2.10}$$

Note that $\mathsf{E}_{\varnothing}$ is a closed interval, so f is a *convex* function.

Assume that the bar is subjected to a differential load capable of producing a differential plastic axial strain. For example, we can continue stretching the bar that is already yielding in tension, or continue compressing it when it is already yielding in compression. Denoting the magnitude of the plastic strain rate by $\dot{\lambda} \geq 0$, we write

$$\dot{\varepsilon}^{\mathrm{p}} = \begin{cases} +\dot{\lambda}\,, & \text{if } \sigma > 0\,, \\ -\dot{\lambda}\,, & \text{if } \sigma < 0\,. \end{cases} \tag{2.11}$$

Alternatively, we can also write

$$\dot{\varepsilon}^{\mathrm{p}} = \dot{\lambda}\,\frac{\partial f}{\partial \sigma}\,, \qquad \frac{\partial f}{\partial \sigma} = \operatorname{sign}(\sigma)\,. \tag{2.12}$$

Equation (2.12) is called the *flow rule* for the plastic strain rate. The flow rule contains the magnitude of the strain rate, $\dot{\lambda}$, as well as its sense, given by the derivative $\partial f/\partial \sigma$. For the one-dimensional bar problem, the flow rule is obtained from the gradient of the yield function f with respect to the axial stress σ.

Assume that the bar continues to yield in tension and that the yield stress changes from the initial value σ_{Y0} to the current value σ_Y. If the yield stress in compression remains equal to the yield stress in tension, then the current yield function changes according to

$$f(\sigma, \sigma_Y) = |\,\sigma\,| - \sigma_Y \leq 0\,. \tag{2.13}$$

In this case, the current elastic region becomes

$$\mathsf{E} = \{\sigma \in \mathbb{R} \,|\, f(\sigma, \sigma_Y) < 0\}\,, \tag{2.14}$$

and its closure is

$$\overline{\mathsf{E}} = \{\sigma \in \mathbb{R} \,|\, f(\sigma, \sigma_Y) \leq 0\}\,. \tag{2.15}$$

A simple law describing the evolution of σ_Y with plastic deformation is given by the *hardening law*

$$\dot{\sigma}_Y = H'\,|\,\dot{\varepsilon}^{\mathrm{p}}\,| = H'\,\dot{\lambda}\,, \tag{2.16}$$

where H' is the plastic modulus. The parameter H' can be interpreted as the slope of the stress-strain curve after removing the elastic component of strain. Since $\dot{\lambda} > 0$, then σ_Y increases or decreases with plastic deformation depending on the sign of H'. By definition, the bar exhibits a hardening response when $H' > 0$, a perfectly plastic response when $H' = 0$, and a softening response when $H' < 0$. Figure 2.1 depicts these three responses.

2.4 Loading/Unloading and Consistency Conditions

The axial stress σ and the parameter $\dot{\lambda}$ are restricted by certain unilateral contraints, namely,

$$\dot{\lambda} \geq 0 \,, \qquad f(\sigma, \sigma_Y) \leq 0 \,. \tag{2.17}$$

Two possibilities exist: if the bar is deforming elastically, then $\dot{\lambda} = 0$ and $f(\sigma, \sigma_Y) < 0$; if it is deforming plastically, then $\dot{\lambda} > 0$ and $f(\sigma, \sigma_Y) = 0$. Either way, the following compact expression holds

$$\dot{\lambda} f(\sigma, \sigma_Y) = 0 \,. \tag{2.18}$$

These conditions are called the Karush-Kuhn-Tucker conditions (Karush 1939; Kuhn and Tucker 1951), or the KKT conditions (also known as the Kuhn-Tucker conditions) representing restrictions that must be satisfied whether the material is yielding or not. The Kuhn-Tucker conditions are first-order necessary conditions for a solution in nonlinear programming to be optimal.

The plastic response is of particular interest and is investigated next. Taking the time-derivative of (2.18) gives

$$\dot{\lambda} \dot{f}(\sigma, \sigma_Y) = 0 \quad \Longrightarrow \quad \dot{f}(\sigma, \sigma_Y) = 0 \,. \tag{2.19}$$

The last equation results from the condition $\dot{\lambda} > 0$, and is called the *consistency condition.* For the one-dimensional bar problem the consistency condition is

$$\dot{f}(\sigma, \sigma_Y) = \operatorname{sign}(\sigma)\, \dot{\sigma} - \dot{\sigma}_Y = 0 \,. \tag{2.20}$$

We now insert (2.6), (2.12), and (2.16) into (2.20), and solve for the plastic multiplier $\dot{\lambda}$. The result is

$$\dot{\lambda} = \frac{\operatorname{sign}(\sigma)\, E\, \dot{\varepsilon}}{E + H'} \quad \Longrightarrow \quad \dot{\varepsilon}^{\mathrm{p}} = \frac{E}{E + H'}\, \dot{\varepsilon} \,. \tag{2.21}$$

Note that the numerator in $(2.21)_1$ is always positive since $E > 0$ and $\operatorname{sign}(\sigma)\, \dot{\varepsilon} > 0$. Since $\dot{\lambda} > 0$, then

$$H' > -E \,, \tag{2.22}$$

which means that H' cannot be more negative than $-E$ or the stress-strain curve will snap back, see Fig. 2.1. There is no upper bound on the values of H', so a positive infinite value of H' gives $\dot{\lambda} = \dot{\varepsilon}^{\mathrm{p}} = 0$, which corresponds to a limiting elastic response.

Inserting (2.21) in (2.6) gives

$$\dot{\sigma} = E\Big[\dot{\varepsilon} - \Big(\frac{E}{E + H'}\Big)\dot{\varepsilon}\Big] = \Big(E - \frac{E^2}{E + H'}\Big)\dot{\varepsilon} \,. \tag{2.23}$$

Comparing the last equation to (2.7), we conclude that the tangential modulus E_T must be related to the plastic modulus H' according to the equation

$$E_T = \frac{H'}{1 + H'/E} \quad \Longleftrightarrow \quad H' = \frac{E_T}{1 - E_T/E} \,. \tag{2.24}$$

The limiting values for the tangential modulus are $E_T \to E$ as $H' \to \infty$, and $E_T \to -\infty$ as $H' \to -E$.

2.5 Isotropic and Kinematic Hardening

The specific hardening law considered in the previous section, in which the yield stress in compression increases by the same amount as the yield stress in tension, pertains to *isotropic hardening.* Essentially, isotropic hardening in one dimension implies that for every incremental change in the yield stress $\Delta\sigma_Y$, the range of the elastic region changes by $2\Delta\sigma_Y$. Therefore, isotropic hardening cannot capture the hysteretic response exhibited by a material subjected to cyclic loading.

To model the hysteresis formed under cyclic loading, we need to introduce a *kinematic hardening.* The simplest form of kinematic hardening is to assume that for every incremental change $\Delta\sigma_Y$ of the yield stress in tension, there is an incremental change $-\Delta\sigma_Y$ of the yield stress in compression, and vice versa. This means that the range of the elastic region does not change, and that the elastic region simply translates. To capture a kinematically hardening response, we need to introduce the notion of a *back stress.*

For the one-dimensional bar problem, assume that yielding is now governed by the translated yield function

$$f(\sigma, \alpha, \sigma_{Y0}) = |\,\sigma - \alpha\,| - \sigma_{Y0} \leq 0 \,. \tag{2.25}$$

Here, α is the back stress ($\alpha = 0$ initially) and is determined as the algebraic mean of the current yield stresses in tension and compression, i.e., the stress coordinate of a point midway between σ_{YT} and $-\sigma_{YC}$ on the stress axis. The current elastic region is then defined by the set

$$\mathsf{E} = \{\sigma \in \mathbb{R} \,|\, f(\sigma, \alpha, \sigma_{Y0}) < 0\} \,, \tag{2.26}$$

with closure

$$\overline{\mathsf{E}} = \{\sigma \in \mathbb{R} \,|\, f(\sigma, \alpha, \sigma_{Y0}) \leq 0\} \,. \tag{2.27}$$

Imposing the consistency condition on the translated yield function gives

$$\dot{f}(\sigma, \alpha, \sigma_{Y0}) = \text{sign}(\sigma - \alpha)(\dot{\sigma} - \dot{\alpha}) - \dot{\sigma}_{Y0} = 0 \quad \Longrightarrow \quad \dot{\sigma} = \dot{\alpha} \tag{2.28}$$

since $\dot{\sigma}_{Y0} = 0$. This last equation states that during plastic yielding the back stress moves by the same amount as the stress point, i.e., the back stress is 'pulled' by the stress point. Therefore, the variation of the back stress may be described by the rate equation

$$\dot{\alpha} = \dot{\gamma}\,\frac{\partial f}{\partial \sigma} \,, \tag{2.29}$$

where $\dot{\gamma} > 0$ is a positive multiplier and $\partial f / \partial \sigma = \text{sign}(\sigma - \alpha)$ defines the sense, either tension or compression, of $\dot{\alpha}$.

To determine the multiplier $\dot{\gamma}$, we argue that since the bar is yielding plastically, then $|\sigma| = \sigma_Y$, where σ_Y is the current yield stress. There are two yield stresses, one in tension and another in compression, so we write

$$\text{sign}(\sigma - \alpha)\,\dot{\sigma} = H'\,|\,\dot{\varepsilon}^{\text{p}}\,| = H'\dot{\lambda}\,, \tag{2.30}$$

where the last part of this equation is obtained from the hardening law (2.16). Substituting (2.29) and (2.30) in $(2.28)_2$ gives

$$\dot{\gamma} = H'\dot{\lambda} > 0\,, \tag{2.31}$$

assuming a hardening response, i.e., $H' > 0$. This means that $\dot{\gamma}$ is not an independent parameter. On substitution of (2.31) into (2.29), we obtain

$$\dot{\alpha} = H'\dot{\lambda}\,\text{sign}(\sigma - \alpha)\,. \tag{2.32}$$

The plastic multiplier $\dot{\lambda}$ can be determined by imposing the consistency condition. Inserting (2.6) and (2.32) into $(2.28)_2$ gives

$$\dot{\lambda} = \frac{\text{sign}(\sigma - \alpha)\,E\,\dot{\varepsilon}}{E + H'} \qquad \Longrightarrow \qquad \dot{\varepsilon}^{\text{p}} = \frac{E}{E + H'}\,\dot{\varepsilon}\,. \tag{2.33}$$

Note that the last equation has the same form as $(2.21)_2$ for isotropic hardening, which means that kinematic hardening produces the same incremental plastic strain as isotropic hardening. However, the two hardening responses differ in the predicted yield stresses on reverse loading, as shown in Fig. 2.2, since kinematic hardening predicts a lower yield stress in compression after yielding in tension. This phenomenon, in which the yield stress in compression decreases with initial yielding in tension, and vice versa, is known as the Bauschinger effect (Bauschinger 1879). The mechanism behind the Bauschinger effect is attributed to a particular kind of residual stresses that form on the grain boundaries of polycrystals, and is related to the dislocation structure in a cold worked metal.

Exercise 2.1. Consider the following yield function for a one-dimensional elastoplastic bar

$$f(\sigma, \alpha, r) = |\,\sigma - \alpha\,| - r \le 0\,,$$

where α is the back stress and r is the distance between the back stress and the yield stresses in tension and compression ($r = \sigma_{Y0}$ initially). A combined isotropic-kinematic hardening can be obtained by setting

$$\dot{r} = \beta H'\dot{\lambda}\,, \qquad \dot{\alpha} = (1 - \beta)H'\dot{\lambda}\,\text{sign}(\sigma - \alpha)\,, \qquad 0 \le \beta \le 1\,.$$

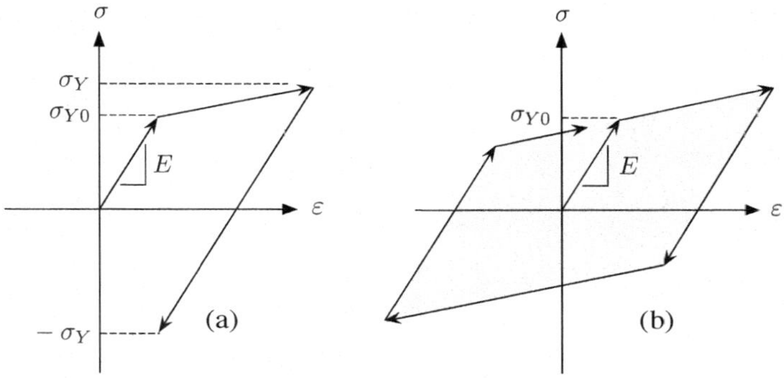

Fig. 2.2 Types of hardening: (a) isotropic, and (b) kinematic

The model reduces to pure kinematic hardening when $\beta = 0$, and to pure isotropic hardening when $\beta = 1$. Using the constitutive equation for $\dot{\sigma}$, the flow rule for $\dot{\varepsilon}^{\mathrm{p}}$, and the consistency condition, show that

$$\dot{\lambda} = \operatorname{sign}(\sigma - \alpha)\frac{E\dot{\varepsilon}}{E + H'} .$$

Determine the final expression for the plastic strain rate $\dot{\varepsilon}^{\mathrm{p}}$ for this combined hardening.

Exercise 2.2. As an illustration of the combined isotropic-kinematic hardening, consider an initially stress-free bar stretched to $\sigma = 2$ kPa, and then compressed to $\sigma = -2$ kPa (ignore buckling). Take $\sigma_{Y0} = 1$ kPa and $E = H' = 100$ kPa. Plot the resulting stress-strain curves for each of the following cases: (a) $\beta = 0$, (b) $\beta = 1/2$, and (c) $\beta = 1$.

2.6 Energy and Plastic Dissipation

This section presents some thermodynamical principles that must be satisfied by the constitutive equations for the one-dimensional bar problem. For simplicity, we assume isotropic hardening and write an expression for the internal energy per unit volume, ψ, herein assumed to take the quadratic form

$$\psi(\varepsilon^{\mathrm{e}}, \xi) = \frac{1}{2}\, E\,(\varepsilon^{\mathrm{e}})^2 + \frac{1}{2}\, H'\,\xi^2 + C\,, \tag{2.34}$$

where C is a constant. The first term represents the recoverable elastic energy per unit volume stored in the bar, whereas the second term represents the effect of plastic deformation. That is, if we stretch the bar until it deforms to twice its initial length, and then compress it back to its initial length, the bar will not be restored to its initial stress state, which gives rise to the second term in the energy function. The quantity ξ is some strain-like plastic internal variable that can be taken for the one-dimensional bar as

$$\xi = \int_0^t \dot{\xi}\,\mathrm{d}t\,, \qquad \dot{\xi} = |\,\dot{\varepsilon}^{\mathrm{p}}\,| = \dot{\lambda}\,. \tag{2.35}$$

Note that ξ is a monotonically increasing function of time.

The reduced dissipation inequality of classical thermodynamics states that the rate of mechanical dissipation must be nonnegative, and for isothermal process we have

$$\mathcal{D} := \sigma\dot{\varepsilon} - \frac{\mathrm{d}\psi}{\mathrm{d}t} \geq 0\,. \tag{2.36}$$

The first term represents the stress power per unit volume, whereas the second term represents the rate of change of internal energy per unit volume. The second term of (2.36) is given by

$$\frac{\mathrm{d}\psi}{\mathrm{d}t} = E\,\varepsilon^{\mathrm{e}}\dot{\varepsilon}^{\mathrm{e}} + H'\xi\dot{\xi} = E\,\varepsilon^{\mathrm{e}}\,(\dot{\varepsilon} - \dot{\varepsilon}^{\mathrm{p}}) + (\sigma_Y - \sigma_{Y0})\,\dot{\xi}\,. \tag{2.37}$$

The second term of (2.37) arises from the assumed form of the hardening law.

Substituting (2.37) into (2.36) and collecting terms gives

$$\mathcal{D} = (\sigma - E\varepsilon^{\mathrm{e}})\dot{\varepsilon} + E\varepsilon^{\mathrm{e}}\dot{\varepsilon}^{\mathrm{p}} - (\sigma_Y - \sigma_{Y0})\dot{\xi} \geq 0\,. \tag{2.38}$$

Since $\dot{\varepsilon}$ can be positive or negative, a necessary condition for the inequality to hold is

$$\sigma - E\varepsilon^{\mathrm{e}} = 0\,. \tag{2.39}$$

Thus, we recover the elastic constitutive equation (2.5) for the bar, see Coleman and Gurtin (1967) for a similar derivation of the stress relation. The dissipation inequality then reduces to the form

$$\mathcal{D} = \sigma\dot{\varepsilon}^{\mathrm{p}} - (\sigma_Y - \sigma_{Y0})\,\dot{\xi} \geq 0\,. \tag{2.40}$$

Using the flow rule, $\dot{\varepsilon}^{\mathrm{p}} = \dot{\lambda}\,\mathrm{sign}(\sigma)$, and noting that $\mathrm{sign}(\sigma)\sigma = |\,\sigma\,|$, we obtain

$$\mathcal{D} = \dot{\lambda}(|\,\sigma\,| - \sigma_Y) + \sigma_{Y0}\dot{\lambda} = \dot{\lambda}f(\sigma,\sigma_Y) + \sigma_{Y0}\dot{\lambda} = \sigma_{Y0}\dot{\lambda} \geq 0\,, \tag{2.41}$$

where the last part of the equation emanates from the Kuhn-Tucker condition. For elastic process, $\dot{\lambda} = 0$ and $\mathcal{D} = 0$, and there is no plastic dissipation (an obvious result). For kinematic hardening a modified form of the internal energy function may be developed to obtain a similar result.

Exercise 2.3. For the isotropically hardening bar of Exercise 2.2, compute the numerical values of the internal energy function ψ at the end of the stretching phase and at the end of the compression phase, assuming $\psi = 0$ initially. What is the significance of having $\mathcal{D} = 0$ when $\sigma_{Y0} > 0$ and $\dot{\lambda} = 0$, or when $\sigma_{Y0} = 0$ and $\dot{\lambda} > 0$?

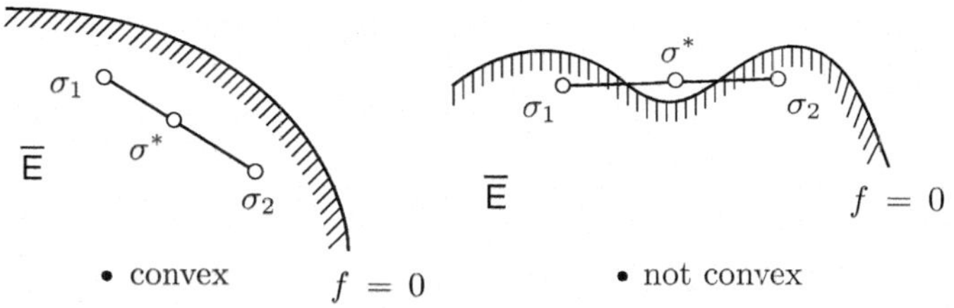

Fig. 2.3 Convex and non-convex yield functions

2.7 Some Properties of the Yield Function

An important property of the yield function is *convexity*, which also implies convexity of the elastic region $\overline{\mathsf{E}}$. For isotropic hardening, convexity of f means

$$f(\sigma^*, \sigma_Y) \leq \eta f(\sigma_1, \sigma_Y) + (1-\eta) f(\sigma_2, \sigma_Y) \tag{2.42}$$

for any σ^* on the line

$$\sigma^* = \eta\sigma_1 + (1-\eta)\sigma_2\,, \quad \eta \in [0,1]\,. \tag{2.43}$$

Convexity of $\overline{\mathsf{E}}$ means that if $f(\sigma_1, \sigma_Y) \leq 0$ and $f(\sigma_2, \sigma_Y) \leq 0$, then $f(\sigma^*, \sigma_Y) \leq 0$. These two relationships are easily proved using the triangle (Minkowski) inequality

$$\begin{aligned} f(\sigma^*, \sigma_Y) &= |\,\sigma^*\,| - \sigma_Y = |\,\eta\sigma_1 + (1-\eta)\sigma_2\,| - \sigma_Y \\ &\leq \eta\,|\,\sigma_1\,| + (1-\eta)\,|\,\sigma_2\,| - \sigma_Y \\ &= \eta f(\sigma_1, \sigma_Y) + (1-\eta) f(\sigma_2, \sigma_Y) \leq 0\,. \end{aligned} \tag{2.44}$$

For kinematic hardening, we obtain a similar result

$$\begin{aligned} f(\sigma^*, \alpha^*) &= |\,\sigma^* - \alpha^*\,| - \sigma_{Y0} = |\,\eta(\sigma_1 - \alpha_1) + (1-\eta)(\sigma_2 - \alpha_2)\,| - \sigma_{Y0} \\ &\leq \eta\,|\,\sigma_1 - \alpha_1\,| + (1-\eta)\,|\,\sigma_2 - \alpha_2\,| - \sigma_{Y0} \\ &= \eta f(\sigma_1, \alpha_1) + (1-\eta) f(\sigma_2, \alpha_2) \leq 0\,. \end{aligned} \tag{2.45}$$

Convex and nonconvex yield functions are shown in Fig. 2.3. Stated in words, $\overline{\mathsf{E}}$ is convex if σ^* is in the elastic region whenever σ_1 and σ_2 are in the elastic region. Convex functionals have many desirable mathematical features as demonstrated throughout this book.

A second important property of the yield function f is that it contains a homogeneous function of degree one. To illustrate this point, we assume isotropic hardening and note that

$$\sigma\frac{\partial f}{\partial \sigma} + \sigma_Y \frac{\partial f}{\partial \sigma_Y} = \sigma\,\mathrm{sign}(\sigma) + \sigma_Y(-1) = |\,\sigma\,| - \sigma_Y = f(\sigma, \sigma_Y)\,. \tag{2.46}$$

Hence, the entire yield function $f(\sigma, \sigma_Y)$ is homogeneous of degree one. On the other hand, for kinematic hardening, we note that

$$\sigma \frac{\partial f}{\partial \sigma} + \alpha \frac{\partial f}{\partial \alpha} = \operatorname{sign}(\sigma - \alpha)(\sigma - \alpha) = |\,\sigma - \alpha\,| = f(\sigma, \alpha) + \sigma_{Y0}\,, \tag{2.47}$$

and so $|\,\sigma - \alpha\,|$ is the relevant homogeneous component of $f(\sigma, \alpha)$. The theorem for homogeneous functions of any degree n is due to Euler (Wylie 1975).

2.8 Algorithm for Isotropic Hardening

A rate form of the constitutive equation is not sufficient to completely define the evolution of the stresses in computer codes. We also need to develop a numerical integration algorithm for describing the evolution of the various state variables over discrete time increments.

Consider the rate equations

$$\dot{\sigma} = E(\dot{\varepsilon} - \dot{\varepsilon}^{\mathrm{p}})\,, \qquad \dot{\varepsilon}^{\mathrm{p}} = \dot{\lambda}\operatorname{sign}(\sigma)\,, \qquad \dot{\sigma}_Y = H'\dot{\lambda}\,. \tag{2.48}$$

Over a typical time interval $t \in [\,t_n, t_{n+1}\,]$, we assume that σ_n and $\sigma_{Y,n}$ are given, and we want to find σ_{n+1} and $\sigma_{Y,n+1}$ for a given strain increment $\Delta\varepsilon$. As a first step, we write the discrete counterpart of $(2.48)_1$ as follows

$$\sigma_{n+1} = \sigma_{n+1}^{\mathrm{tr}} - E\Delta\varepsilon^{\mathrm{p}}\,, \qquad \sigma_{n+1}^{\mathrm{tr}} := \sigma_n + E\Delta\varepsilon\,. \tag{2.49}$$

Here, $\sigma_{n+1}^{\mathrm{tr}}$ is the elastic stress predictor obtained by freezing plastic flow ('tr' = 'trial'), and $E\Delta\varepsilon^{\mathrm{p}}$ is the plastic stress corrector applied to the trial stress. Geometrically, the algorithm maps the trial stress to the actual stress through the plastic corrector stress; hence, it is called a return map (Simo and Hughes 1998).

Next, we write the discrete counterpart of $(2.48)_2$ as

$$\Delta\varepsilon^{\mathrm{p}} = \Delta\lambda\operatorname{sign}(\sigma_{n+1}) = \Delta\lambda\operatorname{sign}(\sigma_{n+1}^{\mathrm{tr}})\,, \tag{2.50}$$

where $\Delta\lambda$ is the discrete counterpart of $\dot{\lambda}$. Inherent in the last equation is the implicit evaluation of $\operatorname{sign}(\sigma_{n+1})$, which is trivially equal to $\operatorname{sign}(\sigma_{n+1}^{\mathrm{tr}})$ for the one-dimensional problem. Finally, the discrete counterpart of $(2.48)_3$ is

$$\sigma_{Y,n+1} = \sigma_{Y,n} + H'\Delta\lambda\,. \tag{2.51}$$

The as yet unknown discrete plastic multiplier $\Delta\lambda$ is now solved by imposing the *discrete consistency condition*

$$f(\sigma_{n+1}, \sigma_{Y,n+1}) = |\,\sigma_{n+1}\,| - \sigma_{Y,n+1} = 0\,, \tag{2.52}$$

which yields the following expression for the discrete plastic multiplier

$$\Delta\lambda = \frac{|\,\sigma_{n+1}^{\text{tr}}\,| - \sigma_{Y,n}}{E + H'} > 0\,. \tag{2.53}$$

Note that $\Delta\lambda > 0$ provided $|\,\sigma_{n+1}^{\text{tr}}\,| > \sigma_{Y,n}$.

The resulting predictor-corrector algorithm is summarized in Box 2.1 and elaborated geometrically in Fig. 2.4. Note that the algorithm does not require that the initial stress σ_n be on f: if the initial stress is inside the elastic region, then the algorithm still calculates the correct amount of plastic strain necessary to map the trial stress to f. An attribute of the algorithm is that it is very easy to implement.

Step 1. Compute $\sigma_{n+1}^{\text{tr}} = \sigma_n + E\Delta\varepsilon$
Step 2. $|\,\sigma_{n+1}^{\text{tr}}\,| - \sigma_{Y,n} > 0$?
No, set $\sigma_{n+1} = \sigma_{n+1}^{\text{tr}}$, $\sigma_{Y,n+1} = \sigma_{Y,n}$ and exit.
Step 3. Yes, compute $\Delta\lambda = (|\,\sigma_{n+1}^{\text{tr}}\,| - \sigma_{Y,n})/(E + H')$
Step 4. Set $\sigma_{n+1} = \sigma_{n+1}^{\text{tr}} - E\Delta\lambda\text{sign}(\sigma_{n+1}^{\text{tr}})$ and
$\sigma_{Y,n+1} = \sigma_{Y,n} + H'\Delta\lambda$ and exit.

Box 2.1 Algorithm for isotropically hardening elastoplastic bar problem

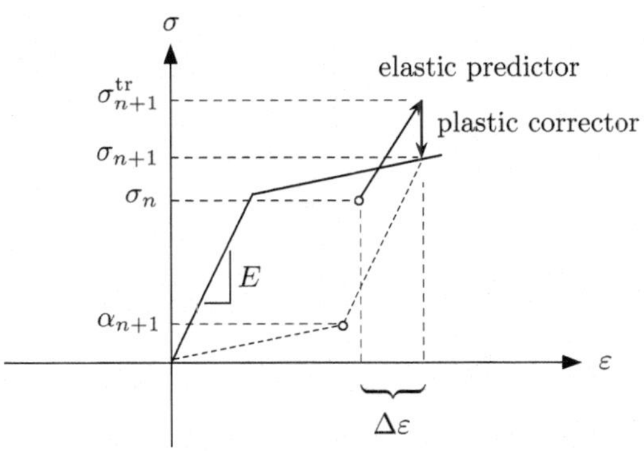

Fig. 2.4 Algorithm for isotropic and kinematic hardening: Beginning at σ_n inside the elastic region, the incremental strain $\Delta\varepsilon$ leads to an elastic stress predictor σ_{n+1}^{tr}. The plastic corrector maps the predictor stress to the final stress σ_{n+1} on the loading branch. The back stress α_{n+1} is always zero for isotropic hardening.

2.9 Algorithm for Kinematic Hardening

We now consider the rate equations

$$\dot{\sigma} = E(\dot{\varepsilon} - \dot{\varepsilon}^{\mathrm{p}}), \qquad \dot{\varepsilon}^{\mathrm{p}} = \dot{\lambda}\, \mathrm{sign}(\sigma - \alpha), \qquad \dot{\alpha} = H'\dot{\lambda}\, \mathrm{sign}(\sigma - \alpha), \quad (2.54)$$

and take the same time interval of interest, $t \in [t_n, t_{n+1}]$. Here, we assume that σ_n, α_n and $\Delta\varepsilon$ are given, and the desired quantities are σ_{n+1} and α_{n+1}. The discrete counterparts of (2.54) are

$$\sigma_{n+1} = \sigma_{n+1}^{\mathrm{tr}} - E\Delta\varepsilon^{\mathrm{p}}, \qquad \sigma_{n+1}^{\mathrm{tr}} := \sigma_n + E\Delta\varepsilon \quad (2.55)$$

for the first part,

$$\Delta\varepsilon^{\mathrm{p}} = \Delta\lambda\, \mathrm{sign}(\sigma_{n+1} - \alpha_{n+1}) = \Delta\lambda\, \mathrm{sign}(\sigma_{n+1}^{\mathrm{tr}} - \alpha_n) \quad (2.56)$$

for the second part, and

$$\alpha_{n+1} = \alpha_n + H'\Delta\lambda\, \mathrm{sign}(\sigma_{n+1} - \alpha_{n+1}) = \alpha_n + H'\Delta\lambda\, \mathrm{sign}(\sigma_{n+1}^{\mathrm{tr}} - \alpha_n) \quad (2.57)$$

for the third part. The algorithm makes use of the simple substitution

$$\mathrm{sign}(\sigma_{n+1} - \alpha_{n+1}) \equiv \mathrm{sign}(\sigma_{n+1}^{\mathrm{tr}} - \alpha_n), \quad (2.58)$$

since both $\sigma_{n+1}^{\mathrm{tr}}$ and α_n are known. In other words, there is no need to know the values of σ_{n+1} and α_{n+1} to determine the correct sign of their difference. The geometric meaning of the algorithm is as follows: the back stress is fixed during the elastic predictor phase, and then it catches up with the stress point during the plastic corrector phase. For this reason, it is called the 'catching-up' algorithm (Moreau 1977).

We now use the discrete consistency condition to solve for the discrete plastic multiplier $\Delta\lambda$,

$$f(\sigma_{n+1}, \alpha_{n+1}, \sigma_{Y0}) = |\,\sigma_{n+1} - \alpha_{n+1}\,| - \sigma_{Y0} = 0, \quad (2.59)$$

where σ_{Y0} is the fixed distance from the back stress to the yield points. This gives

$$\Delta\lambda = \frac{|\,\sigma_{n+1}^{\mathrm{tr}} - \alpha_n\,| - \sigma_{Y0}}{E + H'} > 0, \quad (2.60)$$

where the inequality is true provided $|\,\sigma_{n+1}^{\mathrm{tr}} - \alpha_n\,| > \sigma_{Y0}$. The algorithm is summarized in Box 2.2 and also depicted geometrically in Fig. 2.4.

Exercise 2.4. Develop an algorithm similar to Boxes 2.1 and 2.2 for the combined isotropic-kinematic hardening considered in Exercise 2.1. Identify where the parameter β enters into the algorithm.

Step 1. Compute $\sigma_{n+1}^{\text{tr}} = \sigma_n + E\Delta\varepsilon$
Step 2. $|\,\sigma_{n+1}^{\text{tr}} - \alpha_n\,| > \sigma_{Y0}$?
No, set $\sigma_{n+1} = \sigma_{n+1}^{\text{tr}},\ \ \alpha_{n+1} = \alpha_n$ and exit.
Step 3. Yes, compute $\Delta\lambda = (|\,\sigma_{n+1}^{\text{tr}} - \alpha_n\,| - \sigma_{Y0})/(E + H')$
Step 4. Set $\sigma_{n+1} = \sigma_{n+1}^{\text{tr}} - E\Delta\lambda \text{sign}(\sigma_{n+1}^{\text{tr}} - \alpha_n)$ and
$\alpha_{n+1} = \alpha_n + H'\Delta\lambda\,\text{sign}(\sigma_{n+1}^{\text{tr}} - \alpha_n)$ and exit.

Box 2.2 Algorithm for kinematically hardening elastoplastic bar problem

2.10 Algorithm for Nonlinear Hardening

We now consider the case where H' is a function of the accumulated plastic strain ξ, which means that the internal energy function ψ is no longer quadratic in ξ. From the point of view of the integration algorithm, this means that an iterative strategy must now be employed to determine the discrete plastic multiplier $\Delta\lambda$.

To illustrate how the additional nonlinearity impacts the algorithm, we consider isotropic hardening and write the discrete counterpart of $(2.48)_3$ as

$$\sigma_{Y,n+1} = \sigma_{Y,n} + H'_{n+1}\Delta\lambda\,, \qquad H'_{n+1} = H'|_{\xi_{n+1}}\,. \tag{2.61}$$

Substituting into (2.52) gives

$$f(\Delta\lambda) = |\,\sigma_{n+1}^{\text{tr}}\,| - \sigma_{Y,n} - E\Delta\lambda - H'_{n+1}\Delta\lambda\,. \tag{2.62}$$

We need to calculate the root $\Delta\lambda^*$ that gives $f(\Delta\lambda^*) = 0$. This is easily obtained from a local Newton iteration, where the consistent tangent operator is

$$f'(\Delta\lambda) = -E - H'_{n+1} - \Delta\lambda\,\frac{\mathrm{d}H'}{\mathrm{d}\xi}\bigg|_{n+1}\,. \tag{2.63}$$

Once $\Delta\lambda^*$ is obtained, the algorithm can proceed essentially as in Box 2.1 with the Newton iteration loop now replacing Step 3.

Exercise 2.5. A hyperbolic stress-strain curve is commonly used to fit uniaxial experimental data for some materials like soils (Kondner 1963). The hyperbola takes the form $\sigma = \varepsilon(a + b\varepsilon)^{-1}$, where a and b are material parameters, which have the physical significance that $1/a$ represents the initial tangent modulus at $\varepsilon = 0$, and $1/b$ represents the asymptotic value of σ as $\varepsilon \to \infty$. Cast the hyperbolic model in the framework of an isotropically hardening plasticity model with nonlinear hardening and determine the expression for H' as a function of ξ. Hint: Assume $1/a = E$ and a vanishing initial elastic region, $\mathsf{E}_\varnothing = \varnothing$.

Exercise 2.6. A three-parameter stress-strain curve has been proposed by Ramberg and Osgood (1945) for aluminum alloy and carbon steel. The analytical expression is $\varepsilon = \sigma/E + K(\sigma/E)^n$, where E, K, and n are constant.

Determine expressions for the tangential modulus, $E_T = \mathrm{d}\sigma/\mathrm{d}\varepsilon$, as a function of ε, and the plastic modulus, H', as a function of ξ, assuming $\mathsf{E}_\varnothing = \varnothing$ and an instantaneous elastic response at $\sigma = 0$.

2.11 Uniqueness of Solution

Suppose that a specimen of a certain material is deformed by a set of applied surface forces and/or prescribed surface displacements, and that infinitesimal increments of stresses and/or displacements are subsequently applied on the surface of this same material. We want to determine the corresponding incremental stresses and strains throughout the specimen, as well as the circumstances in which they are uniquely determined by the given boundary conditions and the stress-strain relation for the material.

To illustrate the method of analysis for uniqueness of the solution, we again consider the one-dimensional elastoplastic bar problem and write the differential equation of equilibrium as

$$\frac{\partial \sigma}{\partial x} + \rho \mathrm{g} = 0\,, \qquad x \in (0, L)\,, \tag{2.64}$$

where ρ is the mass density of the bar, g is the gravity acceleration constant, and L is the length of the bar. We assume that the boundary conditions are prescribed at $x = 0$ and at $x = L$ either in the form of load or displacement. Now, consider two possibly distinct stress increments, herein expressed in rate form, $\dot{\sigma}$ and $\dot{\sigma}^*$, arising from two possibly distinct strain increments $\dot{\varepsilon}$ and $\dot{\varepsilon}^*$. For dead loading, the stress rates satisfy the equations

$$\frac{\partial \dot{\sigma}}{\partial x} = 0\,, \qquad \frac{\partial \dot{\sigma}^*}{\partial x} = 0\,, \qquad x \in (0, L)\,. \tag{2.65}$$

Furthermore, the strain rates satisfy the compatibility conditions

$$\dot{\varepsilon} = \frac{\partial v}{\partial x}\,, \qquad \dot{\varepsilon}^* = \frac{\partial v^*}{\partial x}\,, \qquad x \in (0, L)\,, \tag{2.66}$$

where v and v^* are two possibly distinct velocity distributions.

Following Hill (1950), we consider the integral

$$\begin{aligned} \int_L (\dot{\sigma} - \dot{\sigma}^*)(\dot{\varepsilon} - \dot{\varepsilon}^*)\,\mathrm{d}x \;&=\; \int_L \frac{\partial}{\partial x}\Big[(\dot{\sigma} - \dot{\sigma}^*)(v - v^*)\Big]\,\mathrm{d}x \\ &- \int_L (v - v^*)\,\frac{\partial}{\partial x}\,(\dot{\sigma} - \dot{\sigma}^*)\,\mathrm{d}x\,. \end{aligned} \tag{2.67}$$

The first integral on the right-hand side can be converted into a point integral by Gauss theorem, which vanishes, i.e.,

$$\int_L \frac{\partial}{\partial x}\Big[(\dot{\sigma} - \dot{\sigma}^*)(v - v^*)\Big]\,\mathrm{d}x = (\dot{\sigma} - \dot{\sigma}^*)(v - v^*)\Big|_{x=0,L} = 0\,, \tag{2.68}$$

since the prescribed boundary conditions are the same for both solutions. Furthermore, the second integral on the right-hand side also vanishes since the two stress rates are required to satisfy the same equilibrium condition.

Therefore, using the constitutive equations $\dot{\sigma} = E_T\dot{\varepsilon}$ and $\dot{\sigma}^* = E_T\dot{\varepsilon}^*$, the remaining integral on the left-hand side of (2.67) simplifies to the form

$$\int_L (\dot{\sigma} - \dot{\sigma}^*)(\dot{\varepsilon} - \dot{\varepsilon}^*)\, \mathrm{d}x = \int_L E_T\, (\dot{\varepsilon} - \dot{\varepsilon}^*)^2\, \mathrm{d}x = \int_L E_T \dot{\phi}^2\, \mathrm{d}x = 0\,, \tag{2.69}$$

where $\dot{\phi} = \dot{\varepsilon} - \dot{\varepsilon}^*$ is a possible discontinuity, or jump, in the strain rate. In the elastic regime, $E_T = E > 0$, and so $\dot{\phi}$ must necessarily be zero, which means that there is only one possible solution to the boundary value problem. If the tangential modulus E_T varies in a smooth fashion from its initial elastic value E, then multiple solutions are possible once the tangential modulus vanishes, i.e., $E_T = 0$. This critical condition is the *bifurcation point* and signals the loss of uniqueness of the solution to the boundary-value problem.

2.12 Solved Problem

We consider an elastoplastic bar with combined isotropic and kinematic hardening. The material parameters are $E = 100$ MPa, $E_T = 0.2E$, and the initial uniaxial yield stress is $\sigma_{Y0} = 1$ MPa. The bar is subjected to cyclic uniaxial stress of $\sigma = \pm 2.0$ MPa starting from an initial back stress $\alpha = 0$. Figure 2.5 shows the resulting cyclic stress-strain curves.

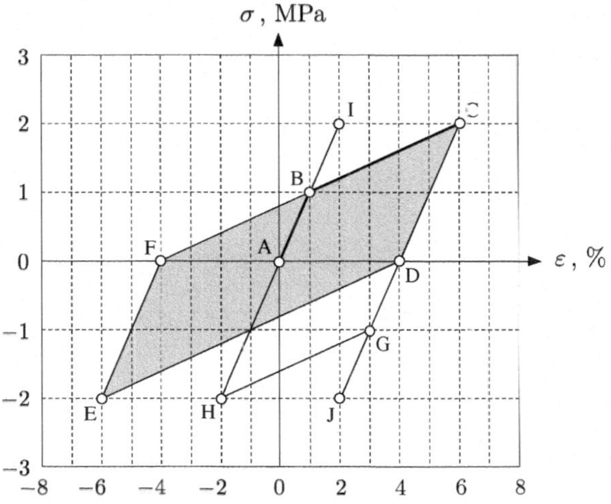

Fig. 2.5 Stress-strain curves for elastoplastic bar subjected to cyclic uniaxial stress of $\sigma = \pm 2.0$ MPa. Heavier line ABC is the initial response at the beginning of loading; shaded region CDEF defines a fixed hysteretic loop for $\beta = 0$.

From Exercise 2.1, the yield criterion used to generate the stress-strain curves of Figure 2.5 is

$$f(\sigma, \alpha, r) = |\,\sigma - \alpha\,| - r = 0\,,$$

where $r = \sigma_{Y0}$ initially. Combined isotropic-kinematic hardening is obtained from the evolution equations

$$\dot{r} = \beta H'\dot{\lambda}\,, \qquad \dot{\alpha} = (1-\beta)H'\dot{\lambda}\,\text{sign}(\sigma - \alpha)\,, \qquad 0 \le \beta \le 1\,,$$

where $\beta = 0$ for pure kinematic hardening, and $\beta = 1$ for pure isotropic hardening. At the beginning of loading the modeled response is given by the heavier line ABC shown in Fig. 2.5, independent of β. However, the unloading/reloading responses are dependent on the value of β.

For $\beta = 0$ (kinematic hardening), r remains constant at 1 MPa. So, when the bar is unloaded at C the yield point on reverse loading is encountered at D. Thereafter, the steady-state stress-strain path settles into a fixed loop CDEF (shaded region). For $\beta = 1/2$ (combined hardening), the accumulated plastic strain at C is $\Delta\lambda = 0.04$, which gives $\Delta r = \Delta\alpha = 0.5$ MPa. Thus, r increases to 1.5 MPa while α increases to 0.5 MPa at point C. On reverse loading the yield point is encountered at G, and at point H the variable r increases further to 2 MPa while α goes back to zero. The stress-strain path makes one complete loop before settling into a purely elastic response HI passing through the origin. Finally, for $\beta = 1$ (isotropic hardening), r expands to 2 MPa at point C, and so we observe a long-term purely elastic response CJ, but this time with a residual plastic strain of 4%.

2.13 Summary

This chapter introduces the following main ingredients of the mathematical theory of plasticity in the context of a one-dimensional bar problem: elastic region, yield function, yield stress, yield criterion, consistency condition, isotropic hardening, kinematic hardening, back stress, loading and unloading conditions, and the flow rule. From a computational standpoint, the chapter also introduces the notion of return mapping, also called the predictor-corrector algorithm, that will prove useful once the theory is extended to multidimensional problems.

3 J2 Plasticity

3.1 Introduction

During his studies on plastic yielding of metals, Tresca (1864) conducted experiments on punching and extrusion through dies of various shapes. He concluded that yielding in metals would occur when the maximum shear stress reaches a certain yield value. This maximum shear stress is the difference between the major principal stress σ_1 and minor principal stress σ_3. Tresca stated his yield criterion in the form

$$(\sigma_1 - \sigma_3)_{\max} = Y\,, \tag{3.1}$$

where $Y > 0$ is the yield stress. Almost fifty years later, von Mises (1913) suggested that the yielding of metals is governed by the second invariant of the deviatoric stress tensor, denoted as J_2. Compared to the Tresca yield criterion, the von Mises yield criterion is known to be in better agreement with experiments for ductile metals such as copper, nickel, aluminum, and alloy steels.

Figure 3.1 shows the results of combined torsion-tension tests conducted by Taylor and Quinney (1931) on three different types of metal that have been work-hardened to the same state. Under this stress condition the Tresca yield criterion is $\sigma^2 + 4\tau^2 = \sigma_Y^2$, whereas the von Mises yield criterion is $\sigma^2 + 3\tau^2 = \sigma_Y^2$, where σ is the axial stress, τ is the shear stress, and σ_Y is the tensile yield stress. For these three metals, yielding has been predicted more accurately by the von Miss criterion particularly under pure shear. This observation is typical and is not specific to these tests. In addition to the accuracy of the von Mises criterion in predicting yielding of ductile materials such as metals, it is also defined by a smooth yield function that can be implemented more easily into nonlinear finite element codes. This chapter thus focuses on the theoretical formulation and numerical implementation of the von Mises criterion, which we shall simply refer to as "J_2 plasticity" throughout this work.

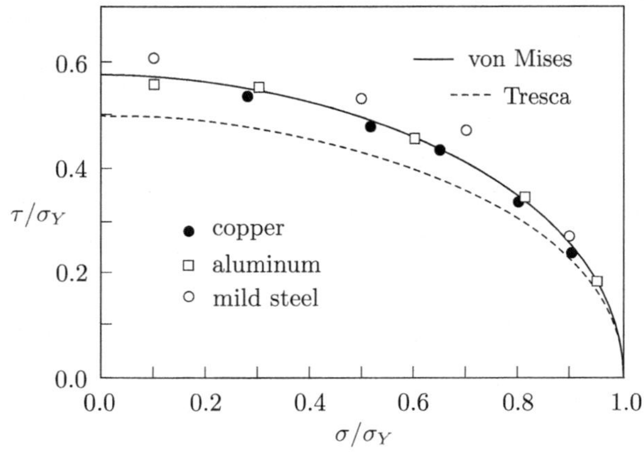

Fig. 3.1 Experimental results of Taylor and Quinney (1931) from combined tension and torsion tests. Each metal was work-hardened to the same state for all tests.

3.2 The J2 Yield Criterion

We denote the symmetric Cauchy stress tensor by $\boldsymbol{\sigma}$ with components σ_{ij}. An additive decomposition of this tensor into volumetric and deviatoric parts gives

$$\boldsymbol{\sigma} = p\mathbf{1} + \boldsymbol{s}\,, \tag{3.2}$$

where $p = \mathrm{tr}(\boldsymbol{\sigma})/3$ is the mean normal stress, $\mathbf{1}$ is the second-rank identity tensor with components δ_{ij}, the Kronecker delta, $\boldsymbol{s}$ is the deviatoric stress tensor satisfying the condition $\mathrm{tr}(\boldsymbol{s}) = 0$, and 'tr' is the trace operator. A similar additive decomposition of the infinitesimal strain tensor $\boldsymbol{\epsilon}$ gives

$$\boldsymbol{\epsilon} = \frac{1}{3}\,\epsilon_v\mathbf{1} + \boldsymbol{e}\,, \tag{3.3}$$

where $\epsilon_v = \mathrm{tr}(\boldsymbol{\epsilon})$ is the volumetric strain, and $\boldsymbol{e}$ is the deviatoric strain tensor satisfying the condition $\mathrm{tr}(\boldsymbol{e}) = 0$.

For isotropic linearly elastic material, the elastic constitutive equations are

$$p = K\epsilon_v\,, \qquad \boldsymbol{s} = 2\mu\boldsymbol{e}\,, \tag{3.4}$$

where K and μ are the elastic bulk and shear moduli, respectively. This gives

$$\boldsymbol{\sigma} = K\epsilon_v\mathbf{1} + 2\mu\boldsymbol{e} = \boldsymbol{c}^{\mathrm{e}} : \boldsymbol{\epsilon}\,, \tag{3.5}$$

where

$$\boldsymbol{c}^{\mathrm{e}} = K\mathbf{1}\otimes\mathbf{1} + 2\mu\Big(\boldsymbol{I} - \frac{1}{3}\,\mathbf{1}\otimes\mathbf{1}\Big) \tag{3.6}$$

is the rank-four tensor of elastic moduli. Here, we have used the symbol $\otimes$ to form a dyad such that $(\boldsymbol{a}\otimes\boldsymbol{b})_{ijkl} = a_{ij}b_{kl}$ for any second-order symmetric tensors $\boldsymbol{a}$ and $\boldsymbol{b}$. The symbol : denotes an inner product $(\boldsymbol{c} : \boldsymbol{\epsilon})_{ij} = c_{ijkl}\epsilon_{kl}$, while the symbol $\boldsymbol{I}$ denotes the rank-four symmetric identity operator with components $(\boldsymbol{I})_{ijkl} = (\delta_{ik}\delta_{jl} + \delta_{il}\delta_{jk})/2$.

Throughout this chapter we shall refer to J_2 as the second invariant of the deviatoric stress tensor $\boldsymbol{s}$,

$$J_2 = \frac{1}{2}\,\boldsymbol{s} : \boldsymbol{s} = \frac{1}{2}\,s_{ij}s_{ij}\,. \tag{3.7}$$

In principal stress axes,

$$J_2 = \frac{1}{6}\,[\,(\sigma_1 - \sigma_2)^2 + (\sigma_2 - \sigma_3)^2 + (\sigma_1 - \sigma_3)^2\,]\,. \tag{3.8}$$

For uniaxial loading, $J_2 = \sigma^2/3$, where σ is the uniaxial stress; for biaxial loading, $J_2 = (\sigma_1 - \sigma_3)^2/3$, where σ_1 is the axial stress and σ_3 is the radial stress.

Consider the yield function

$$f(\boldsymbol{\sigma}, \kappa) = \sqrt{2J_2} - \kappa \le 0\,, \tag{3.9}$$

where $\sqrt{2J_2} = \|\boldsymbol{s}\|$ and $\kappa > 0$ is a stress-like plastic internal variable (Lubliner 2008). The elastic region is defined by the set

$$\mathsf{E} = \{(\boldsymbol{\sigma}, \kappa) \in \mathbb{S} \times \mathbb{R}^1) \,|\, f(\boldsymbol{\sigma}, \kappa) < 0\}\,, \tag{3.10}$$

with closure

$$\overline{\mathsf{E}} = \{(\boldsymbol{\sigma}, \kappa) \in \mathbb{S} \times \mathbb{R}^1 \,|\, f(\boldsymbol{\sigma}, \kappa) \le 0\}\,, \tag{3.11}$$

where $\mathbb{S}$ is the space of linear, second-order symmetric tensors. Setting $f(\boldsymbol{\sigma}, \kappa) = 0$ defines the *yield surface* for J_2 plasticity.

Let $\boldsymbol{\sigma}_1$ and $\boldsymbol{\sigma}_2$ be any two stress points in $\overline{\mathsf{E}}$, and let $\boldsymbol{\sigma}^* = \eta\boldsymbol{\sigma}_1 + (1-\eta)\boldsymbol{\sigma}_2$, where $\eta \in [0, 1]$. It follows that $\boldsymbol{s}^* = \eta\boldsymbol{s}_1 + (1-\eta)\boldsymbol{s}_2$. Using the triangle inequality,

$$\begin{aligned} f(\boldsymbol{\sigma}^*, \kappa) &= \|\boldsymbol{s}^*\| - \kappa = \|\eta\boldsymbol{s}_1 + (1-\eta)\boldsymbol{s}_2\| - \kappa \\ &\le \eta\,\|\boldsymbol{s}_1\| + (1-\eta)\,\|\boldsymbol{s}_2\| - \kappa \\ &= \eta f(\boldsymbol{\sigma}_1, \kappa) + (1-\eta) f(\boldsymbol{\sigma}_2, \kappa) \le 0\,, \end{aligned} \tag{3.12}$$

since $f(\boldsymbol{\sigma}_1, \kappa) \le 0$ and $f(\boldsymbol{\sigma}_2, \kappa) \le 0$, by hypothesis. Therefore, f is a convex function.

In principal stress space, the yield function can be written in the form

$$f(\sigma_1, \sigma_2, \sigma_3, \kappa) = \frac{1}{\sqrt{3}}\,[\,(\sigma_1 - \sigma_2)^2 + (\sigma_2 - \sigma_3)^2 + (\sigma_1 - \sigma_3)^2\,]^{1/2} - \kappa \le 0\,. \tag{3.13}$$

Since $\kappa > 0$, then $f < 0$ at any point on the space diagonal defined by the equation

$$\sigma_1 = \sigma_2 = \sigma_3\,. \tag{3.14}$$

This means that the space diagonal is contained in the elastic region, see Fig. 3.2. Furthermore, let $f(\sigma_1, \sigma_2, \sigma_3, \kappa) = 0$; then $f(\sigma_1 + \alpha, \sigma_2 + \alpha, \sigma_3 + \alpha, \kappa) = 0$. This means that a translation of a stress point initially on the yield surface in the direction of the space diagonal does not make it move away from the yield surface.

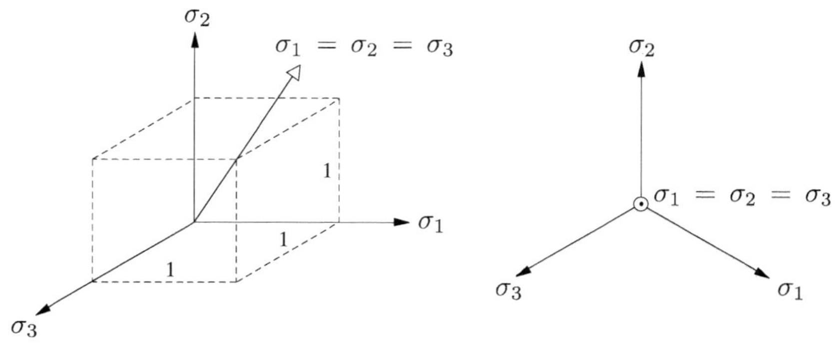

Fig. 3.2 Space diagonal and the π-plane

Next, we rotate the coordinate axes according to the sequence shown in Fig. 3.3. The sequence consists of a positive rotation of $\pi/4$ about the σ_2-axis creating the transformation $(\sigma_1, \sigma_2, \sigma_3) \to (\sigma_1', \sigma_2', \sigma_3')$, followed by a negative rotation of $\theta = \cos^{-1}\sqrt{2/3}$ about the σ_1'-axis producing the transformation $(\sigma_1', \sigma_2', \sigma_3') \to (\sigma_1'', \sigma_2'', \sigma_3'')$. This results in the rotated σ_3''-axis coinciding with the space diagonal. This sequence of rotation is described by a multiplicative matrix equation of the form

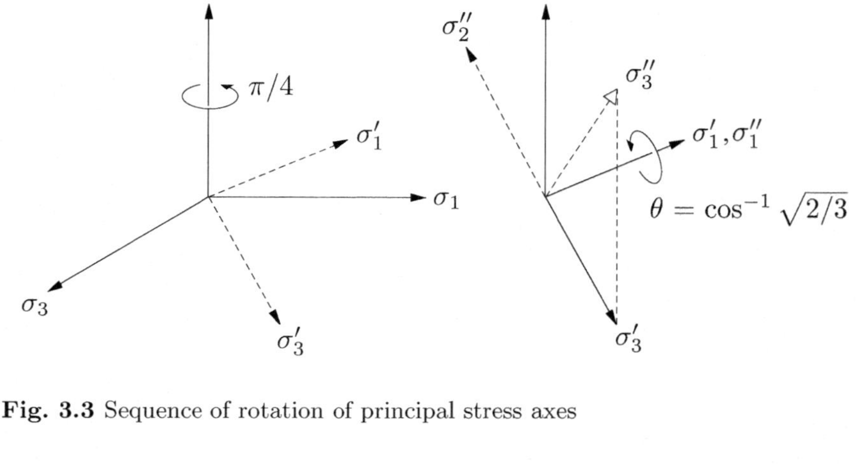

Fig. 3.3 Sequence of rotation of principal stress axes

$$\begin{Bmatrix} \sigma_1 \\ \sigma_2 \\ \sigma_3 \end{Bmatrix} = \begin{bmatrix} \sqrt{2}/2 & 0 & \sqrt{2}/2 \\ 0 & 1 & 0 \\ -\sqrt{2}/2 & 0 & \sqrt{2}/2 \end{bmatrix} \begin{bmatrix} 1 & 0 & 0 \\ 0 & \sqrt{2/3} & 1/\sqrt{3} \\ 0 & -1/\sqrt{3} & \sqrt{2/3} \end{bmatrix} \begin{Bmatrix} \sigma_1'' \\ \sigma_2'' \\ \sigma_3'' \end{Bmatrix} . \tag{3.15}$$

With respect to the rotated stress frame, the principal stress differences of the Tresca yield criterion can be written as

$$\left.\begin{aligned} \sigma_1 - \sigma_2 &= \frac{\sqrt{2}}{2}\sigma_1'' - \sqrt{\frac{3}{2}}\sigma_2'' \\ \sigma_1 - \sigma_3 &= \sqrt{2}\,\sigma_1'' \\ \sigma_2 - \sigma_3 &= \frac{\sqrt{2}}{2}\sigma_1'' + \sqrt{\frac{3}{2}}\sigma_2'' \end{aligned}\right\} . \tag{3.16}$$

With respect to the same rotated stress frame, the J_2 yield function takes the form

$$f(\sigma_1'', \sigma_2'', \sigma_3'', \kappa) = [\,(\sigma_1'')^2 + (\sigma_2'')^2\,]^{1/2} - \kappa \le 0 . \tag{3.17}$$

We recall that neither the Tresca nor the J_2 yield criterion is a function of the mean normal stress, so the term σ_3'' does not appear in any of the above expressions. In principal stress space, the yield surface $f = 0$ is a right circular cylinder of radius $R = \kappa$ and centered about the space diagonal, see Figs. 3.4 and 3.5.

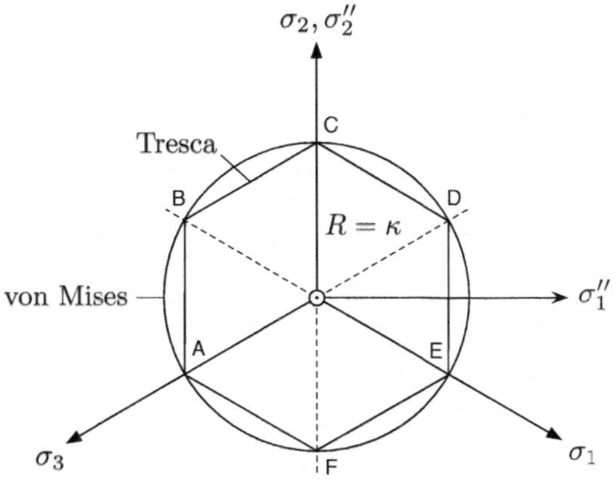

Fig. 3.4 Tresca and J_2 yield surfaces on deviatoric plane

The plane defined by the equation $\sigma_3'' = 0$, or by the equation

$$\sigma_1 + \sigma_2 + \sigma_3 = 0 , \tag{3.18}$$

is called the π-plane. It is a plane perpendicular to the space diagonal and passing through the origin of the principal stress axes. On the π-plane the

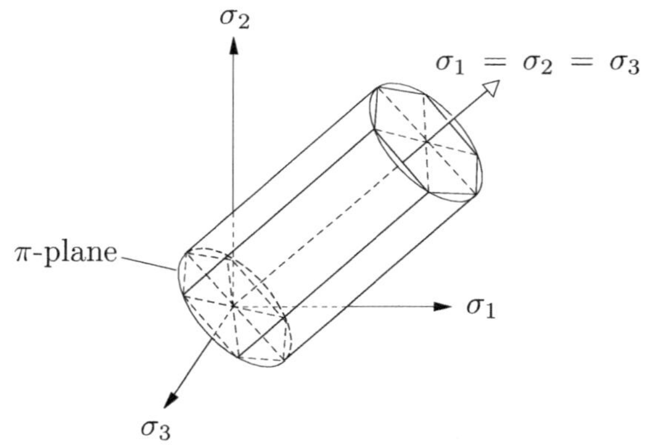

Fig. 3.5 Tresca and J_2 yield surfaces in principal stress space

J_2 yield surface plots as a circle of radius κ. Principal stress coordinates are shortened by a factor of $\sqrt{2/3}$ when projected onto the π-plane.

For the Tresca yield criterion the elastic region is defined by the following inequalities

$$\left.\begin{aligned} |\,\sigma_1 - \sigma_2\,| &= \left|\,\frac{\sqrt{2}}{2}\,\sigma_1'' - \sqrt{\frac{3}{2}}\,\sigma_2''\,\right| \leq Y \\ |\,\sigma_1 - \sigma_3\,| &= \left|\,\sqrt{2}\,\sigma_1''\,\right| \leq Y \\ |\,\sigma_2 - \sigma_3\,| &= \left|\,\frac{\sqrt{2}}{2}\,\sigma_1'' + \sqrt{\frac{3}{2}}\,\sigma_2''\,\right| \leq Y \end{aligned}\right\} . \tag{3.19}$$

The geometric region on the plane (σ_1'', σ_2'') defined by these inequalities is a right hexagonal cylinder whose generating planes are all parallel to the space diagonal, see Figs. 3.4 and 3.5. The Tresca yield stress Y has been chosen in these figures so that the outer corners of the hexagon intersect the circle generated by the J_2 yield surface, i.e., $Y = \sqrt{3/2}\,\kappa$.

Exercise 3.1. Sketch two-dimensional representations of the Tresca and J_2 yield criteria on the plane defined by the axes $(\sigma_1 - \sigma_3)$ and $(\sigma_2 - \sigma_3)$. Note that when a yield function depends on the second and third invariants alone (see Chapter 4 for further discussion), it can be written in terms of the above stress differences.

Exercise 3.2. Express the Tresca and J_2 yield parameters Y and κ in terms of the uniaxial yield stress σ_Y. Then, develop expressions for the two yield criteria in terms of the normal stress σ and shear stress τ for a combined tension/torsion test.

Exercise 3.3. The J_2 yield criterion is equivalent to yielding when the recoverable elastic energy of distortion, U^D, reaches a critical value (Hencky

1924). Show that if this distortional energy is $U^D = s_{ij}e_{ij}/2 = \boldsymbol{s} : \boldsymbol{e}/2$, the critical value of U^D is $U^D_{\mathrm{cr}} = \kappa^2/4\mu$, where μ is the elastic shear modulus and κ is the J_2 yield parameter. Derive an expression for the elastic energy of expansion/compression.

Exercise 3.4. Determine the minimum value of the Tresca yield parameter Y as a function of the J_2 yield parameter κ so that the circular cylinder defined by the J_2 yield criterion is fully inscribed in the hexagonal cylinder defined by the Tresca yield criterion.

Exercise 3.5. Show that the elastic region defined by the Tresca yield criterion is convex.

3.3 Perfect Plasticity

Some historical developments of the perfectly plastic J_2 theory date back to the works by de Saint-Venant (1870) and Lévy (1871), who proposed a flow rule for rigid-plastic materials. It was later extended to elastoplastic materials by Prandtl (1924) for two-dimensional problems, and Reuss (1930) for three-dimensional problems. In what follows, we summarize the theoretical formulations of the elastic-perfectly plastic J_2 theory.

Consider an additive decomposition of the infinitesimal strain tensor into elastic and plastic parts

$$\boldsymbol{\epsilon} = \boldsymbol{\epsilon}^{\mathrm{e}} + \boldsymbol{\epsilon}^{\mathrm{p}} . \tag{3.20}$$

We can write the elastic constitutive equation as

$$\boldsymbol{\sigma} = \boldsymbol{c}^{\mathrm{e}} : \boldsymbol{\epsilon}^{\mathrm{e}} = \boldsymbol{c}^{\mathrm{e}} : (\boldsymbol{\epsilon} - \boldsymbol{\epsilon}^{\mathrm{p}}) , \tag{3.21}$$

and the J_2 yield surface as

$$f(\boldsymbol{\sigma}, \kappa) = \sqrt{2J_2} - \kappa = \|\boldsymbol{s}\| - \kappa = 0 . \tag{3.22}$$

In the three-dimensional case, the flow rule determines the plastic flow direction. For example, consider the case where the yield function f defines the *plastic potential function*. Then,

$$\dot{\boldsymbol{\epsilon}}^{\mathrm{p}} = \dot{\lambda}\, \frac{\partial f}{\partial \boldsymbol{\sigma}} = \dot{\lambda} \widehat{\boldsymbol{n}} , \qquad \widehat{\boldsymbol{n}} = \boldsymbol{s}/\|\boldsymbol{s}\| = \boldsymbol{s}/\kappa . \tag{3.23}$$

This expression for the plastic flow is due to Prandtl and Reuss (Hill 1950). We recall from Chapter 2 that for one-dimensional problem the plastic flow direction is simply a sign function (± 1) defining the sense of the plastic flow. In the three-dimensional case, the flow rule now defines a symmetric tensor of plastic strain rate, with unit direction $\widehat{\boldsymbol{n}}$ and norm $\dot{\lambda}$. Note that $\mathrm{tr}(\dot{\boldsymbol{\epsilon}}^{\mathrm{p}}) = \dot{\lambda}\, \mathrm{tr}(\widehat{\boldsymbol{n}}) = 0$, which means that the plastic flow defines an isochoric deformation, i.e., there is no plastic volume change.

As before, loading and unloading are governed by the Kuhn-Tucker conditions

$$\dot{\lambda} \geq 0\,, \qquad f(\boldsymbol{\sigma}, \kappa) \leq 0\,, \qquad \dot{\lambda} f(\boldsymbol{\sigma}, \kappa) = 0\,. \tag{3.24}$$

Differentiating the last equation gives the consistency condition

$$\dot{f}(\boldsymbol{\sigma}, \kappa) = 0\,. \tag{3.25}$$

Plastic loading is defined by $\dot{\lambda} > 0$ and $f(\boldsymbol{\sigma}, \kappa) = \dot{f}(\boldsymbol{\sigma}, \kappa) = 0$, whereas elastic deformation is defined by $f(\boldsymbol{\sigma}, \kappa) < 0$ and $\dot{\lambda} = 0$. There exists a special case, $\dot{\lambda} = 0$ and $f(\boldsymbol{\sigma}, \kappa) = \dot{f}(\boldsymbol{\sigma}, \kappa) = 0$, called neutral loading, where no plastic deformation is produced by the tangential motion of the stress point on the yield surface.

For J_2 plasticity, a constant κ defines a perfectly plastic response for which the consistency condition can be written as

$$\dot{f}(\boldsymbol{\sigma}, \kappa) = \frac{\partial f}{\partial \boldsymbol{\sigma}} : \dot{\boldsymbol{\sigma}} = \widehat{\boldsymbol{n}} : \dot{\boldsymbol{\sigma}} = \widehat{\boldsymbol{n}} : \dot{\boldsymbol{s}} = 0\,. \tag{3.26}$$

Here, we have used the identity $\partial(\|\boldsymbol{s}\|)/\partial\boldsymbol{\sigma} = \boldsymbol{s}/\|\boldsymbol{s}\| = \widehat{\boldsymbol{n}}$. From the elastic constitutive equation $(3.4)_2$, we write

$$\dot{\boldsymbol{s}} = 2\mu\dot{\boldsymbol{e}}^{\mathrm{e}} = 2\mu(\dot{\boldsymbol{e}} - \dot{\boldsymbol{e}}^{\mathrm{p}}) = 2\mu(\dot{\boldsymbol{e}} - \dot{\lambda}\widehat{\boldsymbol{n}})\,, \tag{3.27}$$

where $\boldsymbol{e}$ is the deviatoric component of $\boldsymbol{\epsilon}$. Substituting into (3.26) and noting that $\mathrm{tr}(\widehat{\boldsymbol{n}}) = 0$ gives

$$\dot{\lambda} = \widehat{\boldsymbol{n}} : \dot{\boldsymbol{e}} = \widehat{\boldsymbol{n}} : \dot{\boldsymbol{\epsilon}} \qquad \Longrightarrow \qquad \dot{\boldsymbol{\epsilon}}^{\mathrm{p}} = \dot{\lambda}\widehat{\boldsymbol{n}} = (\widehat{\boldsymbol{n}} \otimes \widehat{\boldsymbol{n}}) : \dot{\boldsymbol{\epsilon}}\,. \tag{3.28}$$

Inserting back into the elastic constitutive equation gives

$$\dot{\boldsymbol{\sigma}} = \boldsymbol{c}^{\mathrm{e}} : (\dot{\boldsymbol{\epsilon}} - \dot{\boldsymbol{\epsilon}}^{\mathrm{p}}) = \boldsymbol{c}^{\mathrm{ep}} : \dot{\boldsymbol{\epsilon}}\,, \tag{3.29}$$

where $\boldsymbol{c}^{\mathrm{ep}}$ is the rank-four elastic-perfectly plastic tangent tensor given by

$$\boldsymbol{c}^{\mathrm{ep}} = K\mathbf{1} \otimes \mathbf{1} + 2\mu\Big(\boldsymbol{I} - \frac{1}{3}\mathbf{1} \otimes \mathbf{1} - \widehat{\boldsymbol{n}} \otimes \widehat{\boldsymbol{n}}\Big)\,. \tag{3.30}$$

Comparing with the one-dimensional formulation of Chapter 2, $\boldsymbol{c}^{\mathrm{e}}$ may be analogized to Young's modulus of elasticity E, whereas $\boldsymbol{c}^{\mathrm{ep}}$ may be viewed as the counterpart of the tangential modulus E_T. Note that whereas $E_T = 0$ for perfect plasticity, $\boldsymbol{c}^{\mathrm{ep}} \neq \mathbf{0}$ in two or three dimensions, even if the elastic region does not change. Perfect plasticity in two or three dimensions simply constrains the stress point to move on a stationary yield surface.

3.4 Radial Return Algorithm

Motivated by the shape of the yield surface, which is a right circular cylinder in principal stress space, Wilkins (1964) proposed a radial return algorithm to numerically integrate the rate constitutive equations for J_2 plasticity. The format of calculations is the same as in Chapter 2, i.e., the stress tensor $\boldsymbol{\sigma}_n$ and the strain increment $\Delta\boldsymbol{\epsilon}$ are given, and the goal is to calculate the stress tensor $\boldsymbol{\sigma}_{n+1}$. The rate equations to be integrated are

$$\dot{\boldsymbol{\sigma}} = \boldsymbol{c}^{\mathrm{e}} : (\dot{\boldsymbol{\epsilon}} - \dot{\boldsymbol{\epsilon}}^{\mathrm{p}}) , \qquad \dot{\boldsymbol{\epsilon}}^{\mathrm{p}} = \dot{\lambda}\widehat{\boldsymbol{n}} , \tag{3.31}$$

where κ = constant (perfect plasticity).

A first step in the numerical integration is to write the stress tensor in the predictor-corrector format,

$$\boldsymbol{\sigma}_{n+1} = \boldsymbol{\sigma}_{n+1}^{\mathrm{tr}} - \boldsymbol{c}^{\mathrm{e}} : \Delta\boldsymbol{\epsilon}^{\mathrm{p}} , \qquad \boldsymbol{\sigma}_{n+1}^{\mathrm{tr}} = \boldsymbol{\sigma}_n + \boldsymbol{c}^{\mathrm{e}} : \Delta\boldsymbol{\epsilon} . \tag{3.32}$$

Using the backward implicit scheme for integrating the incremental plastic flow, we get

$$\Delta\boldsymbol{\epsilon}^{\mathrm{p}} = \int_{t_n}^{t_{n+1}} \dot{\lambda}\widehat{\boldsymbol{n}}\,\mathrm{d}t \approx \Delta\lambda\,\widehat{\boldsymbol{n}}_{n+1} , \qquad \widehat{\boldsymbol{n}}_{n+1} = \boldsymbol{s}_{n+1}/\|\boldsymbol{s}_{n+1}\| , \tag{3.33}$$

where $\Delta\lambda$ is the discrete plastic multiplier. Whereas $\dot{f} = 0$ is the consistency condition for the rate equation, the discrete consistency condition is now written as

$$f_{n+1} = \|\boldsymbol{s}_{n+1}\| - \kappa = 0 , \tag{3.34}$$

where $\boldsymbol{s}_{n+1}$ is the final deviatoric stress tensor. Enforcing the consistency condition at the end of the loading step ensures that the stress point would not drift away from the yield surface.

Next we take the trace of both sides of equation (3.32) to obtain

$$\mathrm{tr}(\boldsymbol{\sigma}_{n+1}) = \mathrm{tr}(\boldsymbol{\sigma}_{n+1}^{\mathrm{tr}}) \quad \Longrightarrow \quad p_{n+1} = p_{n+1}^{\mathrm{tr}} , \tag{3.35}$$

where $p = \mathrm{tr}\,\boldsymbol{\sigma}/3$ is the mean normal stress. This last equation states that the final value of the mean normal stress is equal to its elastic predictor value. In other words, there is no plastic correction needed for the mean normal stress. Now, taking the deviatoric part of (3.32) and imposing (3.34) gives

$$\boldsymbol{s}_{n+1} = \boldsymbol{s}_{n+1}^{\mathrm{tr}} - 2\mu\Delta\lambda\,\boldsymbol{s}_{n+1}/\kappa . \tag{3.36}$$

Therefore, we obtain

$$\widehat{\boldsymbol{n}}_{n+1} = \boldsymbol{s}_{n+1}/\kappa = \boldsymbol{s}_{n+1}^{\mathrm{tr}}/\|\boldsymbol{s}_{n+1}^{\mathrm{tr}}\| \quad \Longrightarrow \quad \boldsymbol{s}_{n+1} = (\kappa/a)\,\boldsymbol{s}_{n+1}^{\mathrm{tr}} , \tag{3.37}$$

where $a = \|\boldsymbol{s}_{n+1}^{\mathrm{tr}}\|$. In other words, the tensor $\boldsymbol{s}_{n+1}$ must be in the same direction as $\boldsymbol{s}_{n+1}^{\mathrm{tr}}$, but its norm is scaled to the radius κ. Geometrically, the plastic corrector phase can be interpreted as a simple radial mapping of the elastic predictor stress $\boldsymbol{s}_{n+1}^{\mathrm{tr}}$ onto the yield surface $f(\boldsymbol{\sigma}_{n+1}, \kappa) = 0$, hence, the name of the algorithm. The final stress tensor may be obtained from the sum of volumetric and deviatoric parts,

$$\boldsymbol{\sigma}_{n+1} = p_{n+1}\mathbf{1} + \boldsymbol{s}_{n+1} \,. \tag{3.38}$$

Box 3.1 summarizes the radial return algorithm for the elastic-perfectly plastic J_2 theory. Figure 3.6, on the other hand, portrays the geometric meaning of the predictor-corrector scheme. Here, $\boldsymbol{\sigma}_n$ is assumed to lie within the elastic region $\overline{\mathsf{E}}$, and the incremental strain $\Delta\boldsymbol{\epsilon}$ is assumed to be sufficiently large that the predictor stress $\boldsymbol{\sigma}_{n+1}^{\text{tr}}$ overshoots the yield surface at a contact point $\boldsymbol{\sigma}^*$. The algorithm does not need to know this particular contact point; instead, it simply returns the elastic stress predictor radially onto $f = 0$ to determine the final stress $\boldsymbol{\sigma}_{n+1}$.

Step 1. Compute $\boldsymbol{\sigma}_{n+1}^{\text{tr}} = \boldsymbol{\sigma}_n + \boldsymbol{c}^{\text{e}} : \Delta\boldsymbol{\epsilon}$,
$p = \frac{1}{3}\operatorname{tr}(\boldsymbol{\sigma}_{n+1}^{\text{tr}})$, $\boldsymbol{s}_{n+1}^{\text{tr}} = \boldsymbol{\sigma}_{n+1}^{\text{tr}} - p\mathbf{1}$
Step 2. Check $a = \|\boldsymbol{s}_{n+1}^{\text{tr}}\| > \kappa$?
No, set $\boldsymbol{\sigma}_{n+1} = \boldsymbol{\sigma}_{n+1}^{\text{tr}}$ and exit.
Step 3. Yes, set $\boldsymbol{\sigma}_{n+1} = p\mathbf{1} + (\kappa/a)\boldsymbol{s}_{n+1}^{\text{tr}}$ and exit.

Box 3.1 Radial return algorithm for elastic-perfectly plastic J_2 theory

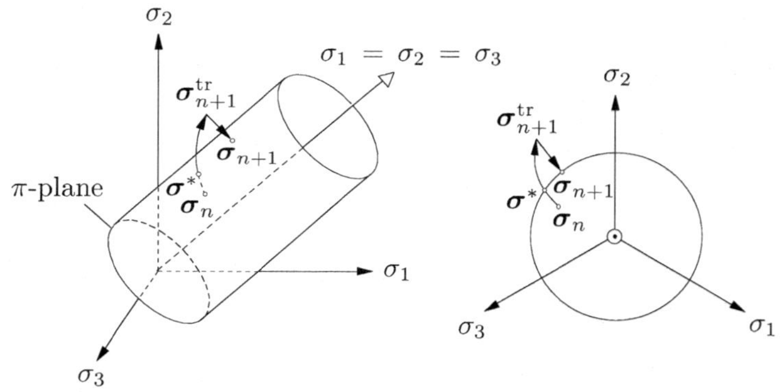

Fig. 3.6 Geometric meaning of the radial return algorithm for elastic-perfectly plastic J_2 theory

3.5 Isotropic Hardening

Hardening (alternatively, softening) refers to an increase (alternatively, decrease) in the value of the yield stress with plastic deformation. For J_2 plasticity the relevant constitutive variable that determines whether it is a hardening or softening response is the plastic internal variable κ. It has been observed that the J_2 yield criterion remains valid even after a significant amount of plastic straining for some metals such as copper and aluminum. For other metals such as mild steel, the shape of the yield surface could change slightly (Hill 1950), but in general, the J_2 yield criterion still remains fairly accurate over a wide range of plastic deformation.

We can describe the evolution of the plastic internal variable κ using either a work-hardening or a strain-hardening hypothesis. In the work-hardening hypothesis, it is assumed that κ varies with plastic work,

$$\kappa = \kappa(W^{\mathrm{p}})\,, \tag{3.39}$$

where

$$W^{\mathrm{p}} = \int_t \boldsymbol{\sigma} : \dot{\boldsymbol{\epsilon}}^{\mathrm{p}}\, \mathrm{d}t > 0 \tag{3.40}$$

is the plastic work. Note that W^{p} is nonnegative because plastic deformation is an irreversible process, and so, plastic dissipation cannot be less than zero. To establish an appropriate work-hardening law, an experimental program must be designed to provide the specific form of the function W^{p}.

A more frequently used alternative framework for hardening relates the plastic internal variable κ to a certain measure of cumulative plastic strain,

$$\kappa = \kappa(\lambda)\,. \tag{3.41}$$

For J_2 plasticity this cumulative plastic strain is taken as the path-dependent integral

$$\lambda = \int_t \dot{\lambda}\, \mathrm{d}t\,, \qquad \dot{\lambda} = \|\dot{\boldsymbol{\epsilon}}^{\mathrm{p}}\|\,. \tag{3.42}$$

The cumulative plastic strain λ is also a nonnegative function, and in fact, it is a monotonically increasing function of time since $\dot{\lambda}$ cannot be less than zero. In developing the specific hardening law, an experimental program must also be designed to provide the specific variation of the hardening law $\kappa(\lambda)$.

For a non-constant κ the consistency condition can be written as

$$\dot{f}(\boldsymbol{\sigma}, \kappa) = \frac{\partial f}{\partial \boldsymbol{\sigma}} : \dot{\boldsymbol{\sigma}} + \frac{\partial f}{\partial \kappa}\dot{\kappa} = 0\,, \tag{3.43}$$

where

$$\dot{\kappa}/\dot{\lambda} = \begin{cases} \kappa\kappa'(W^{\mathrm{p}})\,, & \text{if work hardening;} \\ \kappa'(\lambda)\,, & \text{if strain hardening.} \end{cases} \tag{3.44}$$

Therefore, the consistency condition can be written in the more general form

$$\frac{\partial f}{\partial \boldsymbol{\sigma}} : \dot{\boldsymbol{\sigma}} - H\dot{\lambda} = 0\,, \qquad H = -\frac{\partial f}{\partial \kappa}(\dot{\kappa}/\dot{\lambda})\,, \tag{3.45}$$

where H is the *generalized plastic modulus.* Note that H is not the same as the plastic modulus H' introduced in Chapter 2; however, there is a relation between the two plastic moduli. Under uniaxial loading, $\|\boldsymbol{s}\| = \sqrt{2/3}\,\sigma$, and so, $\dot{\kappa} = \sqrt{2/3}\,\dot{\sigma}_Y$, where σ_Y is the uniaxial yield stress. Also, $\dot{\lambda} = \|\dot{\boldsymbol{\epsilon}}^{\mathrm{p}}\| = \sqrt{3/2}\,\dot{\epsilon}^{\mathrm{p}}$, where $\dot{\epsilon}^{\mathrm{p}}$ is the axial plastic strain rate. Since the hardening law states that $\dot{\sigma}_Y = H'\dot{\epsilon}^{\mathrm{p}}$, then

$$\sqrt{\frac{2}{3}}\,\dot{\sigma}_Y = H\sqrt{\frac{3}{2}}\,\dot{\epsilon}^{\mathrm{p}} \quad \Longrightarrow \quad H = \frac{2}{3}\,H' \,. \tag{3.46}$$

Therefore, if H' is constant, the evolution of κ with plastic strain must be of the form

$$\kappa = \frac{2}{3}H'\lambda + \kappa_0 \,, \tag{3.47}$$

where κ_0 is the reference value of κ when $\lambda = 0$.

3.6 Combined Isotropic-Kinematic Hardening

When a material yields in tension and the loading is reversed, one typically observes a reduction in the compressive yield stress. A similar observation is true when a material yields in compression and is then stretched, i.e., the yield stress on reverse loading decreases. This phenomenon is known as the Bauschinger effect (Bauschinger 1879) and pertains to a property where the material's stress-strain characteristics change as a result of the microscopic stress distribution during yielding. The Bauschinger effect cannot be captured by an isotropically hardening J_2 theory. In order to capture this effect, some form of kinematic hardening must be introduced. For the J_2 model, kinematic hardening may be captured by allowing the elastic region to translate on the deviatoric plane in such as way that its generating axis remains parallel to the space diagonal.

Introduction of a deviatoric *back stress* tensor $\boldsymbol{\alpha}$, with $\mathrm{tr}(\boldsymbol{\alpha}) = 0$, facilitates the translation of the yield surface. A translated J_2 yield function is of the form

$$f(\boldsymbol{\sigma}, \boldsymbol{\alpha}, \kappa) = \|\boldsymbol{\xi}\| - \kappa \leq 0 \,, \qquad \boldsymbol{\xi} = \boldsymbol{s} - \boldsymbol{\alpha} \,. \tag{3.48}$$

Here, $\boldsymbol{\xi}$ is called the *effective stress*, with $\mathrm{tr}(\boldsymbol{\xi}) = \mathrm{tr}(\boldsymbol{s}) - \mathrm{tr}(\boldsymbol{\alpha}) \equiv 0$. This gives the following modified expression for the elastic region

$$\overline{\mathsf{E}} = \{(\boldsymbol{\sigma}, \boldsymbol{\alpha}, \kappa) \in \mathbb{S} \times \mathbb{S} \times \mathbb{R} \,|\, f(\boldsymbol{\sigma}, \boldsymbol{\alpha}, \kappa) \leq 0\} \,. \tag{3.49}$$

Figure 3.7 compares the two types of hardening and portrays the geometric meaning of the back stress tensor $\boldsymbol{\alpha}$.

In the presence of the back stress tensor $\boldsymbol{\alpha}$, the flow rule takes the form

$$\dot{\boldsymbol{\epsilon}}^{\mathrm{p}} = \dot{\lambda}\,\frac{\partial f}{\partial \boldsymbol{\sigma}} = \dot{\lambda}\widehat{\boldsymbol{n}} \,, \qquad \widehat{\boldsymbol{n}} = \boldsymbol{\xi}/\|\boldsymbol{\xi}\| \,, \tag{3.50}$$

where the back stress enters through the effective stress $\boldsymbol{\xi}$. Similarly, the form of the consistency condition changes as a result of the back stress,

$$\dot{f}(\boldsymbol{\sigma}, \boldsymbol{\alpha}, \kappa) = \widehat{\boldsymbol{n}} : (\dot{\boldsymbol{s}} - \dot{\boldsymbol{\alpha}}) - \dot{\kappa} = 0 \,. \tag{3.51}$$

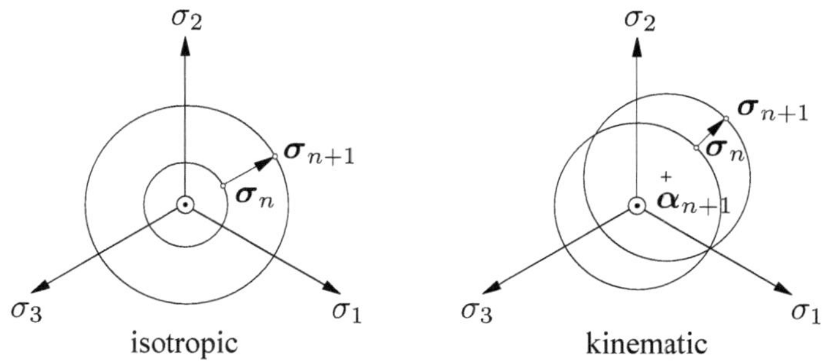

Fig. 3.7 Isotropic and kinematic hardening

In the following discussions, we present a formulation of a simple J_2 theory that accommodates both isotropic and kinematic hardening.

The constitutive J_2 theory of interest consists of a *linear combination* of isotropic and kinematic hardening (Hughes 1984). Consider a scalar parameter $\beta \in [0,1]$, and let the radius and center of the yield surface evolve according to the following expressions

$$\dot{\kappa} = \beta H \dot{\lambda}\,, \qquad \dot{\boldsymbol{\alpha}} = (1-\beta) H \dot{\boldsymbol{\epsilon}}^{\mathrm{p}}\,. \tag{3.52}$$

Isotropic hardening is represented by the evolution of the radius κ, whereas kinematic hardening is represented by the translation of the back stress tensor $\boldsymbol{\alpha}$. This particular expression for kinematic hardening, in which $\boldsymbol{\alpha}$ moves in the direction of the plastic strain rate, is called Prager's translation rule (Prager 1955). If we set $\beta = 0$, then we get $\dot{\kappa} = 0$ (no isotropic hardening), whereas if we set $\beta = 1$, then we get $\dot{\boldsymbol{\alpha}} = \mathbf{0}$ (no kinematic hardening). Thus, by specifying a value of β between zero and one, we can combine the two types of hardening linearly.

From the flow rule, $\dot{\boldsymbol{\epsilon}}^{\mathrm{p}} = \dot{\lambda}\widehat{\boldsymbol{n}}$, and from the fact that $\widehat{\boldsymbol{n}} : \dot{\boldsymbol{s}} = \widehat{\boldsymbol{n}} : \dot{\boldsymbol{\sigma}}$ since $\widehat{\boldsymbol{n}}$ is a deviatoric tensor, we can write the consistency condition for combined hardening as

$$\dot{f} = \frac{\partial f}{\partial \boldsymbol{\sigma}} : \dot{\boldsymbol{\sigma}} - H\dot{\lambda} = 0\,, \tag{3.53}$$

where H is the same generalized plastic modulus introduced earlier. Solving the consistency condition for $\dot{\lambda}$ gives

$$\dot{\lambda} = \frac{\widehat{\boldsymbol{n}} : \dot{\boldsymbol{\epsilon}}}{1 + H/2\mu}\,. \tag{3.54}$$

Note that $\dot{\lambda}$ does not depend on the parameter β, which means that the amount of plastic deformation is the same irrespective of the type of hardening.

Having determined the consistency parameter $\dot{\lambda}$ for a given plastic strain rate $\dot{\boldsymbol{\epsilon}}$, the plastic strain rate tensor can now be calculated as follows

$$\dot{\boldsymbol{\epsilon}}^{\mathrm{p}} = \frac{\widehat{\boldsymbol{n}} \otimes \widehat{\boldsymbol{n}}}{1 + H/2\mu} : \dot{\boldsymbol{\epsilon}} \,. \tag{3.55}$$

We can now also determine the evolution of the hardening variables in terms of the same plastic strain rate,

$$\dot{\kappa} = \frac{\beta H}{1 + H/2\mu} \widehat{\boldsymbol{n}} : \dot{\boldsymbol{\epsilon}} \,, \qquad \dot{\boldsymbol{\alpha}} = \frac{(1-\beta)H}{1 + H/2\mu} (\widehat{\boldsymbol{n}} \otimes \widehat{\boldsymbol{n}}) : \dot{\boldsymbol{\epsilon}} \,. \tag{3.56}$$

Substituting these back into the rate constitutive equation gives the rank-four elastoplastic tangent constitutive tensor

$$\boldsymbol{c}^{\mathrm{ep}} = K \mathbf{1} \otimes \mathbf{1} + 2\mu \Big(\boldsymbol{I} - \frac{1}{3} \mathbf{1} \otimes \mathbf{1} - \frac{\widehat{\boldsymbol{n}} \otimes \widehat{\boldsymbol{n}}}{1 + H/2\mu} \Big) \,. \tag{3.57}$$

Note that $\boldsymbol{c}^{\mathrm{ep}}$ does not contain the parameter β either, which means that the tangential stress-strain response does not depend on the type of hardening response.

3.7 Algorithm for Combined Hardening

With some modifications of the algorithm for the perfectly plastic J_2 theory, Krieg and Key (1976) accommodated the combined isotropic-kinematic hardening within the framework of the radial return algorithm. The format of calculations is the same as in the perfectly plastic case: $\boldsymbol{\sigma}_n$, κ_n, $\boldsymbol{\alpha}_n$, and $\Delta\boldsymbol{\epsilon}$ are given, and $\boldsymbol{\sigma}_{n+1}$, κ_{n+1}, and $\boldsymbol{\alpha}_{n+1}$ are calculated.

With reference to Fig. 3.8, and using the backward difference scheme, the predictor-corrector expression for $\boldsymbol{\sigma}_{n+1}$ is written as

$$\boldsymbol{\sigma}_{n+1} = \boldsymbol{\sigma}^{\mathrm{tr}}_{n+1} - 2\mu\Delta\lambda\, \boldsymbol{\xi}_{n+1}/\kappa_{n+1} \,. \tag{3.58}$$

Taking the trace of both sides of the equation gives

$$\mathrm{tr}(\boldsymbol{\sigma}_{n+1}) = \mathrm{tr}(\boldsymbol{\sigma}^{\mathrm{tr}}_{n+1}) \quad \Longrightarrow \quad p_{n+1} = p^{\mathrm{tr}}_{n+1} \,. \tag{3.59}$$

Taking the deviatoric part gives

$$\boldsymbol{s}_{n+1} = \boldsymbol{s}^{\mathrm{tr}}_{n+1} - 2\mu\Delta\lambda\, \boldsymbol{\xi}_{n+1}/\kappa_{n+1} \,. \tag{3.60}$$

Using the same backward difference scheme for integrating $\dot{\kappa}$ and $\dot{\boldsymbol{\alpha}}$, we obtain the expressions

$$\kappa_{n+1} = \kappa_n + \beta H \Delta\lambda \tag{3.61}$$

$$\boldsymbol{\alpha}_{n+1} = \boldsymbol{\alpha}_n + (1-\beta) H \Delta\lambda\, \boldsymbol{\xi}_{n+1}/\kappa_{n+1} \,. \tag{3.62}$$

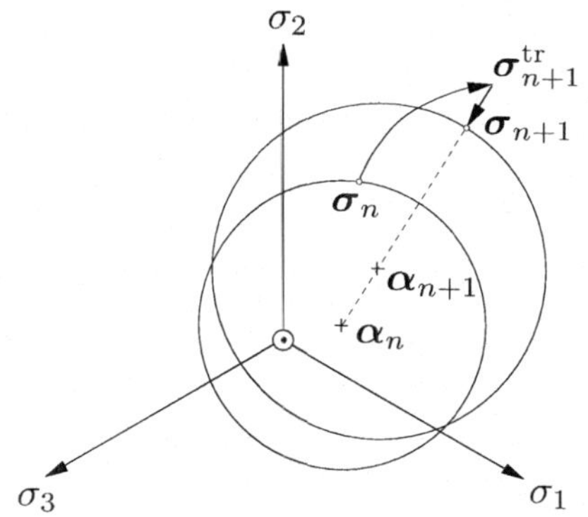

Fig. 3.8 Return mapping algorithm for J_2 theory with combined isotropic-kinematic hardening

Hence,

$$\boldsymbol{\xi}_{n+1} = \boldsymbol{s}_{n+1} - \boldsymbol{\alpha}_{n+1} = \boldsymbol{\xi}^{\mathrm{tr}}_{n+1} - \left[1 + (1-\beta)\,\frac{H}{2\mu}\right]\frac{2\mu\Delta\lambda}{\kappa_{n+1}}\,\boldsymbol{\xi}_{n+1}\,, \tag{3.63}$$

where $\boldsymbol{\xi}^{\mathrm{tr}}_{n+1} = \boldsymbol{s}^{\mathrm{tr}}_{n+1} - \boldsymbol{\alpha}_n$. This last equation shows that $\boldsymbol{\xi}_{n+1} \propto \boldsymbol{\xi}^{\mathrm{tr}}_{n+1}$, so we can simply take the norms of (3.63) to obtain

$$\|\boldsymbol{\xi}_{n+1}\| = \|\boldsymbol{\xi}^{\mathrm{tr}}_{n+1}\| - 2\mu\Delta\lambda\left[1 + (1-\beta)\,\frac{H}{2\mu}\right]. \tag{3.64}$$

Combining (3.61) and (3.64) yields

$$\Delta\lambda = \frac{\|\boldsymbol{\xi}^{\mathrm{tr}}_{n+1}\| - \kappa_n}{2\mu + H}\,. \tag{3.65}$$

Box 3.2 summarizes the algorithm for J_2 plasticity with combined hardening.

Step 1. Compute $\boldsymbol{\sigma}^{\mathrm{tr}}_{n+1} = \boldsymbol{\sigma}_n + \boldsymbol{c}^{\mathrm{e}} : \Delta\boldsymbol{\epsilon}, \quad p = \frac{1}{3}\,\mathrm{tr}(\boldsymbol{\sigma}^{\mathrm{tr}}_{n+1}),$
$\boldsymbol{s}^{\mathrm{tr}}_{n+1} = \boldsymbol{\sigma}^{\mathrm{tr}}_{n+1} - p\mathbf{1}, \quad \boldsymbol{\xi}^{\mathrm{tr}}_{n+1} = \boldsymbol{s}^{\mathrm{tr}}_{n+1} - \boldsymbol{\alpha}_n.$

Step 2. Check $a = \|\,\boldsymbol{\xi}^{\mathrm{tr}}_{n+1}\,\| > \kappa_n$?
No, set $\boldsymbol{\sigma}_{n+1} = \boldsymbol{\sigma}^{\mathrm{tr}}_{n+1}$ and exit.

Step 3. Yes, set $\widehat{\boldsymbol{n}}_{n+1} = \boldsymbol{\xi}^{\mathrm{tr}}_{n+1}/a, \quad \Delta\lambda = (a - \kappa_n)/(2\mu + H),$
$\boldsymbol{\sigma}_{n+1} = \boldsymbol{\sigma}^{\mathrm{tr}}_{n+1} - 2\mu\Delta\lambda\,\widehat{\boldsymbol{n}}_{n+1},$
$\kappa_{n+1} = \kappa_n + \beta H\Delta\lambda,$
$\boldsymbol{\alpha}_{n+1} = \boldsymbol{\alpha}_n + (1-\beta)H\Delta\lambda\,\widehat{\boldsymbol{n}}_{n+1},$ and exit.

Box 3.2 Radial return algorithm for J_2 theory with a linear combination of isotropic and kinematic hardening

3.8 Algorithmic Tangent Operator

In Chapter 1 we emphasized the importance of an exact algorithmic tangent operator to realize the optimal convergence properties of Newton's method. With this idea in mind, we now develop the consistent tangent operator for J_2 plasticity with a linear combination of isotropic and kinematic hardening.

As a background for discussion, we first consider the following boundary-value problem: Let $\mathcal{B}$ denote a body with boundary $\partial\mathcal{B} = \overline{\partial\mathcal{B}_{\boldsymbol{t}} \cup \partial\mathcal{B}_{\boldsymbol{u}}}$, where $\partial\mathcal{B}_{\boldsymbol{t}}$ and $\partial\mathcal{B}_{\boldsymbol{u}}$ are non-overlapping portions of $\partial\mathcal{B}$ (i.e., $\partial\mathcal{B}_{\boldsymbol{t}} \cap \partial\mathcal{B}_{\boldsymbol{u}} = \emptyset$) on which surface traction and boundary displacements are prescribed, respectively. Without loss of generality, we assume a quasi-static condition. The variational equation takes the form

$$G(\boldsymbol{\eta}, \boldsymbol{u}) = \int_{\mathcal{B}} \boldsymbol{\nabla}^s\boldsymbol{\eta} : \boldsymbol{\sigma}\,\mathrm{d}V - \int_{\mathcal{B}} \boldsymbol{\eta}\cdot\rho\boldsymbol{g}\,\mathrm{d}V - \int_{\partial\mathcal{B}_t} \boldsymbol{\eta}\cdot\boldsymbol{t}\,\mathrm{d}A = 0\,, \tag{3.66}$$

where $\boldsymbol{u} \in \mathcal{V}$ is the displacement field, $\mathcal{V}$ is the space of trial functions, $\boldsymbol{\eta} \in \mathcal{W}$ is the virtual displacement field, $\mathcal{W}$ is the space of weighting functions, ρ is the mass density, $\boldsymbol{g}$ is the vector of gravity accelerations, and $\boldsymbol{t}$ is the surface traction vector. The variational equation must be satisfied at all times, including at time t_{n+1}, so we can write

$$G(\boldsymbol{\eta}, \boldsymbol{u}_{n+1}) = \int_{\mathcal{B}} \boldsymbol{\nabla}^s\boldsymbol{\eta} : \boldsymbol{\sigma}_{n+1}\,\mathrm{d}V - \int_{\mathcal{B}} \boldsymbol{\eta}\cdot\rho\boldsymbol{g}\,\mathrm{d}V - \int_{\partial\mathcal{B}_t} \boldsymbol{\eta}\cdot\boldsymbol{t}_{n+1}\,\mathrm{d}A = 0\,. \tag{3.67}$$

We assume that the algorithm in Box 3.2 is used to determine $\boldsymbol{\sigma}_{n+1}$ as a function of $\boldsymbol{u}_{n+1}$, with an imposed incremental strain of the form

$$\Delta\boldsymbol{\epsilon} = \boldsymbol{\nabla}^s(\boldsymbol{u}_{n+1} - \boldsymbol{u}_n)\,. \tag{3.68}$$

For dead loading the linearization of (3.67) is given by

$$\mathcal{L}G = G^0 + \int_{\mathcal{B}} \boldsymbol{\nabla}^s\boldsymbol{\eta} : \boldsymbol{c}_{n+1} : \boldsymbol{\nabla}^s\delta\boldsymbol{u}\,\mathrm{d}V - \int_{\partial\mathcal{B}_t} \boldsymbol{\eta}\cdot\delta\boldsymbol{t}_{n+1}\,\mathrm{d}A\,, \tag{3.69}$$

where $G^0 = G(\boldsymbol{\eta}, \boldsymbol{u}^0)$ and $\delta\boldsymbol{u}$ and $\delta\boldsymbol{t}$ are the variations of the vectors $\boldsymbol{u}$ and $\boldsymbol{t}$, respectively. The algorithmic stress-strain operator is then defined by the rank-four tensor

$$\boldsymbol{c}_{n+1} = \frac{\partial\boldsymbol{\sigma}_{n+1}}{\partial\boldsymbol{\epsilon}_{n+1}}\,. \tag{3.70}$$

Note that $\boldsymbol{c}_{n+1}$ must not be confused with the tensor $\boldsymbol{c}^{\mathrm{ep}}$: the latter reflects the material tangent response, whereas the former reflects not only the material tangent response but also all of the artifacts of the numerical algorithm used for calculating $\boldsymbol{\sigma}_{n+1}$ (Simo and Taylor 1985). More specifically, for the J_2 theory the algorithmic tensor $\boldsymbol{c}_{n+1}$ must account for all the computational steps of the algorithm described in Box 3.2.

To obtain the exact algorithmic tangent operator $\boldsymbol{c}_{n+1}$, we first consider the algorithm for computing $\boldsymbol{\sigma}_{n+1}$ given in Step 3 of Box 3.2. Dropping the subscript '$n+1$' for brevity, we write

$$\boldsymbol{\sigma} = \boldsymbol{\sigma}^{\mathrm{tr}} - 2\mu\Delta\lambda\widehat{\boldsymbol{n}}\,, \tag{3.71}$$

where $\widehat{\boldsymbol{n}} = \boldsymbol{\xi}/\kappa = \boldsymbol{\xi}^{\mathrm{tr}}/\|\boldsymbol{\xi}^{\mathrm{tr}}\|$. Using the chain rule, we obtain

$$\boldsymbol{c}_{n+1} = \frac{\partial\boldsymbol{\sigma}^{\mathrm{tr}}}{\partial\boldsymbol{\epsilon}} - 2\mu\Delta\lambda\frac{\partial\widehat{\boldsymbol{n}}}{\partial\boldsymbol{\epsilon}} - 2\mu\widehat{\boldsymbol{n}}\otimes\frac{\partial\Delta\lambda}{\partial\boldsymbol{\epsilon}}\,. \tag{3.72}$$

Since $\boldsymbol{\sigma}^{\mathrm{tr}} = \boldsymbol{\sigma}_n + \boldsymbol{c}^{\mathrm{e}} : (\boldsymbol{\epsilon} - \boldsymbol{\epsilon}_n)$ from Step 1 of the algorithm, where $\boldsymbol{\sigma}_n$ and $\boldsymbol{\epsilon}_n$ are fixed stress and strain tensors from the previous load calculations, it follows that

$$\frac{\partial\boldsymbol{\sigma}^{\mathrm{tr}}}{\partial\boldsymbol{\epsilon}} = \boldsymbol{c}^{\mathrm{e}}\,. \tag{3.73}$$

This is the *linearization* of Step 1.

The linearization of $\mathrm{tr}(\boldsymbol{\sigma}^{\mathrm{tr}})/3$ is

$$\frac{1}{3}\frac{\partial\mathrm{tr}(\boldsymbol{\sigma}^{\mathrm{tr}})}{\partial\boldsymbol{\epsilon}} = K\frac{\partial\mathrm{tr}(\boldsymbol{\epsilon})}{\partial\boldsymbol{\epsilon}} = K\mathbf{1}\,. \tag{3.74}$$

The linearization of $\boldsymbol{\xi}^{\mathrm{tr}}$ is

$$\frac{\partial\boldsymbol{\xi}^{\mathrm{tr}}}{\partial\boldsymbol{\epsilon}} = \frac{\partial\boldsymbol{s}^{\mathrm{tr}}}{\partial\boldsymbol{\epsilon}} = \boldsymbol{c}^{\mathrm{e}} - K\mathbf{1}\otimes\mathbf{1} = 2\mu\Big(\boldsymbol{I} - \frac{1}{3}\mathbf{1}\otimes\mathbf{1}\Big)\,. \tag{3.75}$$

Note that $\boldsymbol{\alpha}_n$ is fixed, so its variation is zero.

From Step 3 of the algorithm, we get

$$\frac{\partial\|\boldsymbol{\xi}^{\mathrm{tr}}\|}{\partial\boldsymbol{\epsilon}} = \widehat{\boldsymbol{n}} : \frac{\partial\boldsymbol{\xi}^{\mathrm{tr}}}{\partial\boldsymbol{\epsilon}} = \widehat{\boldsymbol{n}} : (\boldsymbol{c}^{\mathrm{e}} - K\mathbf{1}\otimes\mathbf{1}) = 2\mu\widehat{\boldsymbol{n}}\,, \tag{3.76}$$

since $\widehat{\boldsymbol{n}} : \boldsymbol{c}^{\mathrm{e}} = 2\mu\widehat{\boldsymbol{n}}$ and $\widehat{\boldsymbol{n}} : \mathbf{1} = 0$. Hence,

$$\frac{\partial\widehat{\boldsymbol{n}}}{\partial\boldsymbol{\epsilon}} = \frac{2\mu}{\|\boldsymbol{\xi}^{\mathrm{tr}}\|}\Big(\boldsymbol{I} - \frac{1}{3}\mathbf{1}\otimes\mathbf{1} - \widehat{\boldsymbol{n}}\otimes\widehat{\boldsymbol{n}}\Big)\,. \tag{3.77}$$

Also from Step 3, we write

$$\frac{\partial\Delta\lambda}{\partial\boldsymbol{\epsilon}} = \frac{\widehat{\boldsymbol{n}}}{1 + H/2\mu}\,. \tag{3.78}$$

Note that κ_n is also fixed, so its variation is zero.

Substituting into (3.72) and simplifying yields the following closed-form expression for the algorithmic tangent operator

$$\boldsymbol{c}_{n+1} = \boldsymbol{c}^{\mathrm{ep}} - 2\mu\gamma\Big(\boldsymbol{I} - \frac{1}{3}\mathbf{1}\otimes\mathbf{1} - \widehat{\boldsymbol{n}}\otimes\widehat{\boldsymbol{n}}\Big)\,, \tag{3.79}$$

where $\gamma = 2\mu\Delta\lambda/\|\boldsymbol{\xi}^{\text{tr}}\|$ and $\boldsymbol{c}^{\text{ep}}$ is elastoplastic constitutive operator given in (3.57). The additional terms that are factors of $2\mu\gamma$ result from the artifacts of the algorithm and would be zero if the algorithm was exact. This can be seen by noting that as $\Delta\lambda \to 0$, $\boldsymbol{c}_{n+1} \to \boldsymbol{c}^{\text{ep}}$, and the algorithmic contributions to the tangent operator disappear, i.e., the algorithmic tangent operator $\boldsymbol{c}_{n+1}$ approaches the elastoplastic constitutive operator $\boldsymbol{c}^{\text{ep}}$. In general, however, the algorithmic tangent operator $\boldsymbol{c}_{n+1}$ is not the same as the elastoplastic constitutive operator $\boldsymbol{c}^{\text{ep}}$ because the numerical algorithm is not exact. Since we are dealing with a numerically integrated boundary-value problem, the algorithmic tangent operator must be used to realize the desirable convergence properties of Newton's method.

Exercise 3.6. A material behavior is described by the J_2 theory with combined isotropic-kinematic hardening, where $E = 40$ MPa, $\nu = 0.3$, $E_T = 0.1E$, and initial uniaxial yield stress $\sigma_{Y0} = 1$ MPa. A bar specimen of this material is subjected to cyclic axial stress of $\sigma = \pm 2.0$ MPa. Sketch steady-state axial stress-axial strain responses assuming the following types of hardening: (a) purely kinematic, $\beta = 0$; (b) purely isotropic, $\beta = 1$; and (c) combined isotropic-kinematic, $\beta = 0.50$.

Exercise 3.7. Repeat Exercise 3.6 if instead of subjecting the specimen to cyclic axial stress it is now subjected to cyclic axial strain of $\epsilon = \pm 5\%$.

3.9 Maximum Plastic Dissipation

The principle of maximum plastic dissipation is central to the variational formulation of plasticity theory. It states that among all possible stress points satisfying the yield criterion, the maximum plastic dissipation occurs at the actual stress point.

Consider the following Helmholtz free energy function

$$\psi(\boldsymbol{\epsilon}^{\text{e}}, \boldsymbol{\xi}) = \psi^{\text{e}}(\boldsymbol{\epsilon}^{\text{e}}) + \psi^{\text{p}}(\boldsymbol{\xi}) , \tag{3.80}$$

where $\boldsymbol{\xi}$ is a vector of strain-like plastic internal variables analogous to the scalar cumulative plastic shear strain λ, although it could also contain other measures of plastic strains, such as the volumetric plastic strain. The dissipation inequality of thermodynamics can be written as

$$\mathcal{D} := \boldsymbol{\sigma} : \dot{\boldsymbol{\epsilon}} - \frac{\mathrm{d}\psi}{\mathrm{d}t} \geq 0 . \tag{3.81}$$

By the additive decomposition of the strain tensor, we have $\dot{\boldsymbol{\epsilon}} = \dot{\boldsymbol{\epsilon}}^{\text{e}} + \dot{\boldsymbol{\epsilon}}^{\text{p}}$. Therefore, the last equation can be expanded to yield

$$\mathcal{D} = \left(\boldsymbol{\sigma} - \frac{\partial \psi^{\text{e}}}{\partial \boldsymbol{\epsilon}^{\text{e}}}\right) : \dot{\boldsymbol{\epsilon}}^{\text{e}} + \boldsymbol{\sigma} : \dot{\boldsymbol{\epsilon}}^{\text{p}} - \frac{\partial \psi^{\text{p}}}{\partial \boldsymbol{\xi}} \cdot \dot{\boldsymbol{\xi}} \geq 0 . \tag{3.82}$$

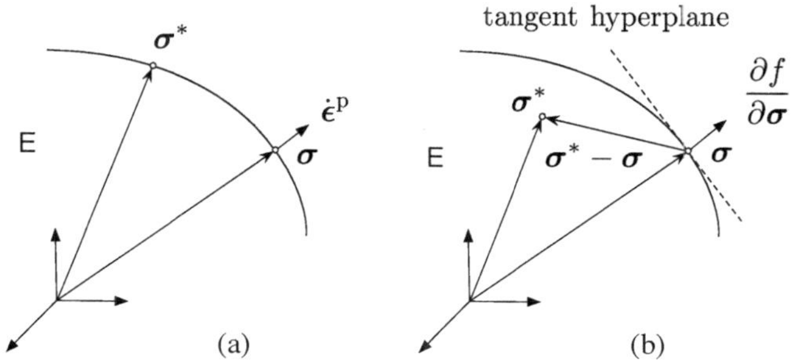

Fig. 3.9 Illustration of the principle of maximum plastic dissipation: (a) plastic dissipation is maximum at $\boldsymbol{\sigma}$; (b) alternative definition of convexity of the yield function

The inequality must hold for any $\dot{\boldsymbol{\epsilon}}^{\mathrm{e}}$, $\dot{\boldsymbol{\epsilon}}^{\mathrm{p}}$, and $\dot{\boldsymbol{\xi}}$; hence, we must have

$$\boldsymbol{\sigma} = \frac{\partial \psi^{\mathrm{e}}}{\partial \boldsymbol{\epsilon}^{\mathrm{e}}}, \tag{3.83}$$

which defines a *hyperelastic constitutive equation.* Thus, the dissipation inequality reduces to the form

$$\mathcal{D}^{\mathrm{p}} = \boldsymbol{\sigma} : \dot{\boldsymbol{\epsilon}}^{\mathrm{p}} + \boldsymbol{q} \cdot \dot{\boldsymbol{\xi}} \geq 0, \qquad \boldsymbol{q} = -\frac{\partial \psi^{\mathrm{p}}}{\partial \boldsymbol{\xi}}, \tag{3.84}$$

where $\boldsymbol{q}$ is a stress-like vector of plastic internal variables conjugate to $\boldsymbol{\xi}$. For example, for J_2 plasticity the vector $\boldsymbol{q}$ is simply the scalar internal variable κ.

Consider any pair $(\boldsymbol{\sigma}^*, \boldsymbol{q}^*)$ satisfying the yield criterion $f(\boldsymbol{\sigma}^*, \boldsymbol{q}^*) = 0$. The principle of maximum plastic dissipation states that the maximum value of $\mathcal{D}^{\mathrm{p}}$ occurs at the actual pair $(\boldsymbol{\sigma}, \boldsymbol{q})$. Therefore,

$$\mathcal{D}^{\mathrm{p}}_{\max} = \boldsymbol{\sigma} : \dot{\boldsymbol{\epsilon}}^{\mathrm{p}} + \boldsymbol{q} \cdot \dot{\boldsymbol{\xi}} \geq \boldsymbol{\sigma}^* : \dot{\boldsymbol{\epsilon}}^{\mathrm{p}} + \boldsymbol{q}^* \cdot \dot{\boldsymbol{\xi}}, \qquad \forall (\boldsymbol{\sigma}^*, \boldsymbol{q}^*) \in \partial \mathsf{E}. \tag{3.85}$$

A geometric interpretation of this statement is shown in Fig. 3.9a for perfect plasticity where $\dot{\boldsymbol{q}} = \boldsymbol{0}$. The principle states that if we compute the value of the scalar function $\boldsymbol{\sigma}^* : \dot{\boldsymbol{\epsilon}}^{\mathrm{p}}$ for all possible values of $\boldsymbol{\sigma}^*$ satisfying the yield criterion, the maximum value occurs at the actual stress point $\boldsymbol{\sigma}$.

We can interpret the principle of maximum plastic dissipation as a problem of constrained minimization in which the functional $-\mathcal{D}^{\mathrm{p}} = -\boldsymbol{\sigma}^* : \dot{\boldsymbol{\epsilon}}^{\mathrm{p}} - \boldsymbol{q}^* \cdot \dot{\boldsymbol{\xi}}$ is minimized subject to the constraint $f(\boldsymbol{\sigma}^*, \boldsymbol{q}^*) = 0$. This is a standard problem in variational calculus (Wiley 1975), which is solved systematically by constructing a Hamiltonian function

$$\mathcal{H} = -\boldsymbol{\sigma}^* : \dot{\boldsymbol{\epsilon}}^{\mathrm{p}} - \boldsymbol{q}^* \cdot \dot{\boldsymbol{\xi}} + \dot{\lambda} f(\boldsymbol{\sigma}^*, \boldsymbol{q}^*), \tag{3.86}$$

where $\dot{\lambda} \geq 0$ is a Lagrange multiplier. For $\mathcal{H}$ to have a stationary value, we must set the following derivatives all equal to zero:

$$\left.\begin{aligned} \frac{\partial \mathcal{H}}{\partial \boldsymbol{\sigma}^*} &= -\dot{\boldsymbol{\epsilon}}^{\mathrm{p}} + \dot{\lambda}\frac{\partial f}{\partial \boldsymbol{\sigma}^*} = \mathbf{0} \\ \frac{\partial \mathcal{H}}{\partial \boldsymbol{q}^*} &= -\dot{\boldsymbol{\xi}} + \dot{\lambda}\frac{\partial f}{\partial \boldsymbol{q}^*} = \mathbf{0} \\ \frac{\partial \mathcal{H}}{\partial \dot{\lambda}} &= f(\boldsymbol{\sigma}^*, \boldsymbol{q}^*) = 0 \end{aligned}\right\} . \tag{3.87}$$

Evaluating at the actual stress point gives

$$\dot{\boldsymbol{\epsilon}}^{\mathrm{p}} = \dot{\lambda}\left.\frac{\partial f}{\partial \boldsymbol{\sigma}^*}\right|_{(\boldsymbol{\sigma},\boldsymbol{q})} = \dot{\lambda}\frac{\partial f}{\partial \boldsymbol{\sigma}}, \qquad \dot{\boldsymbol{\xi}} = \dot{\lambda}\left.\frac{\partial f}{\partial \boldsymbol{q}^*}\right|_{(\boldsymbol{\sigma},\boldsymbol{q})} = \dot{\lambda}\frac{\partial f}{\partial \boldsymbol{q}} . \tag{3.88}$$

The first equation is the *associative flow rule*, a topic explored further in the next section, whereas the second equation is the *associative hardening law*.

Now, suppose we have a pair $(\boldsymbol{\sigma}^*, \boldsymbol{q}^*)$ such that $f(\boldsymbol{\sigma}^*, \boldsymbol{q}^*) < 0$, implying that the stress point $\boldsymbol{\sigma}^*$ lies in the interior of the elastic region. Then, the function $-\mathcal{D}^{\mathrm{p}} = -\boldsymbol{\sigma}^* : \dot{\boldsymbol{\epsilon}}^{\mathrm{p}} - \boldsymbol{q}^* : \dot{\boldsymbol{\xi}}$ must be minimized such that $f(\boldsymbol{\sigma}^*, \boldsymbol{q}^*) < 0$. In this case, we can consider a modified Hamiltonian function

$$\mathcal{H} = -\boldsymbol{\sigma}^* : \dot{\boldsymbol{\epsilon}}^{\mathrm{p}} - \boldsymbol{q}^* : \dot{\boldsymbol{\xi}} + \dot{\lambda}(f(\boldsymbol{\sigma}^*, \boldsymbol{q}^*) + C) , \tag{3.89}$$

where C is a positive real number such that $f(\boldsymbol{\sigma}^*, \boldsymbol{q}^*) + C = 0$. Setting the derivatives equal to zero at the actual stress point leads to similar results as before. In other words, whether $\boldsymbol{\sigma}^*$ lies in the interior or on the boundary of the elastic region, we must have

$$\dot{\lambda} \geq 0 , \qquad f \leq 0 , \qquad \dot{\lambda} f = 0 , \tag{3.90}$$

which are nothing but the Kuhn-Tucker conditions.

Finally, since $\mathcal{D}^{\mathrm{p}}$ is maximum at $(\boldsymbol{\sigma}, \boldsymbol{q})$, then

$$(\boldsymbol{\sigma}^* - \boldsymbol{\sigma}) : \dot{\boldsymbol{\epsilon}}^{\mathrm{p}} + (\boldsymbol{q}^* - \boldsymbol{q}) \cdot \dot{\boldsymbol{\xi}} \leq 0 . \tag{3.91}$$

Substituting (3.88) gives

$$(\boldsymbol{\sigma}^* - \boldsymbol{\sigma}) : \frac{\partial f}{\partial \boldsymbol{\sigma}} + (\boldsymbol{q}^* - \boldsymbol{q}) \cdot \frac{\partial f}{\partial \boldsymbol{q}} \leq 0 . \tag{3.92}$$

This inequality is satisfied by any convex, continuously differentiable yield function f, see Fig. 3.9b. The geometrical meaning of the above relation is that the elastic region is entirely contained in one of two half-spaces delimited by the tangent hyperplane to the boundary at the stress point $\boldsymbol{\sigma}$. If a strict inequality is satisfied, then the elastic region is said to be strictly convex. The principle of maximum plastic dissipation thus implies (i) associative flow rule; (ii) associative hardening; and (iii) convexity of the yield function.

Exercise 3.8. Consider an isotropically hardening J_2 theory whose yield function is given by $f = \|\boldsymbol{s}\| - \kappa = 0$. Assuming a free energy function of the form

$$\psi = \psi^{\mathrm{e}} + \psi^{\mathrm{p}}\,, \qquad \psi^{\mathrm{e}} = \frac{1}{2}\boldsymbol{\epsilon}^{\mathrm{e}} : \boldsymbol{c}^{\mathrm{e}} : \boldsymbol{\epsilon}^{\mathrm{e}}\,, \qquad \psi^{\mathrm{p}} = \frac{1}{2}H\xi^2\,,$$

derive the flow rule for $\dot{\boldsymbol{\epsilon}}^{\mathrm{p}}$ and the hardening law for κ using the principle of maximum plastic dissipation. What elastic constitutive equation results from the requirement of a nonnegative dissipation function?

3.10 Non-associative Plasticity

The preceding section presents the mathematical theory of plasticity from a constrained minimization standpoint. While the formulation exhibits some nice mathematical structure such as symmetry, deviations from reality are to be expected. For example, experimental evidence suggests that a *non-associative plastic flow*, where the plastic strain increment is not generated from the gradient of the yield function, may provide a more realistic description of actual material behavior. Furthermore, a majority of materials exhibit a much more complicated hardening law than the one presented in the previous section.

As an alternative to the associative flow rule where the plastic flow is determined from the yield function f, we can consider a non-associative flow rule where the plastic flow direction for $\dot{\boldsymbol{\epsilon}}^{\mathrm{p}}$ is determined from a certain plastic potential function g that is distinct from the yield function f, i.e.,

$$\dot{\boldsymbol{\epsilon}}^{\mathrm{p}} = \dot{\lambda}\frac{\partial g}{\partial \boldsymbol{\sigma}}\,. \tag{3.93}$$

Furthermore, we can define an evolution for $\boldsymbol{q}$ such that

$$H = -\frac{\partial f}{\partial \boldsymbol{q}} \cdot (\dot{\boldsymbol{q}}/\dot{\lambda})\,, \tag{3.94}$$

where $\dot{\boldsymbol{q}}/\dot{\lambda}$ may be some experimentally derived hardening relation. In other words, we may only need to describe the evolution of the generalized plastic modulus H without having to define the variation of $\boldsymbol{q}$. Plastic response then requires that the following consistency condition be satisfied

$$\dot{f}(\boldsymbol{\sigma}, \kappa) = \frac{\partial f}{\partial \boldsymbol{\sigma}} : \dot{\boldsymbol{\sigma}} - H\dot{\lambda} = 0\,. \tag{3.95}$$

If ψ^{e} is quadratic in $\boldsymbol{\epsilon}^{\mathrm{e}}$, then we have a linear elastic relation

$$\dot{\boldsymbol{\sigma}} = \boldsymbol{c}^{\mathrm{e}} : \left(\dot{\boldsymbol{\epsilon}} - \dot{\lambda}\frac{\partial g}{\partial \boldsymbol{\sigma}}\right). \tag{3.96}$$

Inserting into the consistency condition yields

$$\dot{\lambda} = \frac{1}{\chi}\frac{\partial f}{\partial \boldsymbol{\sigma}} : \boldsymbol{c}^{\mathrm{e}} : \dot{\boldsymbol{\epsilon}} \,, \qquad \chi = \frac{\partial f}{\partial \boldsymbol{\sigma}} : \boldsymbol{c}^{\mathrm{e}} : \frac{\partial g}{\partial \boldsymbol{\sigma}} + H \,. \tag{3.97}$$

Finally, substituting back into the rate constitutive equation gives

$$\dot{\boldsymbol{\sigma}} = \boldsymbol{c}^{\mathrm{ep}} : \dot{\boldsymbol{\epsilon}} \,, \qquad \boldsymbol{c}^{\mathrm{ep}} = \boldsymbol{c}^{\mathrm{e}} - \frac{1}{\chi}\boldsymbol{c}^{\mathrm{e}} : \frac{\partial g}{\partial \boldsymbol{\sigma}} \otimes \frac{\partial f}{\partial \boldsymbol{\sigma}} : \boldsymbol{c}^{\mathrm{e}} \,. \tag{3.98}$$

If $\partial g/\partial \boldsymbol{\sigma} \neq \partial f/\partial \boldsymbol{\sigma}$, then the major symmetry of the constitutive operator $\boldsymbol{c}^{\mathrm{ep}}$ is lost.

3.11 Numerical Simulations

We consider a 1 m × 1 m square plate of thickness 0.1 m containing a 0.5 m-diameter hole. The plate is made of steel supported on frictionless rollers at both its top and bottom surfaces. We use J_2 plasticity to describe the constitutive response of the material, with initial uniaxial yield stress $\sigma_{Y0} = 300$ MPa and tangential modulus $E_T = 0.1E$. The elastic constants are Young's modulus of elasticity $E = 200$ GPa and Poisson's ratio $\nu = 0.45$. Figure 3.10 shows the geometry of the problem. Under a condition of plane stress the plate is first stretched vertically by moving the top and bottom sides away from each other by a distance of 0.02 m, and then compressed vertically by moving them toward each other by −0.02 m relative to the initial configuration. By symmetry, we only model one quarter of the total problem domain.

Figure 3.11 shows the finite element mesh consisting of 706 four-node quadrilateral elements (out-of-plane thickness = 1 m). Because of the volume-preserving plastic flow induced by the theory, many standard elements, including the four-node quadrilateral elements used in this simulation, tend to lock. Therefore, special finite element techniques must be employed to circumvent mesh-locking. Here, we employ the B-bar approach (Hughes 1980) to circumvent this problem. The B-bar approach pertains to the relaxation of the incompressibility constraint in the strain-displacement transformation matrix $\boldsymbol{B}$ that enables the recovery of the strain distribution in the incompressible element without causing the mesh to lock.

Figures 3.12–3.14 show contours of the von Mises stress ($\sqrt{3/2}\|\boldsymbol{s}\|$), current plastic strain ($\sqrt{2/3}\|\boldsymbol{\epsilon}^{\mathrm{p}}\|$), and cumulative plastic strain ($\sqrt{3/2}\lambda$), superimposed on the deformed meshes at times $t = 1$, 2, and 3. To obtain reasonably accurate time histories of the state variables, a total of 120 time increments, or a time step of $\Delta t = 0.025$, was used. We can observe that the solutions are the same at $t = 1$ irrespective of the type of hardening. However, the solutions depend on the type of hardening on unloading and on further loading in compression. Note that unlike the cumulative plastic strain, which increases monotonically, the current plastic strain nearly follows the variation of the magnitude of the total strain.

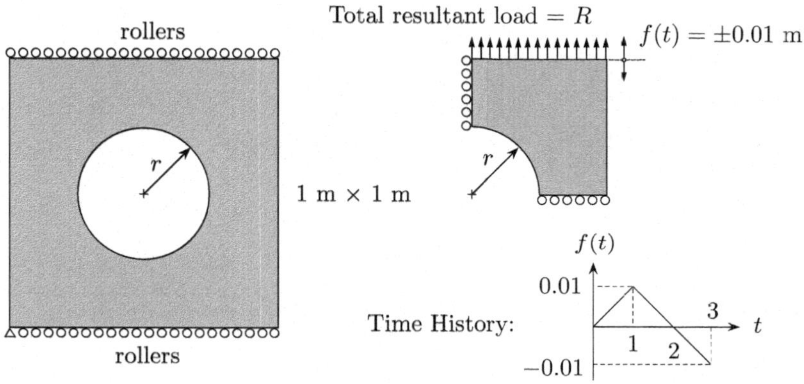

Fig. 3.10 Square steel plate with a circular hole subjected to vertical stretching followed by vertical compression

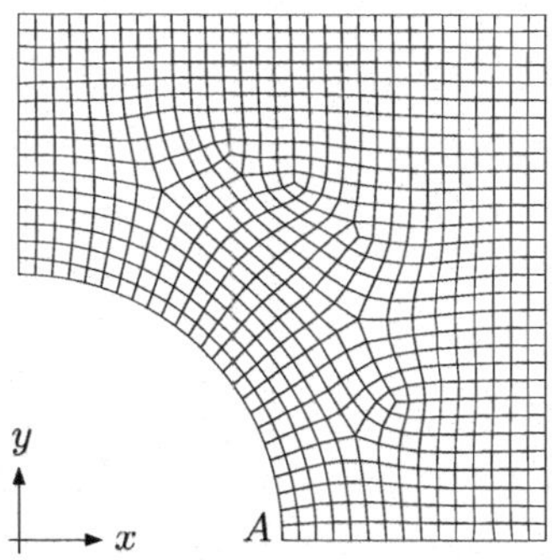

Fig. 3.11 Finite element mesh for square plate with a circular hole. The mesh has 765 nodes and 706 quadrilateral elements numerically integrated with the B-bar method.

Figure 3.15 portrays the evolution of the von Mises stress at a Gauss point in a corner element A of the quarter mesh shown in Fig. 3.11. The figure suggests that this stress point has unloaded well before time $t = 2$ due to 'locked-in' stresses induced by plastic deformation in the structure. Furthermore, the elastic region has clearly expanded in the isotropically hardening simulation causing a sharp V-shaped plot in the von Mises stress, whereas the V-shaped variation is much broader for the kinematically hardening simulation since the elastic region shifted but its size did not change. Figure 3.16 shows time histories of the vertical reactions predicted by the two simulations. Once again, the vertical reactions are the same during the stretching part of the experiment ($0 \leq t \leq 1$), as well as during the early part of elastic unloading. However, the kinematically hardening simulation predicted early yielding on reverse loading.

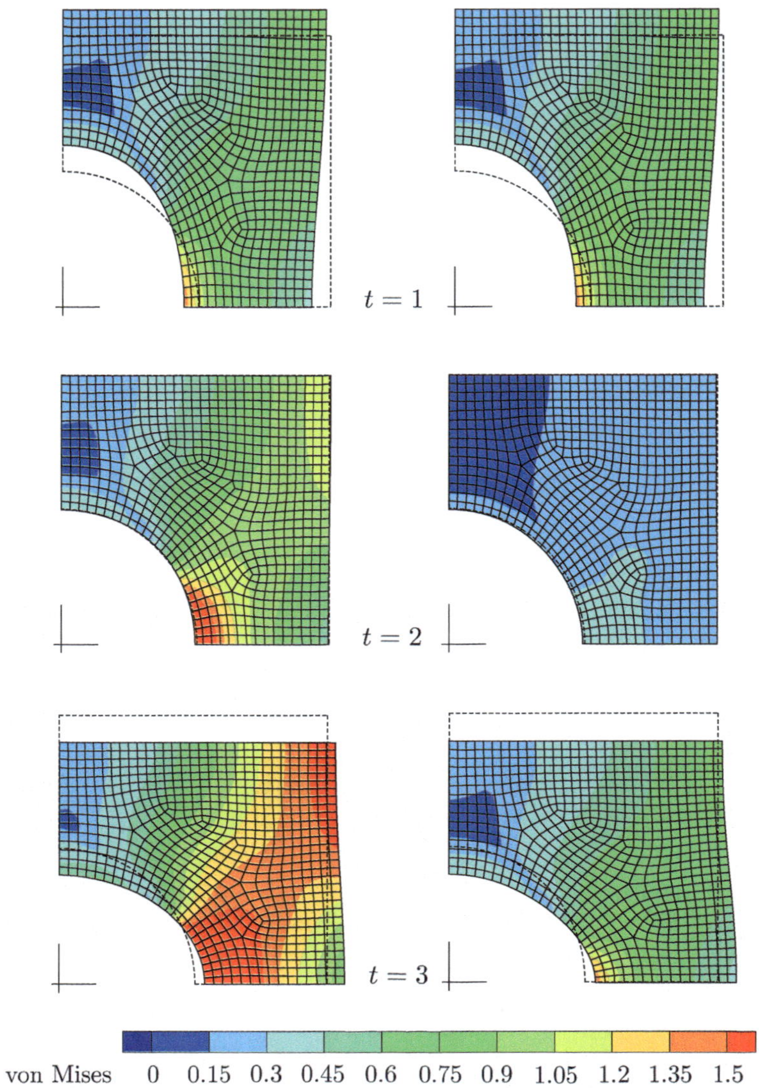

Fig. 3.12 Deformed meshes and contours of the von Mises stress ($\sqrt{3/2}\|\boldsymbol{s}\|$): isotropic hardening (left), and kinematic hardening (right). Color bar in GPa; displacements magnified 5×.

Figure 3.17 shows the optimal convergence rates exhibited by Newton's method. Here we employed the normalized error norm, $\|\boldsymbol{r}^k\|/\|\boldsymbol{r}^0\|$, as the convergence criterion, where $\boldsymbol{r}^k$ is the global out-of-balance force vector with initial value $\boldsymbol{r}^0$. We assumed the converged value $\boldsymbol{d}_n$ from the previous time step as the initial estimate for $\boldsymbol{d}_{n+1}$.

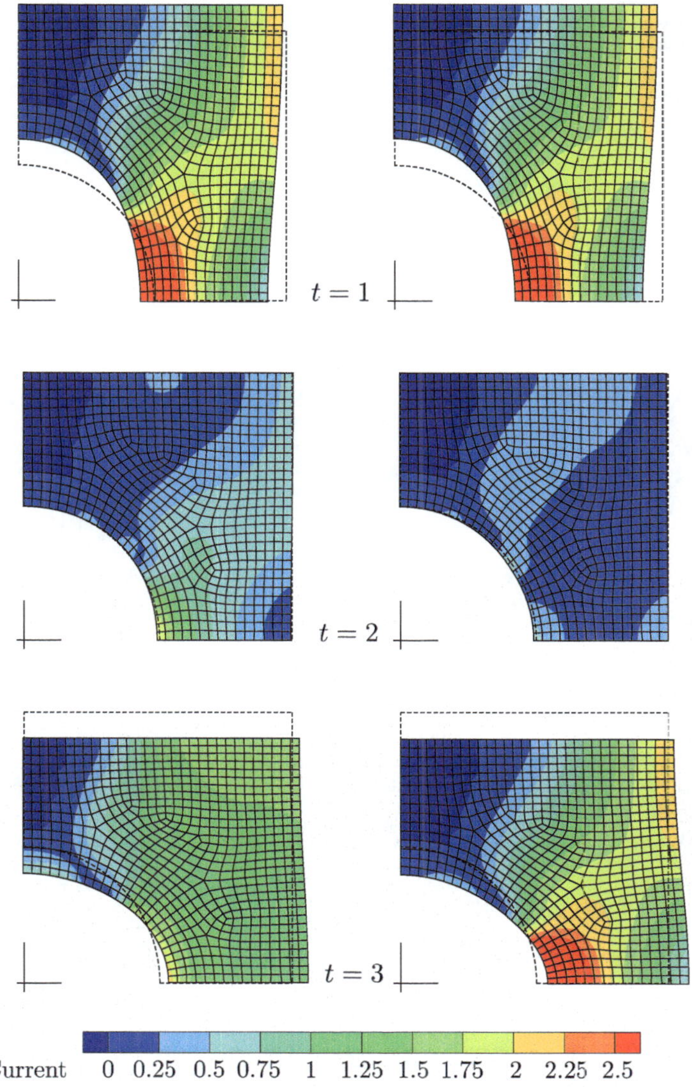

Fig. 3.13 Deformed meshes and contours of current plastic strain ($\sqrt{2/3}\|\boldsymbol{\epsilon}^{\mathrm{p}}\|$): isotropic hardening (left), and kinematic hardening (right). Color bar in percent; displacements magnified 5×.

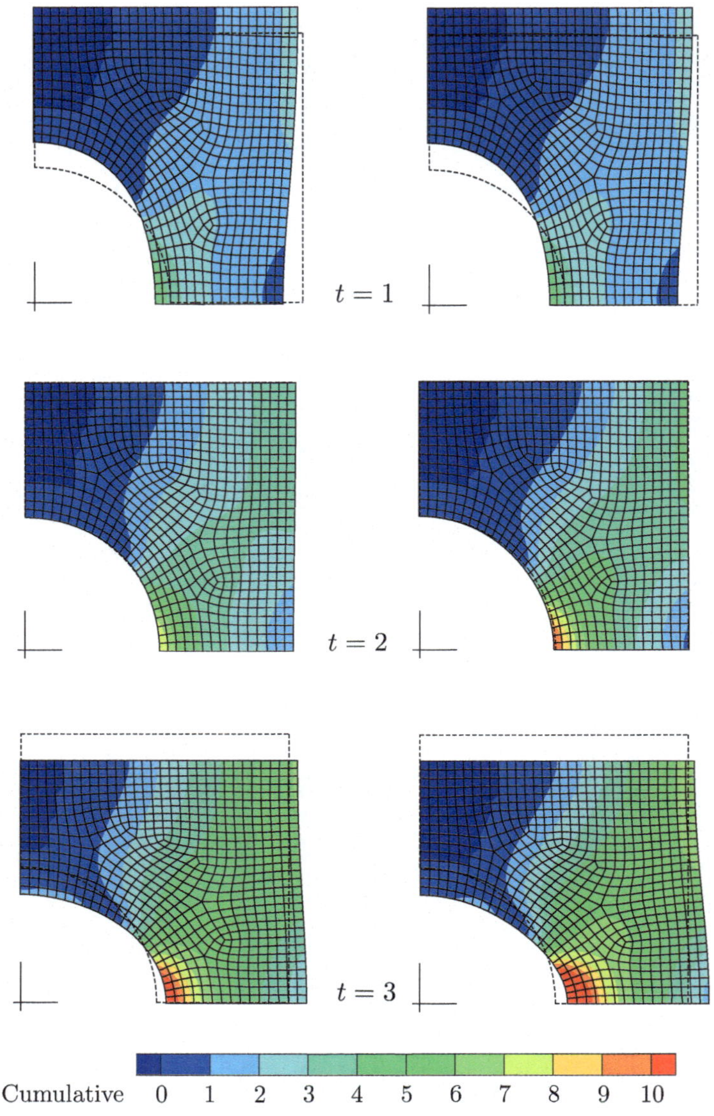

Fig. 3.14 Deformed meshes and contours of cumulative plastic strain ($\int \sqrt{2/3}\|\dot{\boldsymbol{\epsilon}}^{\mathrm{p}}\|\,\mathrm{d}t$): isotropic hardening (left), and kinematic hardening (right). Color bar in percent; displacements magnified 5×.

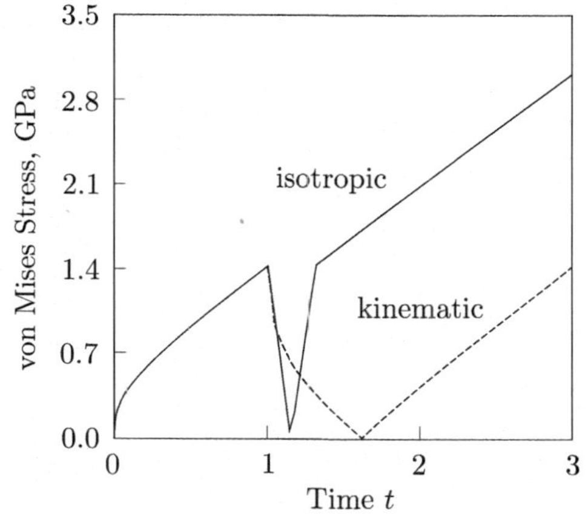

Fig. 3.15 Evolution of the von Mises stress ($\sqrt{3/2}\|\boldsymbol{s}\|$) at a Gauss point in a corner element A, see Fig. 3.11

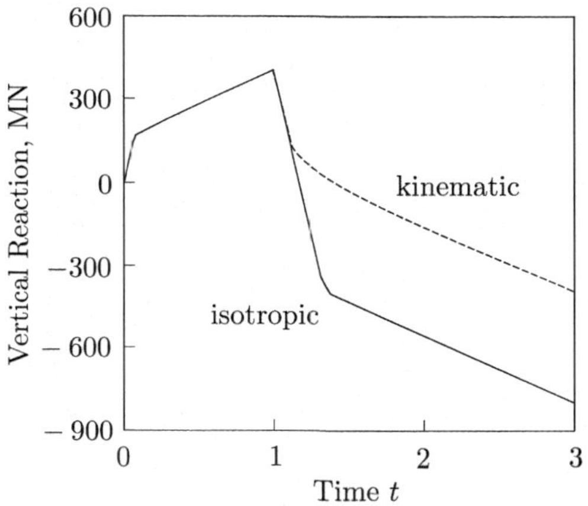

Fig. 3.16 Vertical reactions on the top and bottom sides. Note that the reactions from the quarter mesh are doubled to obtain the forces on the full mesh.

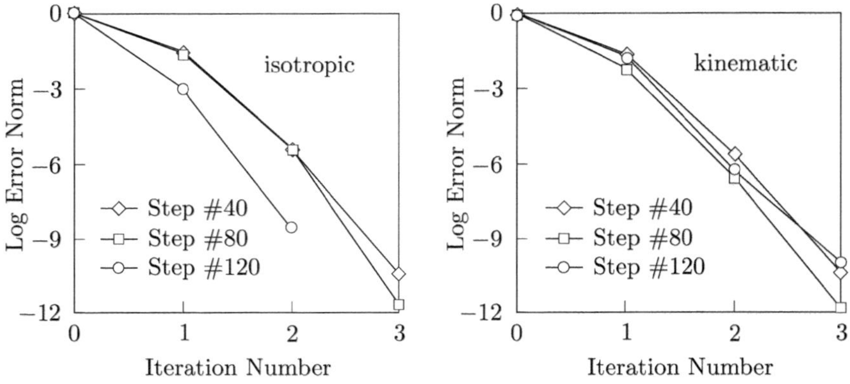

Fig. 3.17 Convergence profiles of Newton iterations for isotropic (left) and kinematic (right) hardening J_2-plasticity simulations

3.12 Summary

Yielding of pressure-insensitive ductile materials such as metals can be described accurately by the J_2 plasticity model. The Bauschinger effect can be captured by the introduction of a deviatoric back stress tensor. An implicit backward stress-point integration scheme is facilitated by the classic radial return algorithm. This technique is suitable for nonlinear finite element analysis, and can be employed for J_2 plasticity with a linear combination of isotropic and kinematic hardening.

4 Isotropic Functions

4.1 Introduction

Many natural and manufactured materials exhibit plastic deformation behavior that cannot be captured by the J_2 theory. They include concrete, rock, and metallic glass, which can compact or dilate inelastically and whose yield strength depends on the confining stress. High-porosity rocks compact under shearing due to collapse of the pores (Baud et al. 2004). Dense sands dilate to enable the moving particles to overcome interlocking (Lambe and Whitman 1969, Reynolds 1885). Bulk metallic glasses produce excess free volumes to accommodate the moving atoms (Spaepen 2006), although other processes such as diffusion and annihilation could also take place concurrently in competition with the free volume production (Flores and Dauskardt 1999, Wright et al. 2001). The J_2 theory, however, predicts no plastic volume change, so it is not appropriate for modeling the inelastic deformation behavior of these materials.

Figure 4.1 illustrates the effect of the mean confining pressure σ_c and the third stress invariant on the yield strength of plain concrete. This structural material consists primarily of aggregate and mortar, and exhibits a plastic deformation response ranging from brittle in tension to ductile at higher confinement. The test data shown in the figure were obtained from testing cylindrical specimens of concrete in triaxial extension (TXE) and triaxial compression (TXC). The test procedure consisted of subjecting the specimens to the same axial and radial stresses ($\sigma_a = \sigma_r \equiv -\sigma_c$), and then either reducing or increasing the axial stress σ_a quasi-statically while holding the radial stress σ_r fixed.

For the same confining pressure, the principal stress difference at failure tends to be greater for TXC than for TXE tests (Fig. 4.1a); however, at higher confining stresses they are nearly the same (Fig. 4.1b). Assuming that deformation in the concrete remained homogeneous and was not dominated by a major fracture or flaw during these tests, the yield strengths may be plotted on the deviatoric plane to show a nearly triangular shape at lower confining pressures and a nearly circular shape at higher confining pressures (Fig. 4.1b). To complement the TXC and TXE tests, and thus better define the shape of the yield surfaces on the deviatoric plane, torsion tests on cylindrical specimens may be conducted to generate a state of simple shear.

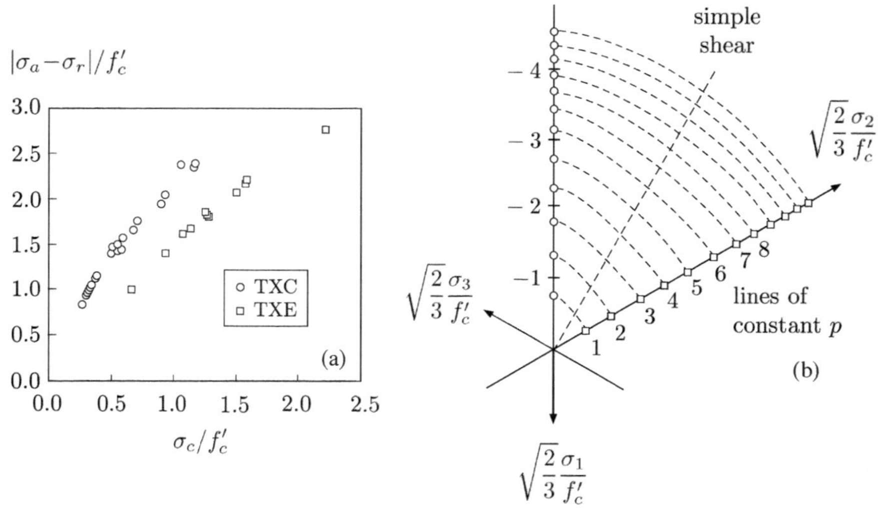

Fig. 4.1 Concrete strength data: (a) maximum principal stress difference versus confining pressure (Mills and Zimmerman 1970); and (b) strength loci on deviatoric plane from TXC and TXE tests (Launay and Gachon 1972). All axes are normalized with respect to ultimate compressive strength of concrete, f_c'.

In this chapter, we consider a fairly general class of plasticity models defined by all three invariants of the stress tensor. The choice of stress invariants as the independent variables is motivated by the restriction that the yield function must be frame-indifferent. Under superposed rigid body motions, the requirement of frame indifference requires that

$$f(\boldsymbol{Q}\cdot\boldsymbol{\sigma}\cdot\boldsymbol{Q}^{\mathsf{T}}) = f(\boldsymbol{\sigma}) \tag{4.1}$$

for any member $\boldsymbol{Q}$ of the proper orthogonal group. Any such function f that is invariant under orthogonal transformation is called an *isotropic function*. The requirement of frame invariance is met by functions that depend only on the invariants of the stress tensor, according to the representation theorem for isotropic functions and the spectral theorem for symmetric tensors (Gurtin 1981). There are at most three independent stress invariants, so the discussions in this chapter will focus on the general formulation and numerical implementation of three-invariant plasticity models.

4.2 Spectral Representation

We recall some useful properties of a symmetric second-order tensor. For any $\boldsymbol{\varphi} \in \mathbb{S}$ the characteristic equation is

$$\det(\boldsymbol{\varphi} - \varphi\mathbf{1}) = -\varphi^3 + \mathcal{I}_1(\boldsymbol{\varphi})\varphi^2 - \mathcal{I}_2(\boldsymbol{\varphi})\varphi + \mathcal{I}_3(\boldsymbol{\varphi}) = 0\,, \tag{4.2}$$

where

$$\left.\begin{aligned}\mathcal{I}_1(\boldsymbol{\varphi}) &= \mathrm{tr}(\boldsymbol{\varphi}) = \varphi_1 + \varphi_2 + \varphi_3 \\ \mathcal{I}_2(\boldsymbol{\varphi}) &= \frac{1}{2}[\mathrm{tr}^2(\boldsymbol{\varphi}) - \mathrm{tr}(\boldsymbol{\varphi}^2)] = \varphi_1\varphi_2 + \varphi_2\varphi_3 + \varphi_1\varphi_3 \\ \mathcal{I}_3(\boldsymbol{\varphi}) &= \frac{1}{6}\mathrm{tr}^3(\boldsymbol{\varphi}) - \frac{1}{2}\mathrm{tr}(\boldsymbol{\varphi})\mathrm{tr}(\boldsymbol{\varphi}^2) + \frac{1}{3}\mathrm{tr}(\boldsymbol{\varphi}^3) = \det\boldsymbol{\varphi} = \varphi_1\varphi_2\varphi_3\end{aligned}\right\} \tag{4.3}$$

are the invariants of $\boldsymbol{\varphi}$, and $\varphi_1, \varphi_2, \varphi_3 \in \mathbb{R}$ are its eigenvalues. By the Cayley-Hamilton theorem, $\boldsymbol{\varphi}$ satisfies its own characteristic equation,

$$-\boldsymbol{\varphi}^3 + \mathcal{I}_1(\boldsymbol{\varphi})\boldsymbol{\varphi}^2 - \mathcal{I}_2(\boldsymbol{\varphi})\boldsymbol{\varphi} + \mathcal{I}_3(\boldsymbol{\varphi})\mathbf{1} = \mathbf{0}\,. \tag{4.4}$$

If all the eigenvalues are distinct, then the tensor $\boldsymbol{\varphi}$ admits the spectral representation

$$\boldsymbol{\varphi} = \sum_{A=1}^{3} \varphi_A \boldsymbol{m}^{(A)}\,, \qquad \boldsymbol{m}^{(A)} = \boldsymbol{n}^{(A)} \otimes \boldsymbol{n}^{(A)}\,, \tag{4.5}$$

where $\boldsymbol{n}^{(A)}$ and $\boldsymbol{m}^{(A)}$ are unit eigenvectors and spectral directions, respectively. The eigenvectors satisfy the orthogonality condition $\boldsymbol{n}^{(A)} \cdot \boldsymbol{n}^{(B)} = \delta_{AB}$ = Kronecker delta. Further, the spectral direction $\boldsymbol{m}^{(A)}$ has the properties

$$\sum_{A=1}^{3} \boldsymbol{m}^{(A)} = \mathbf{1}\,, \qquad \mathrm{tr}(\boldsymbol{m}^{(A)}) = \boldsymbol{m}^{(A)} : \mathbf{1} = 1\,, \tag{4.6}$$

where $(\mathbf{1})_{AB} = \delta_{AB}$. If $\boldsymbol{\varphi}$ has two distinct eigenvalues, φ_1 and φ_2, then its spectral form is

$$\boldsymbol{\varphi} = \varphi_1 \boldsymbol{m}^{(1)} + \varphi_2(\mathbf{1} - \boldsymbol{m}^{(1)})\,, \tag{4.7}$$

where φ_1 is the non-repeated eigenvalue with corresponding spectral direction $\boldsymbol{m}^{(1)} = \boldsymbol{n}^{(1)} \otimes \boldsymbol{n}^{(1)}$, and $\varphi_2 = \varphi_3$ is the repeated eigenvalue. If $\boldsymbol{\varphi}$ has only one eigenvalue, say, φ, then it is an isotropic tensor of the form

$$\boldsymbol{\varphi} = \varphi\mathbf{1}\,. \tag{4.8}$$

If $f(\boldsymbol{\varphi})$ is an isotropic function, then there exists a function $\widetilde{f}$ such that

$$f(\boldsymbol{\varphi}) = \widetilde{f}(\mathcal{I}_1(\boldsymbol{\varphi}), \mathcal{I}_2(\boldsymbol{\varphi}), \mathcal{I}_3(\boldsymbol{\varphi}))\,. \tag{4.9}$$

To prove this assertion, it suffices to show that $f(\boldsymbol{\varphi}) = f(\widetilde{\boldsymbol{\varphi}})$ whenever $\mathcal{I}_A(\boldsymbol{\varphi}) = \mathcal{I}_A(\widetilde{\boldsymbol{\varphi}})$ for any $\boldsymbol{\varphi}, \widetilde{\boldsymbol{\varphi}} \in \mathbb{S}$, where

$$\widetilde{\boldsymbol{\varphi}} = \sum_{A=1}^{3} \varphi_A \boldsymbol{\eta}^{(A)} \otimes \boldsymbol{\eta}^{(A)} \tag{4.10}$$

is a tensor having the same eigenvalues as $\boldsymbol{\varphi}$ but having rotated eigenvectors. To this end, we choose an operator $\boldsymbol{Q}$ satisfying the orthogonality property $\boldsymbol{Q} \cdot \boldsymbol{Q}^{\mathsf{T}} = 1$ such that

$$\boldsymbol{n}^{(A)} = \boldsymbol{Q} \cdot \boldsymbol{\eta}^{(A)}, \qquad A = 1, 2, 3. \tag{4.11}$$

Then,

$$\boldsymbol{\varphi} = \boldsymbol{Q} \cdot \widetilde{\boldsymbol{\varphi}} \cdot \boldsymbol{Q}^{\mathsf{T}}. \tag{4.12}$$

However, by definition of an isotropic function,

$$f(\boldsymbol{\varphi}) = f(\boldsymbol{Q} \cdot \widetilde{\boldsymbol{\varphi}} \cdot \boldsymbol{Q}^{\mathsf{T}}) = f(\widetilde{\boldsymbol{\varphi}}). \tag{4.13}$$

Hence, f admits the invariant representation (4.9).

Alternatively, any isotropic function $f(\boldsymbol{\varphi})$ may be expressed in terms of the principal values of $\boldsymbol{\varphi}$ in the sense that there exists a scalar function $\widehat{f}$ such that

$$f(\boldsymbol{\varphi}) = \widehat{f}(\varphi_1, \varphi_2, \varphi_3) = \widehat{f}(\varphi_1, \varphi_3, \varphi_2) = \widehat{f}(\varphi_2, \varphi_1, \varphi_3) = \ldots \tag{4.14}$$

To prove this assertion, we express the invariants in terms of the principal values, see (4.3), and insert the results into $\widetilde{f}$. Since the invariants of $\boldsymbol{\varphi}$ do not depend on how the principal values are numbered, $\widehat{f}$ does not depend on the order in which the invariants appear in the function.

Exercise 4.1. A tensor $\boldsymbol{a} \in \mathbb{S}$ has the following elements: $a_{11} = 10$, $a_{22} = 20$, $a_{33} = 30$, $a_{12} = a_{21} = -15$, $a_{13} = a_{31} = -25$ and $a_{23} = a_{32} = 0$. Determine the principal values and spectral directions of this tensor and verify that the spectral directions satisfy (4.6).

Exercise 4.2. Repeat Exercise 4.1 for a tensor $\boldsymbol{b} \in \mathbb{S}$ having the following elements: $b_{11} = 5$, $b_{22} = 7$, $b_{33} = 8$, $b_{12} = b_{21} = \sqrt{3}$, $b_{13} = b_{31} = b_{23} = b_{32} = 0$.

4.3 Spin of a Tensor

We now consider the spin of an eigenvector $\boldsymbol{n}^{(A)}$. Here we are interested in the rate at which this vector rotates relative to a fixed Cartesian reference frame. Let $\{\boldsymbol{\eta}^{(1)}, \boldsymbol{\eta}^{(2)}, \boldsymbol{\eta}^{(3)}\}$ denote the basis vectors for the fixed reference frame. Taking the time derivative of the orthogonal transformation (4.11) and eliminating $\boldsymbol{\eta}^{(A)}$, we obtain

$$\dot{\boldsymbol{n}}^{(A)} = \boldsymbol{\omega} \cdot \boldsymbol{n}^{(A)}, \qquad \boldsymbol{\omega} = \dot{\boldsymbol{Q}} \cdot \boldsymbol{Q}^{\mathsf{T}} = -\boldsymbol{Q} \cdot \dot{\boldsymbol{Q}}^{\mathsf{T}}. \tag{4.15}$$

The last expression follows from the relation $\dot{\overline{\boldsymbol{Q} \cdot \boldsymbol{Q}^{\mathsf{T}}}} \equiv \boldsymbol{0}$ and shows that $\boldsymbol{\omega}$ is an anti-symmetric tensor representing the spin of $\boldsymbol{n}^{(A)}$. This spin admits the representation

$$\boldsymbol{\omega} = \sum_{A=1}^{3} \dot{\boldsymbol{n}}^{(A)} \otimes \boldsymbol{n}^{(A)} = -\sum_{A=1}^{3} \boldsymbol{n}^{(A)} \otimes \dot{\boldsymbol{n}}^{(A)}. \tag{4.16}$$

Note that $\dot{\boldsymbol{n}}^{(A)}$ is orthogonal to $\boldsymbol{n}^{(A)}$, as can be seen from the relation

$$\boldsymbol{n}^{(A)} \cdot \dot{\boldsymbol{n}}^{(A)} = \boldsymbol{\omega} : (\boldsymbol{n}^{(A)} \otimes \boldsymbol{n}^{(A)}) \equiv 0 \,, \tag{4.17}$$

since $\boldsymbol{\omega}$ is a skew-symmetric tensor and $\boldsymbol{n}^{(A)} \otimes \boldsymbol{n}^{(A)}$ is a symmetric tensor. Therefore, $\dot{\boldsymbol{n}}^{(A)}$ should lie on the plane defined by the other eigenvectors $\boldsymbol{n}^{(B)}$ and $\boldsymbol{n}^{(C)}$, where $A \neq B \neq C \neq A$ are permutations of 1,2 and 3. In other words,

$$\dot{\boldsymbol{n}}^{(A)} = \omega_{BA}\boldsymbol{n}^{(B)} + \omega_{CA}\boldsymbol{n}^{(C)} = \sum_{B \neq A} \omega_{BA}\boldsymbol{n}^{(B)} \,, \tag{4.18}$$

where ω_{BA} is the component of spin that rotates the basis $\boldsymbol{n}^{(A)}$ to direction $\boldsymbol{n}^{(B)}$, etc.

Next, we use the chain rule to get the derivative of $\boldsymbol{\varphi}$

$$\begin{aligned} \dot{\boldsymbol{\varphi}} &= \sum_{A=1}^{3} \dot{\varphi}_A \boldsymbol{n}^{(A)} \otimes \boldsymbol{n}^{(A)} + \sum_{A=1}^{3} \varphi_A \dot{\boldsymbol{n}}^{(A)} \otimes \boldsymbol{n}^{(A)} + \sum_{A=1}^{3} \varphi_A \boldsymbol{n}^{(A)} \otimes \dot{\boldsymbol{n}}^{(A)} \\ &= \sum_{A=1}^{3} \dot{\varphi}_A \boldsymbol{n}^{(A)} \otimes \boldsymbol{n}^{(A)} + \boldsymbol{\omega} \cdot \boldsymbol{\varphi} - \boldsymbol{\varphi} \cdot \boldsymbol{\omega} \,. \end{aligned} \tag{4.19}$$

Substituting (4.18) into (4.19) gives

$$\begin{aligned} \dot{\boldsymbol{\varphi}} &= \sum_{A=1}^{3} \dot{\varphi}_A \boldsymbol{n}^{(A)} \otimes \boldsymbol{n}^{(A)} + \sum_{A=1}^{3} \varphi_A \Big(\sum_{B \neq A} \omega_{BA} \boldsymbol{n}^{(B)} \Big) \otimes \boldsymbol{n}^{(A)} \\ &\qquad + \sum_{A=1}^{3} \varphi_A \boldsymbol{n}^{(A)} \otimes \Big(\sum_{B \neq A} \omega_{BA} \boldsymbol{n}^{(B)} \Big) \\ &= \sum_{A=1}^{3} \dot{\varphi}_A \boldsymbol{n}^{(A)} \otimes \boldsymbol{n}^{(A)} + \sum_{A=1}^{3} \sum_{B \neq A} \omega_{AB} (\varphi_B - \varphi_A) \boldsymbol{n}^{(A)} \otimes \boldsymbol{n}^{(B)} \,. \end{aligned} \tag{4.20}$$

The geometric meaning of this time derivative is as follows: the first summation is the rate of change of the principal values of $\boldsymbol{\varphi}$ with the principal directions held fixed, while the second pair of summations is the spin of the principal axes with the principal values held fixed (Ogden 1984).

4.4 Constitutive Operators in Spectral Form

For an isotropic linearly elastic material, the Cauchy stress tensor $\boldsymbol{\sigma}$ is coaxial with the elastic infinitesimal strain tensor $\boldsymbol{\epsilon}^{\mathrm{e}}$, i.e., they have the same eigenvectors $\boldsymbol{n}^{(1)}$, $\boldsymbol{n}^{(2)}$, and $\boldsymbol{n}^{(3)}$. Therefore, they admit the spectral representation

$$\boldsymbol{\sigma} = \sum_{A=1}^{3} \sigma_A \boldsymbol{n}^{(A)} \otimes \boldsymbol{n}^{(A)} \,, \qquad \boldsymbol{\epsilon}^{\mathrm{e}} = \sum_{A=1}^{3} \epsilon_A^{\mathrm{e}} \boldsymbol{n}^{(A)} \otimes \boldsymbol{n}^{(A)} \,. \tag{4.21}$$

By letting the isotropic elasticity tensor $\boldsymbol{c}^{\mathrm{e}} = K\mathbf{1} \otimes \mathbf{1} + 2\mu(\boldsymbol{I} - \mathbf{1} \otimes \mathbf{1}/3)$ operate on $\boldsymbol{\epsilon}^{\mathrm{e}}$, the principal values of $\boldsymbol{\sigma}$ and $\boldsymbol{\epsilon}^{\mathrm{e}}$ can be related according to the equation

$$\sigma_A = \sum_{B=1}^{3} a^{\mathrm{e}}_{AB} \epsilon^{\mathrm{e}}_B \,, \qquad \epsilon^{\mathrm{e}}_A = \sum_{B=1}^{3} a^{\mathrm{e}-1}_{AB} \sigma_B \,, \qquad A = 1,2,3 \,, \tag{4.22}$$

where

$$[\,a^{\mathrm{e}}_{AB}\,] = \begin{bmatrix} a & b & b \\ b & a & b \\ b & b & a \end{bmatrix}, \qquad a = K + \frac{4\mu}{3}\,, \qquad b = K - \frac{2\mu}{3} \tag{4.23}$$

is the matrix of elastic moduli in principal axes.

Writing the Cauchy stress rate tensor in spectral form gives

$$\dot{\boldsymbol{\sigma}} = \sum_{A=1}^{3} \dot{\sigma}_A \boldsymbol{n}^{(A)} \otimes \boldsymbol{n}^{(A)} + \sum_{A=1}^{3} \sum_{B \neq A} \omega_{AB} (\sigma_B - \sigma_A) \boldsymbol{n}^{(A)} \otimes \boldsymbol{n}^{(B)} \,. \tag{4.24}$$

Similarly, writing the elastic infinitesimal strain rate tensor in spectral form gives

$$\dot{\boldsymbol{\epsilon}}^{\mathrm{e}} = \sum_{A=1}^{3} \dot{\epsilon}^{\mathrm{e}}_A \boldsymbol{n}^{(A)} \otimes \boldsymbol{n}^{(A)} + \sum_{A=1}^{3} \sum_{B \neq A} \omega_{AB} (\epsilon^{\mathrm{e}}_B - \epsilon^{\mathrm{e}}_A) \boldsymbol{n}^{(A)} \otimes \boldsymbol{n}^{(B)} \,. \tag{4.25}$$

Observe that the spins of the two tensors are the same from the property of coaxiality.

Substituting (4.22) into (4.24) gives

$$\begin{aligned} \dot{\boldsymbol{\sigma}} \;=\; & \sum_{A=1}^{3} \sum_{B=1}^{3} a^{\mathrm{e}}_{AB} \dot{\epsilon}^{\mathrm{e}}_B \boldsymbol{n}^{(A)} \otimes \boldsymbol{n}^{(A)} \\ & + \sum_{A=1}^{3} \sum_{B \neq A} \omega_{AB} (\epsilon^{\mathrm{e}}_B - \epsilon^{\mathrm{e}}_A) \left(\frac{\sigma_B - \sigma_A}{\epsilon^{\mathrm{e}}_B - \epsilon^{\mathrm{e}}_A} \right) \boldsymbol{n}^{(A)} \otimes \boldsymbol{n}^{(B)} \,. \end{aligned} \tag{4.26}$$

This enables the extraction of the isotropic elastic stress-strain tensor in spectral form,

$$\begin{aligned} \boldsymbol{c}^{\mathrm{e}} \;=\; & \sum_{A=1}^{3} \sum_{B=1}^{3} a^{\mathrm{e}}_{AB} \boldsymbol{m}^{(A)} \otimes \boldsymbol{m}^{(B)} \\ + \; & \frac{1}{2} \sum_{A=1}^{3} \sum_{B \neq A} \left(\frac{\sigma_B - \sigma_A}{\epsilon^{\mathrm{e}}_B - \epsilon^{\mathrm{e}}_A} \right) (\boldsymbol{m}^{(AB)} \otimes \boldsymbol{m}^{(AB)} + \boldsymbol{m}^{(AB)} \otimes \boldsymbol{m}^{(BA)}) \,, \end{aligned} \tag{4.27}$$

where $\boldsymbol{m}^{(AB)} = \boldsymbol{n}^{(A)} \otimes \boldsymbol{n}^{(B)}$. This tensor is amenable to the following interpretation: the first pair of summations represents the material elasticity in principal axes, whereas the second represents the spin of the principal axes. It is interesting to note that the spin tensor $\boldsymbol{\omega}$ does not enter into the expression for $\boldsymbol{c}^{\mathrm{e}}$.

The spectral form of $\boldsymbol{c}^{\mathrm{e}}$ is only meaningful when the eigenvalues are all distinct. When two eigenvalues are the same, say, the A and B eigenvalues, then the ratio $(\sigma_B - \sigma_A)/(\epsilon_B - \epsilon_A)$ is indeterminate and thus cannot be evaluated. In this case, it may be necessary to replace this indeterminate expression with the alternative expression $\partial(\sigma_B - \sigma_A)/\partial\epsilon_A^{\mathrm{e}}$ (Ogden 1984). Better still, a simple perturbation of the eigenvalues will do the job for numerical calculation purposes, i.e., the conditions may be slightly perturbed so that all eigenvalues are distinct.

Because of the assumed isotropy in the elastic and plastic responses, we can also develop a spectral representation for the elastoplastic constitutive operator $\boldsymbol{c}^{\mathrm{ep}}$. Consider yield and plastic potential functions of the form

$$f(\boldsymbol{\sigma}, \boldsymbol{\kappa}) = \widehat{f}(\sigma_1, \sigma_2, \sigma_3, \boldsymbol{\kappa}), \qquad g(\boldsymbol{\sigma}, \boldsymbol{\kappa}) = \widehat{g}(\sigma_1, \sigma_2, \sigma_3, \boldsymbol{\kappa}). \tag{4.28}$$

From the chain rule, the gradients with respect to $\boldsymbol{\sigma}$ are

$$\frac{\partial f}{\partial \boldsymbol{\sigma}} = \sum_{A=1}^{3} f_A \frac{\partial \sigma_A}{\partial \boldsymbol{\sigma}}, \qquad \frac{\partial g}{\partial \boldsymbol{\sigma}} = \sum_{A=1}^{3} g_A \frac{\partial \sigma_A}{\partial \boldsymbol{\sigma}}, \tag{4.29}$$

where

$$f_A = \frac{\partial \widehat{f}}{\partial \sigma_A}, \qquad g_A = \frac{\partial \widehat{g}}{\partial \sigma_A}. \tag{4.30}$$

But from the definition of the spectral directions,

$$\dot{\sigma}_C = \frac{\partial \sigma_C}{\partial \boldsymbol{\sigma}} : \dot{\boldsymbol{\sigma}} = \boldsymbol{m}^{(C)} : \dot{\boldsymbol{\sigma}}. \tag{4.31}$$

Since this last equation is true for any arbitrary stress rate $\dot{\boldsymbol{\sigma}}$, it follows that

$$\frac{\partial \sigma_A}{\partial \boldsymbol{\sigma}} = \boldsymbol{m}^{(A)}. \tag{4.32}$$

Thus, the gradients of the yield and plastic potential functions also admit the spectral representation

$$\frac{\partial f}{\partial \boldsymbol{\sigma}} = \sum_{A=1}^{3} f_A \boldsymbol{n}^{(A)} \otimes \boldsymbol{n}^{(A)}, \qquad \frac{\partial g}{\partial \boldsymbol{\sigma}} = \sum_{A=1}^{3} g_A \boldsymbol{n}^{(A)} \otimes \boldsymbol{n}^{(A)}. \tag{4.33}$$

This means that the tensors $\boldsymbol{\sigma}$, $\boldsymbol{\epsilon}^{\mathrm{e}}$, $\partial f/\partial \boldsymbol{\sigma}$, and $\partial g/\partial \boldsymbol{\sigma}$ are all coaxial.

We can now write the elastoplastic constitutive operator in spectral form

$$\begin{aligned}
\boldsymbol{c}^{\mathrm{ep}} &= \boldsymbol{c}^{\mathrm{e}} - \frac{1}{\chi}\boldsymbol{c}^{\mathrm{e}} : \frac{\partial g}{\partial \boldsymbol{\sigma}} \otimes \frac{\partial f}{\partial \boldsymbol{\sigma}} : \boldsymbol{c}^{\mathrm{e}} \\
&= \sum_{A=1}^{3}\sum_{B=1}^{3} a_{AB}^{\mathrm{ep}} \boldsymbol{m}^{(A)} \otimes \boldsymbol{m}^{(B)} \\
&+ \frac{1}{2}\sum_{A=1}^{3}\sum_{B\neq A} \left(\frac{\sigma_B - \sigma_A}{\epsilon_B^{\mathrm{e}} - \epsilon_A^{\mathrm{e}}}\right)(\boldsymbol{m}^{(AB)} \otimes \boldsymbol{m}^{(AB)} + \boldsymbol{m}^{(AB)} \otimes \boldsymbol{m}^{(BA)}) \,, \qquad (4.34)
\end{aligned}$$

where

$$\begin{aligned}
a_{AB}^{\mathrm{ep}} &= a_{AB}^{\mathrm{e}} - \frac{1}{\chi}\Big(\sum_{C=1}^{3} f_C a_{CA}^{\mathrm{e}}\Big)\Big(\sum_{D=1}^{3} a_{BD}^{\mathrm{e}} g_D\Big) \,, \\
\chi &= \frac{\partial f}{\partial \boldsymbol{\sigma}} : \boldsymbol{c}^{\mathrm{e}} : \frac{\partial g}{\partial \boldsymbol{\sigma}} + H = \sum_{A=1}^{3}\sum_{B=1}^{3} f_A a_{AB}^{\mathrm{e}} g_B + H \,. \qquad (4.35)
\end{aligned}$$

Observe that $\boldsymbol{c}^{\mathrm{ep}}$ differs from $\boldsymbol{c}^{\mathrm{e}}$ only through the matrix $[a_{AB}^{\mathrm{ep}}]$. This is due to the lack of plastic spin in the elastoplastic constitutive formulation, resulting in the elastic spin being equal to the total spin.

Exercise 4.3. Let $\boldsymbol{s} = \boldsymbol{\sigma} - \mathrm{tr}(\boldsymbol{\sigma})\mathbf{1}/3$ denote the deviatoric component of the Cauchy stress tensor, and assume that $\boldsymbol{\sigma}$ has the spectral directions $\boldsymbol{m}^{(A)}$, $A = 1, 2, 3$. Show that $\boldsymbol{s}$ has the same spectral directions and express its principal values in terms of the principal values of $\boldsymbol{\sigma}$.

Exercise 4.4. Using the spectral representations for $\dot{\boldsymbol{\sigma}}$, $\dot{\boldsymbol{\epsilon}}^{\mathrm{e}}$, and $\boldsymbol{c}^{\mathrm{e}}$, verify the elastic constitutive equation $\dot{\boldsymbol{\sigma}} = \boldsymbol{c}^{\mathrm{e}} : \dot{\boldsymbol{\epsilon}}^{\mathrm{e}}$.

4.5 Lode's Angle

So far we have considered an isotropic function of the stress tensor of the form $f(\mathcal{I}_1, \mathcal{I}_2, \mathcal{I}_3)$ or $f(\sigma_1, \sigma_2, \sigma_3)$. Isotropic functions in principal stress space always have the hydrostatic axis as their generating axis. When plotted on the deviatoric plane, the cross section of an isotropic function of the first two stress invariants has a circular shape, which permits the representation of this function in terms of $(\mathcal{I}_1, J_2)$. If the third stress invariant is also present, then the cross section of an isotropic function on the deviatoric plane is no longer circular, but it still exhibits symmetry with respect to the three generating principal stress axes (six, if positive and negative principal stress axes are counted separately). In this case, a convenient alternative representation is through the invariants $(\mathcal{I}_1, J_2, \theta)$, where θ is an angle sweeping a sector of $2\pi/6$ on the deviatoric plane, called Lode's angle (Lode 1926).

Consider the deviatoric component $\boldsymbol{s} \in \mathbb{S}$ of the Cauchy stress tensor. From (4.2) and (4.3), the characteristic equation of this tensor is

$$s^3 - J_2 s - J_3 = 0\,, \tag{4.36}$$

where s is a principal value of $\boldsymbol{s}$, and

$$J_2 = -\mathcal{I}_2(\boldsymbol{s}) = \frac{1}{2}\mathrm{tr}(\boldsymbol{s}^2)\,, \qquad J_3 = \mathcal{I}_3(\boldsymbol{s}) = \frac{1}{3}\mathrm{tr}(\boldsymbol{s}^3)\,. \tag{4.37}$$

The cubic polynomial (4.36) has a particular form that allows an explicit determination of the three roots (Nayak and Zienkiewicz 1972).

Consider the identity

$$\cos^3\theta - \frac{3}{4}\cos\theta - \frac{1}{4}\cos 3\theta = 0\,. \tag{4.38}$$

Writing s in polar coordinates yields

$$s = r\cos\theta\,. \tag{4.39}$$

Substituting into the characteristic equation and re-arranging gives

$$\cos^3\theta - \frac{J_2}{r^2}\cos\theta - \frac{J_3}{r^3} = 0\,. \tag{4.40}$$

By comparing (4.38) and (4.40), we conclude that

$$\frac{J_2}{r^2} = \frac{3}{4}\,, \qquad \frac{J_3}{r^3} = \frac{1}{4}\cos 3\theta\,. \tag{4.41}$$

Thus, the polar radius r and angle θ are obtained as

$$r = 2\sqrt{J_2/3}\,, \qquad \cos 3\theta = \frac{3\sqrt{3}J_3}{2J_2^{3/2}} = \sqrt{6}\frac{\mathrm{tr}(\boldsymbol{s}^3)}{[\mathrm{tr}(\boldsymbol{s}^2)]^{3/2}}\,. \tag{4.42}$$

We need to impose the restriction $0 \le 3\theta \le \pi$ for the primary root so that we can cover the range $-1 \le \cos 3\theta \le 1$. Hence, θ should cover the range

$$0 \le \theta \le \pi/3\,. \tag{4.43}$$

The cosine function has a cyclic nature $\cos 3\theta = \cos(3\theta+2\pi) = \cos(3\theta+4\pi) =$ and so on. Therefore, the three roots of the cubic polynomial must be

$$s_A = r\cos\theta\,, \qquad s_B = r\cos(\theta + 4\pi/3)\,, \qquad s_C = r\cos(\theta + 2\pi/3)\,, \tag{4.44}$$

where A, B, and C are permutations of 1, 2 and 3, and where $s_A > s_B > s_C$ according to restriction (4.43) on the values of θ, see Fig. 4.2.

Alternatively, consider the identity

$$\sin^3\overline{\theta} - \frac{3}{4}\sin\overline{\theta} + \frac{1}{4}\sin 3\overline{\theta} = 0\,. \tag{4.45}$$

Writing s in polar coordinates gives

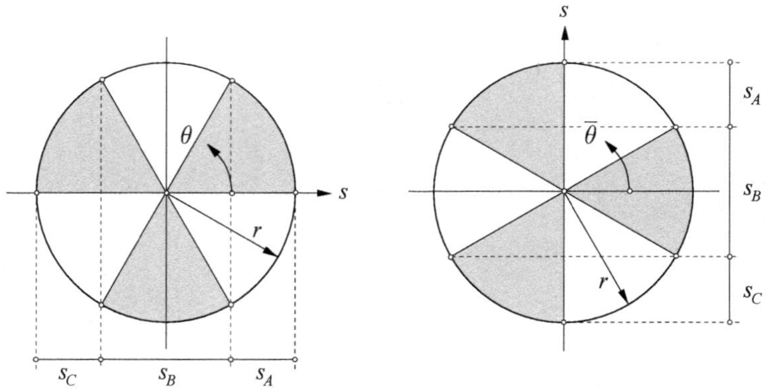

Fig. 4.2 Range of values of s_A, s_B, and s_C. All sectors are $\pi/3$

$$s = r\sin\overline{\theta}\,. \tag{4.46}$$

Substituting into the characteristic equation yields

$$\sin^3\overline{\theta} - \frac{J_2}{r^2}\sin\overline{\theta} - \frac{J_3}{r^3} = 0\,. \tag{4.47}$$

Therefore,

$$\frac{J_2}{r^2} = \frac{3}{4}\,, \qquad -\frac{J_3}{r^3} = \frac{1}{4}\sin 3\overline{\theta}\,. \tag{4.48}$$

Thus, the polar radius r and angle $\overline{\theta}$ are obtained as

$$r = 2\sqrt{J_2/3}\,, \qquad \sin 3\overline{\theta} = -\frac{3\sqrt{3}J_3}{2J_2^{3/2}} = -\sqrt{6}\frac{\mathrm{tr}(\boldsymbol{s}^3)}{[\mathrm{tr}(\boldsymbol{s}^2)]^{3/2}}\,. \tag{4.49}$$

We need to impose the restriction $-\pi/2 \le 3\overline{\theta} \le \pi/2$ for the primary root so that we can cover the range $-1 \le \sin 3\overline{\theta} \le 1$; thus,

$$-\pi/6 \le \overline{\theta} \le \pi/6\,. \tag{4.50}$$

In a similar vein, by noting the cyclic nature of the sine function we can determine the three roots of the cubic polynomial as

$$s_A = r\sin(\overline{\theta} + 2\pi/3)\,, \qquad s_B = r\sin\overline{\theta}\,, \qquad s_C = r\sin(\overline{\theta} + 4\pi/3)\,, \tag{4.51}$$

where $s_A > s_B > s_C$ according to restriction (4.50) on $\overline{\theta}$, see Fig. 4.2. Note that the polar radius r is the same, but the two polar angles are now shifted by an angular distance of $\pi/6$, i.e.,

$$\theta = \overline{\theta} + \pi/6\,. \tag{4.52}$$

To identify the relevant polar axes for the polar angles θ and $\overline{\theta}$, we note that at $\theta = \pi/6$ (or at $\overline{\theta} = 0$), we have $\mathrm{tr}(\boldsymbol{s}) = \mathcal{I}_3(\boldsymbol{s}) = s_1 s_2 s_3 = 0$. If all of the roots are distinct, then one of these roots must be zero. Thus, if $s_B = 0$, then the line $\overline{\theta} = 0$ must be perpendicular to the σ_B axis on the deviatoric plane—this is the line of pure shear. The 'most positive' principal value of $\boldsymbol{s}$ occurs at $\theta = 0$ (or at $\overline{\theta} = -\pi/6$), which is the line of pure extension, while the 'most negative' principal value occurs at $\theta = \pi/3$ (or at $\overline{\theta} = \pi/6$), which is the line of pure compression.

4.6 The Mohr-Coulomb Yield Criterion

Early studies on the failure of granular materials can be traced to Coulomb (1773) and Mohr (1900). The Mohr-Coulomb (MC) failure criterion states that the maximum shear stress τ_{max} on a failure plane varies linearly with the compressive normal stress σ_{n} acting on this plane, as shown in Fig. 4.3 and according to

$$\tau_{\mathrm{max}} = c - \sigma_{\mathrm{n}} \tan\phi\,, \tag{4.53}$$

where c is the cohesion and ϕ is the friction angle. Because there are two parameters, c and ϕ, any material obeying this failure criterion is called a cohesive-frictional material. The notion of 'yielding' is sometimes assumed to be analogous to 'failure' (Owen and Hinton 1980). If we take this point of view, then (4.53) becomes the MC yield criterion. To apply this yield criterion to a general three-dimensional continuum, it is necessary to reformulate (4.53) in the context of an isotropic yield function.

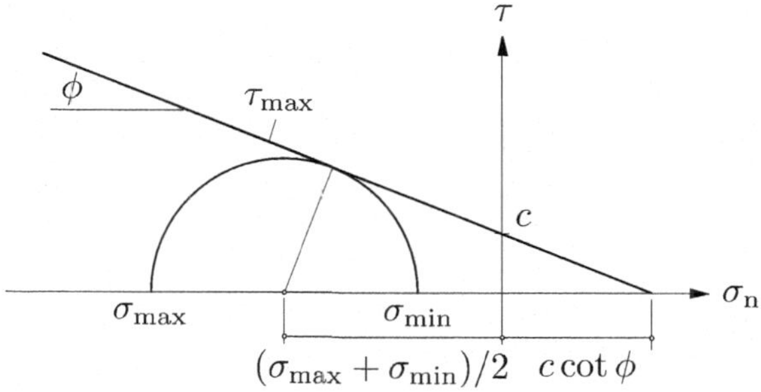

Fig. 4.3 Mohr-Coulomb failure envelope

Using the continuum mechanics sign convention where tensile normal stress is considered positive, the MC yield criterion can be expressed in terms of the principal stresses as

$$-(\sigma_{\mathrm{max}} - \sigma_{\mathrm{min}}) = 2c\cos\phi - (\sigma_{\mathrm{max}} + \sigma_{\mathrm{min}})\sin\phi\,, \tag{4.54}$$

where σ_{max} and σ_{min} are the major and minor principal stresses in compression, respectively, i.e., $-\sigma_{\mathrm{max}} > -\sigma_{\mathrm{min}}$. If $\phi = 0$, then the MC yield criterion coincides with the classic Tresca yield criterion with $2c = Y$, see (3.1).

Consider a yield function in principal stress space of the form

$$\widehat{f}(\sigma_1, \sigma_2, \sigma_3, \kappa) = |\sigma_A - \sigma_B| - \kappa \le 0\,, \tag{4.55}$$

where

$$\kappa = 2c\cos\phi - (\sigma_A + \sigma_B)\sin\phi \tag{4.56}$$

for $A, B = 1, 2, 3$, with $A \neq B$ (for a total of three permutations). With reference to the rotated system of Sec. 3.2, we write

$$\left.\begin{aligned} \sigma_1 + \sigma_2 &= \tfrac{\sqrt{2}}{2}\,\sigma_1'' + \tfrac{1}{2}\sqrt{\tfrac{2}{3}}\,\sigma_2'' + \tfrac{2}{\sqrt{3}}\,\sigma_3'' \\ \sigma_1 + \sigma_3 &= -\sqrt{\tfrac{2}{3}}\,\sigma_2'' + \tfrac{2}{\sqrt{3}}\,\sigma_3'' \\ \sigma_2 + \sigma_3 &= -\tfrac{\sqrt{2}}{2}\,\sigma_1'' + \tfrac{1}{2}\sqrt{\tfrac{2}{3}}\,\sigma_2'' + \tfrac{2}{\sqrt{3}}\,\sigma_3'' \end{aligned}\right\}. \tag{4.57}$$

Combining these equations with the corresponding principal stress differences gives

$$\left.\begin{aligned} &\left|\tfrac{\sqrt{2}}{2}\,\sigma_1'' - \sqrt{\tfrac{3}{2}}\,\sigma_2''\right| \le 2c\cos\phi - \left(\tfrac{\sqrt{2}}{2}\,\sigma_1'' + \tfrac{1}{2}\sqrt{\tfrac{2}{3}}\,\sigma_2'' + \tfrac{2}{\sqrt{3}}\,\sigma_3''\right)\sin\phi \\ &\left|\sqrt{2}\,\sigma_1''\right| \le 2c\cos\phi - \left(-\sqrt{\tfrac{2}{3}}\,\sigma_2'' + \tfrac{2}{\sqrt{3}}\,\sigma_3''\right)\sin\phi \\ &\left|\tfrac{\sqrt{2}}{2}\,\sigma_1'' + \sqrt{\tfrac{3}{2}}\,\sigma_2''\right| \le 2c\cos\phi - \left(-\tfrac{\sqrt{2}}{2}\,\sigma_1'' + \tfrac{1}{2}\sqrt{\tfrac{2}{3}}\,\sigma_2'' + \tfrac{2}{\sqrt{3}}\,\sigma_3''\right)\sin\phi \end{aligned}\right\}. \tag{4.58}$$

In principal stress space, the yield surface $\widehat{f} = 0$ defines a pyramid bounded by six non-parallel planes with a vertex at $\sigma_1'' = \sigma_2'' = 0$ and $\sigma_3'' = \sqrt{3}\,c\cot\phi$ on the tension side. This pyramid has the space diagonal (σ_3''-axis) as its generating axis; the cross section is an irregular hexagon that expands with increasing pressure, as shown in Figs. 4.4 and 4.5. For the same mean normal stress, the hexagon predicts a higher yield stress in compression than in tension. The elastic region lies on the compression side defined by $\sigma_3'' < \sqrt{3}\,c\cot\phi$, or on the side $\sigma_1, \sigma_2, \sigma_3 < c\cot\phi$.

Exercise 4.5. Evaluate relations (4.58) at $\sigma_3'' = 0$ (i.e., the π-plane) and show that the boundaries of the elastic region are defined by six straight lines. Then, match each of these lines with the corresponding sides of the irregular hexagon shown in Fig. 4.4 and show that the σ_2''-coordinates of points C and F are $6\sqrt{2/3}\,c\cos\phi/(3+\sin\phi)$ and $-6\sqrt{2/3}\,c\cos\phi/(3-\sin\phi)$, respectively.

Exercise 4.6. Assume that $\phi = 0$, and set $\sigma_{\max} = p + s_C$ and $\sigma_{\min} = p + s_A$. The MC yield function reduces to the Tresca yield function, $|\sigma_{\max} - \sigma_{\min}| = s_A - s_C \le Y$, where s_A and s_C are defined in Fig. 4.2. Write the Tresca yield function in the invariant form

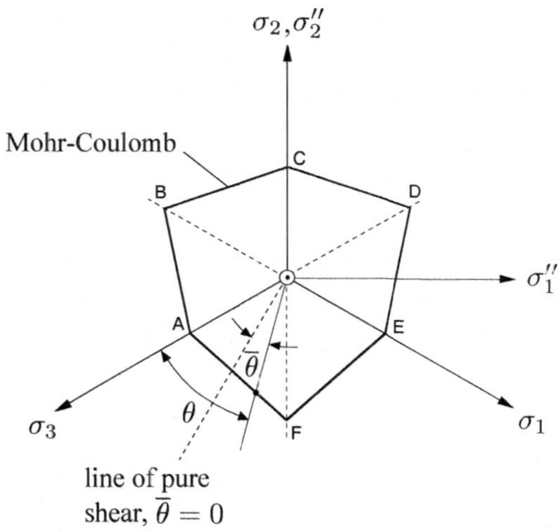

Fig. 4.4 Mohr-Coulomb yield function on deviatoric plane

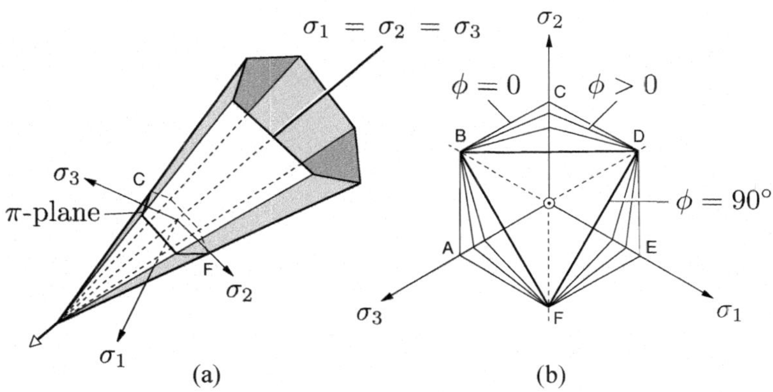

Fig. 4.5 Mohr-Coulomb yield surface in principal stress space: (a) isometric view; (b) cross section on deviatoric plane viewed from positive hydrostatic axis

$$\widetilde{f} = 2\sqrt{J_2}\cos(\theta - \pi/6) - Y \leq 0\,,$$

where θ is Lode's angle. Note that this function does not depend on $\mathcal{I}_1$.

Exercise 4.7. Write the MC yield function in the invariant form

$$\widetilde{f} = (\mathcal{I}_1/3)\sin\phi + \sqrt{J_2}\cos\overline{\theta} - \sqrt{J_2/3}\,\sin\phi\sin\overline{\theta} - c\cos\phi \leq 0\,,$$

where $\overline{\theta} = \theta - \pi/6$. Note that the yield function depends on all three stress invariants.

4.7 Smooth Approximations of the MC Yield Surface

Like the Tresca yield criterion, the MC yield criterion is convenient to use for two-dimensional problems where the corners of the yield surface are not engaged. However, for three-dimensional problems the MC yield criterion is more tricky to implement because the yield surface is not smooth and the stresses tend to get 'attracted' to the corners most of the time. Smooth approximations of the MC yield criterion are, therefore, more desirable in this case. An example of a smooth approximation of the MC yield surface is shown in Fig. 4.6.

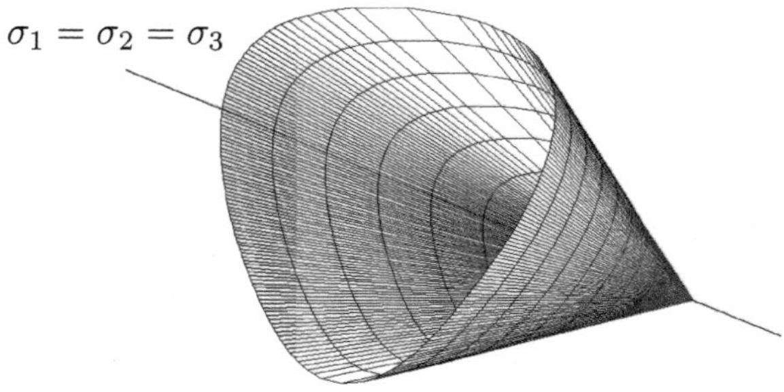

Fig. 4.6 Smooth approximation of the MC yield surface

In developing a class of smooth conical yield functions resembling the MC yield function, it is more convenient to first eliminate the effect of cohesion by translating the principal stresses such that the vertex of the cone coincides with the origin of the translated stress axes, i.e.,

$$\overline{\boldsymbol{\sigma}} = \boldsymbol{\sigma} - \alpha \mathbf{1}\,, \qquad \alpha = c\cot\phi\,. \tag{4.59}$$

It can be shown that if f is an isotropic function of $\boldsymbol{\sigma}$, then it must be an isotropic function of $\overline{\boldsymbol{\sigma}}$. Furthermore, it can be shown that the translated stress tensor $\overline{\boldsymbol{\sigma}}$ also admits the spectral representation

$$\overline{\boldsymbol{\sigma}} = \sum_{A=1}^{3} \overline{\sigma}_A \boldsymbol{n}^{(A)} \otimes \boldsymbol{n}^{(A)}\,, \qquad \overline{\sigma}_A = \sigma_A - \alpha\,. \tag{4.60}$$

We define the following invariants of the translated stress tensor:

$$\overline{\mathcal{I}}_1 = \overline{\sigma}_1 + \overline{\sigma}_2 + \overline{\sigma}_3 \,, \qquad \overline{\mathcal{I}}_2 = \overline{\sigma}_1\overline{\sigma}_2 + \overline{\sigma}_2\overline{\sigma}_3 + \overline{\sigma}_1\overline{\sigma}_3 \,, \qquad \overline{\mathcal{I}}_3 = \overline{\sigma}_1\overline{\sigma}_2\overline{\sigma}_3 \,. \quad (4.61)$$

In terms of these invariants, a number of smooth approximations of the MC yield surface have been proposed.

Matsuoka and Nakai (1974) proposed a smooth approximation of the MC yield function of the form

$$\widetilde{f} = -(\overline{\mathcal{I}}_1\overline{\mathcal{I}}_2)^{1/3} + (\beta\overline{\mathcal{I}}_3)^{1/3} \leq 0 \,. \quad (4.62)$$

The yield surface is defined by the equation $\overline{\mathcal{I}}_1\overline{\mathcal{I}}_2/\overline{\mathcal{I}}_3 = \beta$, see Fig. 4.7a. Since $\overline{\mathcal{I}}_1, \overline{\mathcal{I}}_3 < 0$ and $\overline{\mathcal{I}}_2 > 0$, the cube roots are all nonpositive numbers.

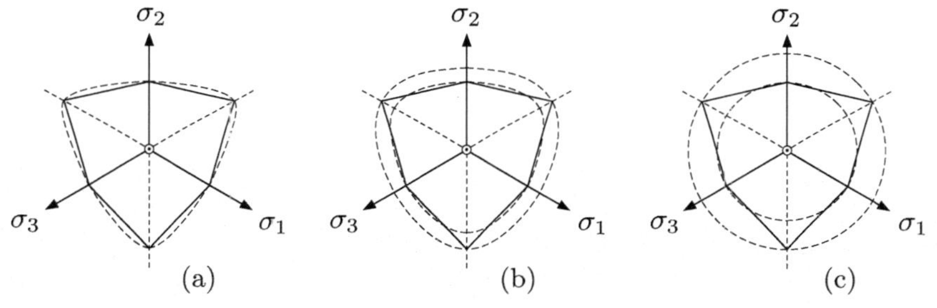

Fig. 4.7 Smooth approximations for the MC yield surface: (a) Matsuoka-Nakai; (b) Lade-Duncan; and (c) Drucker-Prager models.

The parameter β of the Matsuoka-Nakai model can be determined by forcing the yield surface $\widetilde{f} = 0$ to pass through all six corners of the MC yield surface. On the meridian plane $\sigma_1'' = 0$, which is defined with respect to the translated stresses, the tension and compression corners are given by the stress ratios

$$\eta = \frac{\sigma_2''}{\sigma_3''} = \begin{cases} -2\sqrt{2}\sin\phi/(3+\sin\phi) & \text{for tension corner}\,; \\ 2\sqrt{2}\sin\phi/(3-\sin\phi) & \text{for compression corner,} \end{cases} \quad (4.63)$$

where $\sigma_3'' < 0$. With respect to the rotated coordinates, the relation $\beta\overline{\mathcal{I}}_3 - \overline{\mathcal{I}}_1\overline{\mathcal{I}}_2 = 0$ can be rewritten as

$$\begin{aligned} 3(\sigma_1'')^2[(\beta-3)\sigma_3'' + \sqrt{2}\beta\sigma_2''] - 3(3-\beta)(\sigma_2'')^2\sigma_3'' & \\ -2(\beta-9)(\sigma_3'')^3 - \sqrt{2}\beta(\sigma_2'')^3 &= 0 \,. \end{aligned} \quad (4.64)$$

Setting $\sigma_1'' = 0$ and dividing through by $(\sigma_3'')^3$ gives

$$\beta = \frac{18 - 9\eta^2}{\sqrt{2}\eta^3 - 3\eta^2 + 2} = \frac{9\sqrt{2} + 9\eta}{-\sqrt{2}\eta^2 + \eta + \sqrt{2}} \,. \quad (4.65)$$

Substituting (4.63) into (4.65) gives

$$\beta = \frac{9 - \sin^2\phi}{1 - \sin^2\phi}\,. \tag{4.66}$$

Since ϕ cannot be negative, it follows that $\beta \geq 9$. Therefore, the minimum value of β is 9, which corresponds to $\phi = 0$.

Lade and Duncan (1975) proposed a yield function of a slightly different form,

$$\widetilde{f} = -\overline{\mathcal{I}}_1 + (\beta\overline{\mathcal{I}}_3)^{1/3} \leq 0\,. \tag{4.67}$$

The yield surface is given by the equation $\overline{\mathcal{I}}_1^3/\overline{\mathcal{I}}_3 = \beta$. In the transformed coordinates the relation $\beta\overline{\mathcal{I}}_3 - \overline{\mathcal{I}}_1^3 = 0$ can be written as

$$3(\sigma_1'')^2(\sqrt{2}\sigma_2'' + \sigma_3'') - (\sigma_2'')^2(\sqrt{2}\sigma_2'' - 3\sigma_3'') - 2(\sigma_3'')^3\left(\frac{\beta - 27}{\beta}\right) = 0\,. \tag{4.68}$$

Setting $\sigma_1'' = 0$ and dividing by $(\sigma_3'')^3$ gives

$$\beta = \frac{27}{1 - \eta^2(3 - \sqrt{2}\eta)/2}\,, \qquad \eta = \frac{\sigma_2''}{\sigma_3''}\,, \qquad \sigma_3'' \neq 0\,. \tag{4.69}$$

The stress ratio η can be chosen to force the yield surface to pass through either the tension or compression corners of the MC yield surface, see Fig. 4.7b, yielding

$$\beta = \frac{(3 \pm \sin\phi)^3}{1 \pm \sin\phi - \sin^2\phi \mp \sin^3\phi}\,, \tag{4.70}$$

where the superior sign (i.e., + in $\pm$ and − in $\mp$) pertains to a yield surface passing through the tension corners, and the subordinate sign pertains to yield surface passing through the compression corners of the underlying MC yield function. Condition $\phi \geq 0$ implies that $\beta \geq 27$ for the Lade-Duncan model.

Drucker and Prager (1952) proposed a two-invariant yield function of the form

$$\widetilde{f} = -\overline{\mathcal{I}}_1 - (\beta\overline{\mathcal{I}}_2)^{1/2} \leq 0\,. \tag{4.71}$$

The yield surface is defined by the equation $\overline{\mathcal{I}}_1^2/\overline{\mathcal{I}}_2 = \beta$, and has a circular cross section on the deviatoric plane, see Fig. 4.7c. In the transformed coordinates the relation $\beta\overline{\mathcal{I}}_2 - \overline{\mathcal{I}}_1^2 = 0$ is equivalent to

$$(\beta - 3)(\sigma_3'')^2 - \beta[\,(\sigma_1'')^2 + (\sigma_2'')^2\,]/2 = 0\,. \tag{4.72}$$

Setting $\sigma_1'' = 0$ and dividing through by $(\sigma_3'')^2$ gives

$$\beta = \frac{3}{1 - \eta^2/2}\,, \qquad \eta = \frac{\sigma_2''}{\sigma_3''}\,, \qquad \sigma_3'' \neq 0\,. \tag{4.73}$$

The stress ratio η can be chosen to force the yield surface to pass through either the tension or compression corners of the MC yield surface. This implies that β can have two possible values:

$$\beta = \frac{(3 \pm \sin\phi)^2}{3 \pm 2\sin\phi - \sin^2\phi}, \tag{4.74}$$

where the plus and minus signs are for tension and compression approximations, respectively. Since $\phi \geq 0$, then $\beta \geq 3$ for the Drucker-Prager model.

A more convenient representation of the Drucker-Prager yield function is provided by the stress invariants $(\mathcal{I}_1, J_2)$, or, alternatively, by the invariants (p, q), where $p = \mathcal{I}_1/3$ is the mean normal stress and $q = \sqrt{3J_2}$ is the von Mises stress. Noting that $\sqrt{(\sigma_1'')^2 + (\sigma_2'')^2} = \|\boldsymbol{s}\| = \sqrt{2/3}\,q$ and $\sigma_3'' = \sqrt{3}(p - c\cot\phi)$, equation (4.72) can be written in the equivalent form

$$\widetilde{f} = \sqrt{\frac{2}{3}}\,q - (A - Bp) \leq 0\,. \tag{4.75}$$

The two material parameters are

$$A = \frac{2\sqrt{6}\,c\cos\phi}{3 \pm \sin\phi}, \qquad B = \frac{2\sqrt{6}\,\sin\phi}{3 \pm \sin\phi}, \tag{4.76}$$

where the plus and minus signs are for circles passing through the tension and compression corners of the MC yield surface, respectively.

Exercise 4.8. Show from equation (4.75) that the Drucker-Prager yield function is convex.

Exercise 4.9. Consider the following yield surface

$$\widetilde{f} = q - [c_1 - c_2\exp(c_3 p)] = 0\,,$$

where c_1, c_2 and c_3 are positive constants, and $c_1 > c_2$. Describe the shape of this surface in principal stress space.

Exercise 4.10. A rectangular specimen of rock tested in plane strain compression on the plane (x, y) has a crack oriented at 60° with respect to the horizontal x-axis. The mobilized cohesion on the crack is 5 kPa, and the mobilized friction angle is 45°. Assuming that failure is analogous to yielding on the crack, determine an expression for the yield function in terms of stress components σ_{xx}, σ_{yy}, and σ_{zz} (axis z is perpendicular to the plane of loading). Plot the yield function on the π-plane. Is the yield function an isotropic function of $\boldsymbol{\sigma}$?

Exercise 4.11. Repeat Exercise 4.10 if the specimen has two non-intersecting cracks, one oriented at 60° and another oriented at 45° with respect to the horizontal x-axis. Assume the second crack has the same mobilized cohesion and friction angle as the first crack.

4.8 Pressure-Dependent Friction Angle

When the confining pressure is sufficiently high, the yield surface for certain frictional materials could curve from a conical shape at lower confining pressures into a more cylindrical shape at higher confining pressures. This suggests that there is a decrease in the value of the friction angle with increasing pressure. To capture this feature, it is useful to define dimensionless slopes (Borja et al. 2003)

$$\varphi_1 = \frac{\overline{\mathcal{I}}_1^2}{\overline{\mathcal{I}}_2}, \qquad \varphi_2 = \frac{\overline{\mathcal{I}}_1 \overline{\mathcal{I}}_2}{\overline{\mathcal{I}}_3}, \qquad \varphi_3 = \frac{\overline{\mathcal{I}}_1^3}{\overline{\mathcal{I}}_3}. \tag{4.77}$$

Consider now a yield function of the form

$$\widetilde{f} = \varphi_n - \beta \leq 0, \qquad \beta = \beta_0 + \beta_1 \left(\frac{p_{\mathrm{a}}}{\overline{\mathcal{I}}_1} \right)^{\gamma}, \tag{4.78}$$

where $p_{\mathrm{a}} < 0$ is a normalizing variable, β_1 and γ are material parameters, $\beta_0 \geq 0$ is the asymptotic minimum value to which β tends, and $n = 1, 2$, or 3 depending on which yield function is chosen from equation (4.77).

The yield surface is smooth everywhere except at the vertex, so we can use the following inequality to test for convexity of the yield function

$$(\boldsymbol{\sigma}^* - \boldsymbol{\sigma}) : \frac{\partial f}{\partial \boldsymbol{\sigma}} \leq 0. \tag{4.79}$$

The vertex coincides with the origin when $\overline{\boldsymbol{\sigma}} = \boldsymbol{\sigma}$, so choosing $\boldsymbol{\sigma}^* = \mathbf{0}$ requires that

$$\boldsymbol{\sigma} : \frac{\partial f}{\partial \boldsymbol{\sigma}} = \sum_{A=1}^{3} \sigma_A f_A \geq 0. \tag{4.80}$$

From the chain rule, we have

$$f_A = \frac{\partial \widetilde{f}}{\partial \sigma_A} = \frac{\partial \widetilde{f}}{\partial \overline{\mathcal{I}}_1} \frac{\partial \overline{\mathcal{I}}_1}{\partial \sigma_A} + \frac{\partial \widetilde{f}}{\partial \overline{\mathcal{I}}_2} \frac{\partial \overline{\mathcal{I}}_2}{\partial \sigma_A} + \frac{\partial \widetilde{f}}{\partial \overline{\mathcal{I}}_3} \frac{\partial \overline{\mathcal{I}}_3}{\partial \sigma_A}, \tag{4.81}$$

where

$$\frac{\partial \overline{\mathcal{I}}_1}{\partial \sigma_A} = 1, \qquad \frac{\partial \overline{\mathcal{I}}_2}{\partial \sigma_A} = \overline{\mathcal{I}}_1 - \overline{\sigma}_A, \qquad \frac{\partial \overline{\mathcal{I}}_3}{\partial \sigma_A} = \frac{\overline{\mathcal{I}}_3}{\overline{\sigma}_A}. \tag{4.82}$$

The requirement for convexity becomes

$$\sum_{A=1}^{3} \sigma_A f_A = \frac{\partial \widetilde{f}}{\partial \mathcal{I}_1} \mathcal{I}_1 + 2 \frac{\partial \widetilde{f}}{\partial \mathcal{I}_2} \mathcal{I}_2 + 3 \frac{\partial \widetilde{f}}{\partial \mathcal{I}_3} \mathcal{I}_3 \geq 0. \tag{4.83}$$

This condition is identically satisfied by the family of yield functions considered in (4.78) provided that $\gamma \geq 0$. This is consistent with a decay of β to the limiting value β_0 at higher confining pressures. Figure 4.8 shows two yield surfaces generated from this family of yield functions.

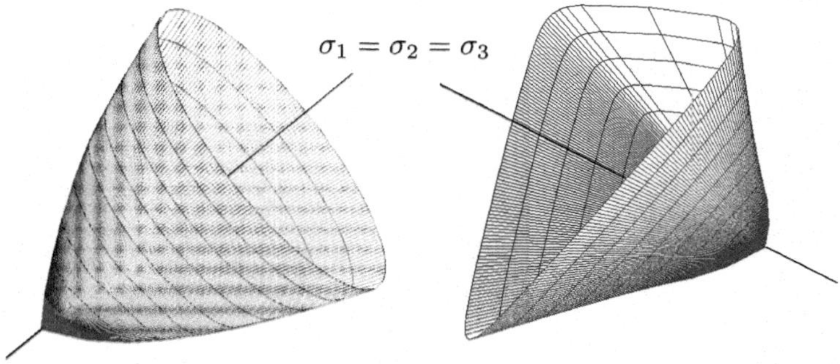

Fig. 4.8 Curved yield surfaces with positive friction angle: two-invariant (left); three-invariant (right)

If $\beta_0 = 0$, then the yield surface converges to the space diagonal to form a 'compression cap' (Lade and Kim 1988). Some authors (Issen and Rudnicki 2000) refer to the compression cap as the zone of negative friction angle; others (DiMaggio and Sandler 1971) even allow the cap to move. The intersection point of the cap with the space diagonal is located at

$$\overline{\mathcal{I}}_1 = p_\mathrm{a}\Big(\frac{\beta_1}{\widetilde{\varphi}_n - \beta_0}\Big)^{1/\gamma}, \tag{4.84}$$

where $\widetilde{\varphi}_n = 3, 9, 27$ for $n = 1, 2, 3$, respectively. Figure 4.9 shows two- and three-invariant yield surfaces generated by setting $\beta_0 = 0$. Yield surfaces with a compression cap are elaborated further in Chapter 6.

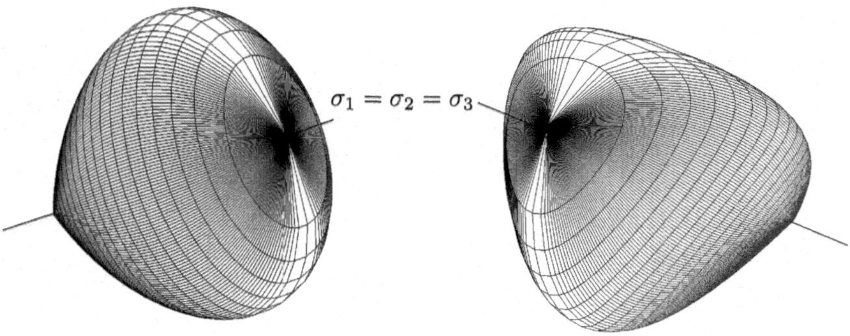

Fig. 4.9 Curved yield surfaces with 'negative' friction angle (cap): two-invariant (left); three-invariant (right)

4.9 Flow Rule and Plastic Dilatancy

Consider a plastic potential function of the form

$$\widehat{g} = |\sigma_A - \sigma_B| - g_0 + (\sigma_A + \sigma_B)\sin\psi\,, \tag{4.85}$$

where g_0 = constant. The flow rule takes the form

$$\left.\begin{aligned} \dot{\epsilon}^{\mathrm{p}}_A &= \dot{\lambda} g_A = \dot{\lambda}\,[\,\mathrm{sign}(\sigma_A - \sigma_B) + \sin\psi\,] \\ \dot{\epsilon}^{\mathrm{p}}_B &= \dot{\lambda} g_B = \dot{\lambda}\,[\,-\mathrm{sign}(\sigma_A - \sigma_B) + \sin\psi\,] \\ \dot{\epsilon}^{\mathrm{p}}_C &= \dot{\lambda} g_C = 0 \end{aligned}\right\}\,, \tag{4.86}$$

where A, B, and C are permutations of 1, 2 and 3. Note that there is no plastic strain in the out-of-plane direction (i.e., $\dot{\epsilon}^{\mathrm{p}}_C = 0$). The volumetric plastic strain rate is given by

$$\dot{\epsilon}^{\mathrm{p}}_v = \mathrm{tr}(\dot{\boldsymbol{\epsilon}}^{\mathrm{p}}) = 2\dot{\lambda}\sin\psi\,. \tag{4.87}$$

If $\psi > 0$, then the plastic flow is dilative. Plastic dilatancy is consistent with observed behavior of dense granular materials (Reynolds 1885). The notion of plastic dilatancy represented by the angle ψ was introduced by Hansen (1958).

Solving for $\dot{\lambda}$ from (4.86) and noting that $\mathrm{sign}(\sigma_A - \sigma_B) = \mathrm{sign}(\dot{\epsilon}^{\mathrm{p}}_A - \dot{\epsilon}^{\mathrm{p}}_B)$ yields

$$\dot{\lambda} = \frac{1}{2}\,\frac{\dot{\epsilon}^{\mathrm{p}}_A - \dot{\epsilon}^{\mathrm{p}}_B}{\mathrm{sign}(\sigma_A - \sigma_B)} = \frac{1}{2}\,|\dot{\epsilon}^{\mathrm{p}}_A - \dot{\epsilon}^{\mathrm{p}}_B| = \frac{1}{2}\,|\dot{\gamma}^{\mathrm{p}}_{AB}|\,, \tag{4.88}$$

where $|\dot{\gamma}^{\mathrm{p}}_{AB}|$ is the rate of plastic distortion (Vermeer and de Borst 1984). Combining the last two equations yields

$$\sin\psi = \dot{\epsilon}^{\mathrm{p}}_v / |\gamma^{\mathrm{p}}_{AB}|\,. \tag{4.89}$$

In order to mobilize the dilatancy angle the plastic distortion must be maximized, which means that the stresses must be chosen such that $\sigma_A > \sigma_C > \sigma_B$.

Figure 4.10 provides a physical interpretation of the dilatancy angle ψ. Consider a homogeneous deformation given by the velocity field

$$v_1 = c_1 y\,, \qquad v_2 = c_2 y\,, \qquad v_3 = 0\,, \tag{4.90}$$

where $c_1, c_2 > 0$. The only nonzero strain rates are

$$\dot{\epsilon}_{22} = c_2\,, \qquad \dot{\epsilon}_{12} = \dot{\epsilon}_{21} = c_1/2\,. \tag{4.91}$$

Assuming perfect plasticity and neglecting the elastic strains, we can take these strain rates as purely plastic. Hence, the Mohr circle for plastic strains shown in Fig. 4.10 gives

$$\dot{\epsilon}^{\mathrm{p}}_v = c_2\,, \qquad |\dot{\gamma}^{\mathrm{p}}_{12}| = \sqrt{c_1^2 + c_2^2}\,, \tag{4.92}$$

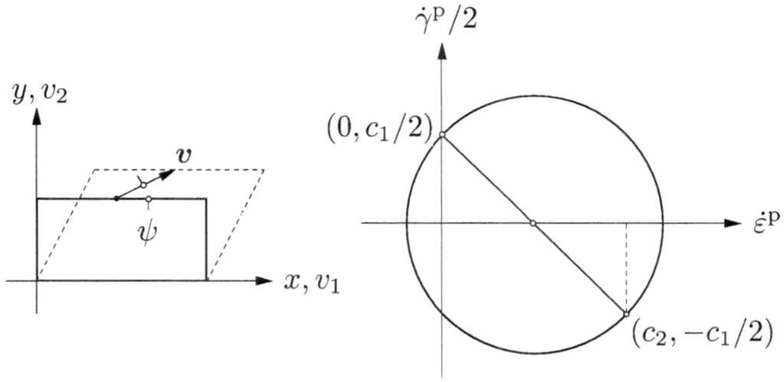

Fig. 4.10 Meaning of dilatancy angle ψ and Mohr circle for plastic strain rates

which implies

$$\sin\psi = \frac{c_2/c_1}{\sqrt{1+(c_2/c_1)^2}} = \frac{v_2/v_1}{\sqrt{1+(v_2/v_1)^2}} . \tag{4.93}$$

Solving for v_2/v_1 gives

$$v_2/v_1 = \tan\psi . \tag{4.94}$$

Thus, ψ is defined from this deformation field as an uplift angle.

Next we investigate the bounds on the values of ψ. For the MC model the requirement of nonnegative plastic work can be written as

$$\dot{W}^p = \boldsymbol{\sigma} : \dot{\boldsymbol{\epsilon}}^p = \dot{\lambda} \sum_{A=1}^{3} \sigma_A g_A = \dot{\lambda}[|\sigma_A - \sigma_B| + (\sigma_A + \sigma_B)\sin\psi] \geq 0 . \tag{4.95}$$

Substituting the yield condition (4.55) gives

$$\dot{W}^p = \dot{\lambda}[2c\cos\phi + (\sigma_A + \sigma_B)(\sin\psi - \sin\phi)] \geq 0 . \tag{4.96}$$

Noting that $\dot{\lambda} > 0$ during plastic flow, we get

$$2c\cos\phi + (\sigma_A + \sigma_B)(\sin\psi - \sin\phi) \geq 0 . \tag{4.97}$$

For cohesionless material $c = 0$ and $\sigma_A + \sigma_B < 0$, and so, for the inequality to be satisfied we must have $\sin\psi \leq \sin\phi$, or

$$\psi \leq \phi . \tag{4.98}$$

In other words, for the plastic work to be nonnegative the dilatancy angle must be less than the friction angle.

4.10 Cohesion and Friction Hardening

Consider the class of two-parameter, smooth yield functions described in Sec. 4.7, where the parameters α and β are related to the cohesion c and friction angle ϕ, respectively. We assume that α and β vary with cumulative plastic strain λ according to the relations

$$\alpha = \alpha(\lambda)\,, \qquad \beta = \beta(\lambda)\,, \qquad \lambda = \int_t \dot{\lambda}\,\mathrm{d}t\,, \tag{4.99}$$

where $\dot{\lambda} \geq 0$. The consistency condition on the yield function $f(\boldsymbol{\sigma}, \alpha, \beta)$ is given by

$$\boldsymbol{f} : \dot{\boldsymbol{\sigma}} + \frac{\partial f}{\partial \alpha}\dot{\alpha} + \frac{\partial f}{\partial \beta}\dot{\beta} = \boldsymbol{f} : \dot{\boldsymbol{\sigma}} - H\dot{\lambda} = 0\,. \tag{4.100}$$

The generalized plastic modulus is

$$H = -\frac{\partial f}{\partial \alpha}\frac{\partial \alpha}{\partial \lambda} - \frac{\partial f}{\partial \beta}\frac{\partial \beta}{\partial \lambda}\,. \tag{4.101}$$

The variation of f with α can be obtained from the chain rule, along with (4.59)

$$\frac{\partial f}{\partial \alpha} = \frac{\partial f}{\partial \overline{\boldsymbol{\sigma}}} : \frac{\partial \overline{\boldsymbol{\sigma}}}{\partial \alpha} = -\frac{\partial f}{\partial \overline{\boldsymbol{\sigma}}} : \mathbf{1} = -\mathrm{tr}\Big(\frac{\partial f}{\partial \boldsymbol{\sigma}}\Big) = -\sum_{A=1}^{3} f_A\,. \tag{4.102}$$

From Sec. 4.8, we find

$$\sum_{A=1}^{3} f_A = 3\frac{\partial \widetilde{f}}{\partial \overline{\mathcal{I}}_1} + 2\overline{\mathcal{I}}_1\frac{\partial \widetilde{f}}{\partial \overline{\mathcal{I}}_2} + \overline{\mathcal{I}}_2\frac{\partial \widetilde{f}}{\partial \overline{\mathcal{I}}_3} > 0 \tag{4.103}$$

for $\phi > 0$, which means that $\partial f/\partial \alpha < 0$. This implies that $\mathrm{sign}(H) = \mathrm{sign}(\partial \alpha/\partial \lambda)$ for a constant β. We require that $\alpha \geq 0$ for the results to be meaningful.

For cohesionless materials where $\alpha = 0$, an increase (alternatively, decrease) in β with λ is called friction hardening (alternatively, softening). This follows from the fact that $\partial f/\partial \beta < 0$, and so $\mathrm{sign}(H) = \mathrm{sign}(\partial \beta/\partial \lambda)$. In particulate materials an increase in the apparent friction angle may be due to increased density of the granular assembly, which enhances interlocking of the particles. The parameters α and β for sands have been documented to vary with plastic deformation (Chen 1982, Stroud 1971, Tatsuoka and Ishihara 1975, Vermeer and de Borst 1984). Furthermore, Roscoe and Burland (1968) and Rowe (1971) have shown that the dilatancy angle ψ can vary with plastic deformation. The specific form of the hardening law can be established from careful laboratory testing on a given material.

Exercise 4.12. Consider a Drucker-Prager plasticity model whose material parameters vary according to the relations (see (4.76))

$$A = A_0 \exp\left[-(\lambda/k_1)^2 \right],$$

and

$$B = \begin{cases} 2B^* \sqrt{k_2 \lambda}/(k_2 + \lambda), & \text{for } \lambda \le k_2; \\ B^*, & \text{for } \lambda \ge k_2, \end{cases}$$

where k_1 and k_2 are positive constants and λ is the cumulative plastic strain. This model describes a cohesion parameter A that softens to zero from an initial value A_0, at the same time that the frictional parameter B hardens from an initial value zero to a final value B^*. For sandstone, cohesion softening represents a decrease in the cementation due to micro-cracking, whereas friction hardening represents conversion of a cemented rock into a particulate material. Determine the expression for the plastic modulus H as a function of λ. If a stress point is loaded at a constant mean normal stress $p = -100$ kPa, at what value of λ would you expect the material to exhibit a perfectly plastic response? Assume $A_0 = 100$ kPa, $B^* = 1.10$, and $k_1 = k_2 = 0.05$. Also, plot A, B, and H as functions of λ.

4.11 Return Mapping in Principal Axes

In this section we present an algorithm for integrating the rate constitutive equation for isotropic plasticity models. The algorithm consists of an elastic predictor followed by plastic corrector in principal stress axes.

As a first step, we write the generalized Hooke's law in rate form as

$$\dot{\boldsymbol{\sigma}} = \boldsymbol{c}^{\mathrm{e}} : \left(\dot{\boldsymbol{\epsilon}} - \dot{\lambda} \frac{\partial g}{\partial \boldsymbol{\sigma}} \right). \tag{4.104}$$

The incremental version of this equation is

$$\boldsymbol{\sigma}_{n+1} = \boldsymbol{\sigma}_{n+1}^{\mathrm{tr}} - \Delta\lambda \boldsymbol{c}_{n+1}^{\mathrm{e}} : \left. \frac{\partial g}{\partial \boldsymbol{\sigma}} \right|_{n+1}, \tag{4.105}$$

where

$$\boldsymbol{\sigma}_{n+1}^{\mathrm{tr}} = \boldsymbol{\sigma}_n + \boldsymbol{c}_{n+1}^{\mathrm{e}} : \Delta\boldsymbol{\epsilon} = \boldsymbol{c}_{n+1}^{\mathrm{e}} : \boldsymbol{\epsilon}_{n+1}^{\mathrm{e\,tr}}, \qquad \boldsymbol{\epsilon}_{n+1}^{\mathrm{e\,tr}} = \boldsymbol{\epsilon}_n^{\mathrm{e}} + \Delta\boldsymbol{\epsilon}. \tag{4.106}$$

The elasticity tensor $\boldsymbol{c}^{\mathrm{e}}$ is constant; however, its spectral representation depends on $\boldsymbol{\sigma}_{n+1}$.

Next, we write the predictor stress and strain tensors in spectral form as

$$\begin{aligned} \boldsymbol{\sigma}_{n+1}^{\mathrm{tr}} &= \sum_{A=1}^{3} \sigma_{A,n+1}^{\mathrm{tr}} \boldsymbol{n}_{n+1}^{\mathrm{tr}(A)} \otimes \boldsymbol{n}_{n+1}^{\mathrm{tr}(A)}, \\ \boldsymbol{\epsilon}_{n+1}^{\mathrm{e\,tr}} &= \sum_{A=1}^{3} \epsilon_{A,n+1}^{\mathrm{e\,tr}} \boldsymbol{n}_{n+1}^{\mathrm{tr}(A)} \otimes \boldsymbol{n}_{n+1}^{\mathrm{tr}(A)}. \end{aligned} \tag{4.107}$$

Omitting the subscript $(n+1)$ for brevity, we can rewrite (4.105) as

$$\sum_{A=1}^{3} \sigma_A \boldsymbol{n}^{(A)} \otimes \boldsymbol{n}^{(A)} = \sum_{A=1}^{3} \sigma_A^{\mathrm{tr}} \boldsymbol{n}^{\mathrm{tr}(A)} \otimes \boldsymbol{n}^{\mathrm{tr}(A)} - \Delta\lambda \sum_{A=1}^{3} \Big(\sum_{B=1}^{3} a_{AB}^{\mathrm{e}} g_B \Big) \boldsymbol{n}^{(A)} \otimes \boldsymbol{n}^{(A)} . \tag{4.108}$$

From this last equation, we can see that

$$\boldsymbol{n}^{(A)} \otimes \boldsymbol{n}^{(A)} = \boldsymbol{n}^{\mathrm{tr}(A)} \otimes \boldsymbol{n}^{\mathrm{tr}(A)} . \tag{4.109}$$

The algorithm thus reduces to a predictor-corrector scheme in principal stress axes, of the form (see Fig. 4.11 for a geometric interpretation of the procedure)

$$\begin{aligned} \sigma_A &= \sigma_A^{\mathrm{tr}} - \Delta\lambda \sum_{B=1}^{3} a_{AB}^{\mathrm{e}} g_B , \\ \epsilon_A^{\mathrm{e}} &= \epsilon_A^{\mathrm{e\,tr}} - \Delta\lambda g_A , \qquad A = 1,2,3 . \end{aligned} \tag{4.110}$$

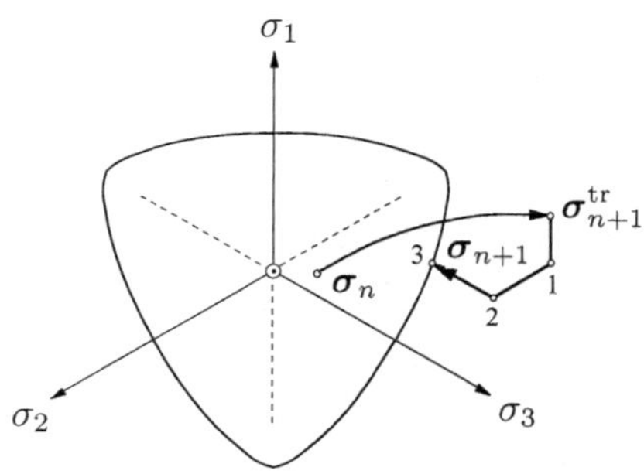

Fig. 4.11 Return mapping in principal stress space. Plastic stress correctors on stress axes σ_1, σ_2 and σ_3 are denoted by vectors 1, 2, 3, respectively

Assuming isotropic hardening, the evolution of the plastic internal variable κ takes the following alternative forms

$$\kappa = \begin{cases} \widehat{\kappa}(W^{\mathrm{p}}) = \widehat{\kappa}(W_n^{\mathrm{p}} + \Delta W^{\mathrm{p}}) , & \text{if work hardening;} \\ \widehat{\kappa}(\lambda) = \widehat{\kappa}(\lambda_n + \Delta\lambda) , & \text{if strain hardening.} \end{cases} \tag{4.111}$$

Here, $\widehat{\kappa}$ is a specific functional form for κ, and λ_n and W_n^{p} are given values at the beginning of the load increment. In residual form, we have

$$\widehat{\ell}(\sigma_1, \sigma_2, \sigma_3, \kappa, \Delta\lambda) = \kappa - \widehat{\kappa}(\sigma_1, \sigma_2, \sigma_3, \Delta\lambda) = 0 , \tag{4.112}$$

while the consistency condition is

$$\widehat{f}(\sigma_1, \sigma_2, \sigma_3, \kappa) = 0 . \tag{4.113}$$

The unknowns must be determined by a local iteration. To this end, a residual vector $\boldsymbol{r}$ and a vector of unknowns $\boldsymbol{x}$ are constructed as

$$\boldsymbol{r}(\boldsymbol{x}) = \left\{ \begin{array}{c} \sum_J a_{1J}^{\mathrm{e}-1}\sigma_J - \epsilon_1^{\mathrm{e\,tr}} + \Delta\lambda g_1 \\ \sum_J a_{2J}^{\mathrm{e}-1}\sigma_J - \epsilon_2^{\mathrm{e\,tr}} + \Delta\lambda g_2 \\ \sum_J a_{3J}^{\mathrm{e}-1}\sigma_J - \epsilon_3^{\mathrm{e\,tr}} + \Delta\lambda g_3 \\ \widehat{\ell}(\sigma_1, \sigma_2, \sigma_3, \kappa, \Delta\lambda) \\ \widehat{f}(\sigma_1, \sigma_2, \sigma_3, \kappa) \end{array} \right\}, \qquad \boldsymbol{x} = \left\{ \begin{array}{c} \sigma_1 \\ \sigma_2 \\ \sigma_3 \\ \kappa \\ \Delta\lambda \end{array} \right\}, \tag{4.114}$$

in which the $\epsilon_A^{\mathrm{e\,tr}}$'s are fixed during the local iteration. Newton's method consists of iterating over the loop

$$\boldsymbol{A}^k \cdot \Delta\boldsymbol{x} = \boldsymbol{r}(\boldsymbol{x}^k), \qquad \boldsymbol{x}^{k+1} \leftarrow \boldsymbol{x}^k - \Delta\boldsymbol{x}, \qquad k \leftarrow k+1 \tag{4.115}$$

until the norm $\|\boldsymbol{r}\|$ reaches an acceptable value.

The local Jacobian is given by

$$\begin{aligned} \boldsymbol{A}^k &= \boldsymbol{r}'(\boldsymbol{x}^k) = \begin{bmatrix} b_{11} & b_{12} & b_{13} & \Delta\lambda\partial g_1/\partial\kappa & g_1 \\ b_{21} & b_{22} & b_{23} & \Delta\lambda\partial g_2/\partial\kappa & g_2 \\ b_{31} & b_{32} & b_{33} & \Delta\lambda\partial g_3/\partial\kappa & g_3 \\ \partial\widehat{\ell}/\partial\sigma_1 & \partial\widehat{\ell}/\partial\sigma_2 & \partial\widehat{\ell}/\partial\sigma_3 & 1 & \partial\widehat{\ell}/\partial\Delta\lambda \\ \partial\widehat{f}/\partial\sigma_1 & \partial\widehat{f}/\partial\sigma_2 & \partial\widehat{f}/\partial\sigma_3 & \partial\widehat{f}/\partial\kappa & 0 \end{bmatrix}, \\ b_{IJ} &= a_{IJ}^{\mathrm{e}-1} + \Delta\lambda\frac{\partial^2\widehat{g}}{\partial\sigma_I\partial\sigma_J}, \qquad I, J = 1, 2, 3. \end{aligned} \tag{4.116}$$

Note that the local Jacobian requires an evaluation of the second derivative of the plastic potential function.

Box 4.1 summarizes the algorithm. First, the elastic strain predictor $\boldsymbol{\epsilon}^{\mathrm{e\,tr}} = \boldsymbol{\epsilon}_n^{\mathrm{e}} + \Delta\boldsymbol{\epsilon}$ is decomposed spectrally and the principal values $\epsilon_A^{\mathrm{e\,tr}}$ and σ_A^{tr} are calculated, along with the spectral directions $\boldsymbol{n}^{\mathrm{tr}\,(A)} \otimes \boldsymbol{n}^{\mathrm{tr}\,(A)}$. Eigenvalue solving can be done either in closed form or by numerical iteration (Borja et al. 2003, Meschke 1996, Tamagnini et al. 2002). If the stress predictor overshoots the yield surface, a return to the yield surface is enforced by mapping

Step 1. Compute $\boldsymbol{\epsilon}^{\mathrm{e\,tr}} = \boldsymbol{\epsilon}_n^{\mathrm{e}} + \Delta\boldsymbol{\epsilon}$
Step 2. Spectrally decompose $\boldsymbol{\epsilon}^{\mathrm{e\,tr}} = \sum_{A=1}^{3} \epsilon_A^{\mathrm{e\,tr}} \boldsymbol{n}^{\mathrm{tr}(A)} \otimes \boldsymbol{n}^{\mathrm{tr}(A)}$
Step 3. Compute $\sigma_A^{\mathrm{tr}} = \sum_{B=1}^{3} a_{AB}^{\mathrm{e}-1} \epsilon_B^{\mathrm{e\,tr}}$,
$\boldsymbol{\sigma}^{\mathrm{tr}} = \sum_{A=1}^{3} \sigma_A^{\mathrm{tr}} \boldsymbol{n}^{\mathrm{tr}(A)} \otimes \boldsymbol{n}^{\mathrm{tr}(A)}$
Step 4. Check $\widehat{f}(\sigma_1^{\mathrm{tr}}, \sigma_2^{\mathrm{tr}}, \sigma_3^{\mathrm{tr}}, \kappa_n) > 0$?
No, set $\boldsymbol{\sigma} = \boldsymbol{\sigma}^{\mathrm{tr}}$ and exit.
Step 5. Yes, solve for σ_1, σ_2, σ_3, and κ such that $\widehat{f}(\sigma_1, \sigma_2, \sigma_3, \kappa) = 0$
Step 6. Compute final stress $\boldsymbol{\sigma} = \sum_{A=1}^{3} \sigma_A \boldsymbol{n}^{\mathrm{tr}(A)} \otimes \boldsymbol{n}^{\mathrm{tr}(A)}$ and exit.

Box 4.1 Return mapping algorithm in principal stress directions.

in the direction of g_B through the metric $a^{\rm e}_{AB}$ (Simo and Hughes 1998). The magnitude of the plastic corrector depends on $\Delta\lambda$, which in turn depends on the stress overshoot and the hardening law. Because the integration algorithm employed is based on implicit scheme, the final values of the principal stresses $\sigma_1, \sigma_2, \sigma_3$, and those of κ and $\Delta\lambda$, are determined by a local iteration.

4.12 Algorithmic Tangent Operator

For numerical calculation purposes, the algorithmic tangent operator $\boldsymbol{c}_{n+1}$ is defined as

$$\boldsymbol{c}_{n+1} = \frac{\partial \boldsymbol{\sigma}_{n+1}}{\partial \boldsymbol{\epsilon}_{n+1}} \equiv \frac{\partial \boldsymbol{\sigma}_{n+1}}{\partial \boldsymbol{\epsilon}^{\rm e\,tr}_{n+1}} . \tag{4.117}$$

According to Section 3.8, this tangent operator is essential to preserve the optimal convergence properties of Newton's method.

Following the procedure of Section 4.4, and omitting the subscript $(n+1)$ again for brevity, we write the time derivatives

$$\dot{\boldsymbol{\sigma}} = \sum_{A=1}^{3} \dot{\sigma}_A \boldsymbol{n}^{(A)} \otimes \boldsymbol{n}^{(A)} + \sum_{A=1}^{3} \sum_{B \neq A} \omega_{AB} (\sigma_B - \sigma_A) \boldsymbol{n}^{(A)} \otimes \boldsymbol{n}^{(B)} \tag{4.118}$$

for the Cauchy stress tensor, and

$$\dot{\boldsymbol{\epsilon}}^{\rm e\,tr} = \sum_{A=1}^{3} \dot{\epsilon}^{\rm e\,tr}_A \boldsymbol{n}^{(A)} \otimes \boldsymbol{n}^{(A)} + \sum_{A=1}^{3} \sum_{B \neq A} \omega_{AB} (\epsilon^{\rm e\,tr}_B - \epsilon^{\rm e\,tr}_A) \boldsymbol{n}^{(A)} \otimes \boldsymbol{n}^{(B)} \tag{4.119}$$

for the elastic strain predictor tensor. Note that the spins of the two tensors are the same from the coaxiality condition (4.109). The algorithmic tangent operator in spectral form is readily obtained as

$$\begin{aligned} \boldsymbol{c}_{n+1} &= \sum_{A=1}^{3} \sum_{B=1}^{3} a_{AB} \boldsymbol{m}^{(A)} \otimes \boldsymbol{m}^{(B)} \\ &+ \frac{1}{2} \sum_{A=1}^{3} \sum_{B \neq A} \left(\frac{\sigma_B - \sigma_A}{\epsilon^{\rm e\,tr}_B - \epsilon^{\rm e\,tr}_A} \right) (\boldsymbol{m}^{(AB)} \otimes \boldsymbol{m}^{(AB)} + \boldsymbol{m}^{(AB)} \otimes \boldsymbol{m}^{(BA)}) , \end{aligned} \tag{4.120}$$

where $\boldsymbol{m}^{(A)} = \boldsymbol{m}^{{\rm tr}(A)}$, etc., and $a_{AB} = \partial \sigma_A / \partial \epsilon^{\rm e\,tr}_B$ is the algorithmic tangent operator in principal stress axes. The algorithmic tangent operator $\boldsymbol{c}_{n+1}$ differs from the constitutive tangent operator $\boldsymbol{c}^{\rm ep}$ given in (4.34) in both the material coefficients ($a^{\rm ep}_{AB}$ versus a_{AB}) and spins ($\epsilon^{\rm e\,tr}_A$ versus $\epsilon^{\rm e}_A$), but their spectral directions remain the same. The new coefficients a_{AB} are evaluated from the procedure described below.

At the locally converged state $\boldsymbol{r}(\boldsymbol{x}^*) = \boldsymbol{0}$, where $\boldsymbol{x}^*$ is the desired local solution. Taking the strain derivative of $\boldsymbol{r}$ at this solution gives

$$\frac{\partial r_I}{\partial \epsilon_B^{\mathrm{e\,tr}}} = \sum_{A=1}^{5} \underbrace{\frac{\partial r_I}{\partial x_A}\bigg|_{\epsilon^{\mathrm{e\,tr}}}}_{A_{IA}^*} \frac{\partial x_A}{\partial \epsilon_B^{\mathrm{e\,tr}}} + \underbrace{\frac{\partial r_I}{\partial \epsilon_B^{\mathrm{e\,tr}}}\bigg|_x}_{-\delta_{IB}} = 0\,, \tag{4.121}$$

for $I = 1, \dots, 5$; $B = 1, 2, 3$. Note that $[A_{IA}^*]$ is simply the Jacobian matrix $\boldsymbol{A}$ defined by (4.116), but evaluated at the converged solution $\boldsymbol{x}^*$. This Jacobian matrix is invertible, so we can solve

$$\frac{\partial x_A}{\partial \epsilon_B^{\mathrm{e\,tr}}} = -\sum_{I=1}^{5} (A_{AI}^*)^{-1} \frac{\partial r_I}{\partial \epsilon_B^{\mathrm{e\,tr}}}\bigg|_x = \frac{\partial \sigma_A}{\partial \epsilon_B^{\mathrm{e\,tr}}} \equiv a_{AB}\,, \qquad A, B = 1, 2, 3\,. \tag{4.122}$$

Thus, the local integration algorithm, through $\boldsymbol{x}^*$, is tied to the global tangent operator, through $\boldsymbol{c}_{n+1}$.

4.13 Return Mapping in Invariant Space

If the yield function is of the form $\widetilde{f}(p, q, \kappa) \le 0$, and if the plastic potential function is of the form $g = \widetilde{g}(p, q)$, where $p = \mathcal{I}_1(\boldsymbol{\sigma})/3$ and $q = \sqrt{3J_2}$, then a return map on the (p, q) space does the same job without having to perform a spectral decomposition. This typically is the case with two-invariant plasticity models. The flow rule is

$$\dot{\boldsymbol{\epsilon}}^{\mathrm{p}} = \dot{\lambda} \frac{\partial g}{\partial \boldsymbol{\sigma}} = \dot{\lambda} \Big(v \frac{\partial p}{\partial \boldsymbol{\sigma}} + e \frac{\partial q}{\partial \boldsymbol{\sigma}} \Big) = \dot{\lambda} \Big(\frac{1}{3} v \boldsymbol{1} + \sqrt{\frac{3}{2}} e \widehat{\boldsymbol{n}} \Big)\,, \tag{4.123}$$

where $v = \partial \widetilde{g}/\partial p$, $e = \partial \widetilde{g}/\partial q$, and $\widehat{\boldsymbol{n}} = \boldsymbol{s}/\|\boldsymbol{s}\|$. The plastic flow is dilative if $v > 0$, compactive if $v < 0$, and isochoric if $v = 0$.

Assuming isotropy in the elastic response,

$$\dot{\boldsymbol{\sigma}} = \boldsymbol{c}^{\mathrm{e}} : \dot{\boldsymbol{\epsilon}} - \dot{\lambda} \Big(K v \boldsymbol{1} + 2\mu \sqrt{\frac{3}{2}} e \widehat{\boldsymbol{n}} \Big)\,, \tag{4.124}$$

where K and μ are the elastic bulk and shear moduli, respectively. The backward implicit scheme yields

$$\boldsymbol{\sigma} = \boldsymbol{\sigma}^{\mathrm{tr}} - \Delta\lambda \Big(K v \boldsymbol{1} + 2\mu \sqrt{\frac{3}{2}} e \widehat{\boldsymbol{n}} \Big)\,. \tag{4.125}$$

On a meridional plane the predictor-corrector algorithm for the mean normal stress is

$$p = p^{\mathrm{tr}} - K \Delta\lambda v\,. \tag{4.126}$$

On the deviatoric plane, the corresponding equation is

$$\boldsymbol{s} = \boldsymbol{s}^{\rm tr} - 2\mu\Delta\lambda\sqrt{\frac{3}{2}}e\frac{\boldsymbol{s}}{\|\boldsymbol{s}\|}\,. \tag{4.127}$$

The last equation suggests that $\boldsymbol{s}$ and $\boldsymbol{s}^{\rm tr}$ have the same directions, so the algorithm reduces to a radial return map on the deviatoric plane. In other words, the last equation reduces to the scalar return map

$$q = q^{\rm tr} - 3\mu\Delta\lambda e\,, \tag{4.128}$$

where $q = \sqrt{3/2}\|\boldsymbol{s}\|$ and $q^{\rm tr} = \sqrt{3/2}\|\boldsymbol{s}^{\rm tr}\|$. The algorithm is essentially the same as the radial return algorithm for J_2 plasticity except that the mean normal stress also contains a plastic corrector term. Box 4.2 summarizes the algorithm.

Step 1. Compute $\boldsymbol{\sigma}^{\rm tr} = \boldsymbol{\sigma}_n + \boldsymbol{c}^{\rm e} : \Delta\boldsymbol{\epsilon}$; $\quad p^{\rm tr} = {\rm tr}(\boldsymbol{\sigma}^{\rm tr})/3$;
$\boldsymbol{s}^{\rm tr} = \boldsymbol{\sigma}^{\rm tr} - p^{\rm tr}\boldsymbol{1}$; $\quad q^{\rm tr} = \sqrt{3/2}\|\boldsymbol{s}^{\rm tr}\|$; $\quad \widehat{\boldsymbol{n}} = \boldsymbol{s}^{\rm tr}/\|\boldsymbol{s}^{\rm tr}\|$
Step 2. Check $\widetilde{f}(p^{\rm tr}, q^{\rm tr}, \kappa_n) > 0$?
No, set $\boldsymbol{\sigma} = \boldsymbol{\sigma}^{\rm tr}$ and exit.
Step 3. Yes, solve for p, q, and κ such that $\widetilde{f}(p, q, \kappa) = 0$
Step 4. Compute final stress $\boldsymbol{\sigma} = p\boldsymbol{1} + \sqrt{2/3}\,q\widehat{\boldsymbol{n}}$ and exit.

Box 4.2 Return mapping algorithm on p, q plane

Exercise 4.13. Consider a Drucker-Prager plasticity theory with constant material parameters A and B as defined in (4.76). Further, assume a plastic potential function of the form

$$\widetilde{g} = \sqrt{\frac{2}{3}}q + bp\,,$$

where b is a constant plastic dilatancy parameter. Using a return-mapping algorithm in stress invariant space, show that the discrete plastic multiplier $\Delta\lambda$ defined in equations (4.126) and (4.128) is given by

$$\Delta\lambda = \frac{\sqrt{2/3}\,q^{\rm tr} - (A - Bp^{\rm tr})}{2\mu + KBb}\,.$$

Show that when $B = 0$, this expression for $\Delta\lambda$ reduces to the perfectly plastic version of the discrete plastic multiplier given in (3.65) for the J_2 theory.

Exercise 4.14. Consider the Drucker-Prager plasticity theory from the previous example, and assume that the final stress tensor $\boldsymbol{\sigma}$ is updated according to Step #4 of Box 4.2. Linearize Step #4 directly and show that the algorithmic stress-strain tangent operator $\boldsymbol{c} = \partial\boldsymbol{\sigma}/\partial\boldsymbol{\epsilon}$ is (see Section 3.8 for further details)

$$\boldsymbol{c} = a_1\boldsymbol{I} + a_2\mathbf{1}\otimes\mathbf{1} + a_3\widehat{\boldsymbol{n}}\otimes\mathbf{1} + a_4\mathbf{1}\otimes\widehat{\boldsymbol{n}} + a_5\widehat{\boldsymbol{n}}\otimes\widehat{\boldsymbol{n}}\,,$$

where

$$a_1 = \frac{2\mu q}{q^{\rm tr}}\,, \qquad a_2 = K - \frac{BbK^2}{2\mu + BbK} - \frac{1}{3}a_1\,, \qquad a_3 = -\frac{2B\mu K}{2\mu + BbK}\,,$$

$$a_4 = -\frac{2b\mu K}{2\mu + BbK}\,, \qquad a_5 = 2\mu - \frac{1}{3}a_1 - \frac{4\mu^2}{2\mu + BbK}\,.$$

Show that when $B = b = 0$, this expression for $\boldsymbol{c}$ reduces to the perfectly plastic version of the algorithmic stress-strain tangent operator given in (3.79) for the J_2 theory.

4.14 Numerical Example

In this example, we simulate the mechanical response of a 100 mm-tall × 30 mm-wide rectangular block with a 28 mm-diameter circular hole at the center. The loading consists of applying a lateral in-plane confining stress of 5 MPa followed by vertical compression in plane strain. The block is modeled with a mesh of constant strain triangular elements, as shown in Fig. 4.12. The following material parameters are used in the simulations: Young's modulus

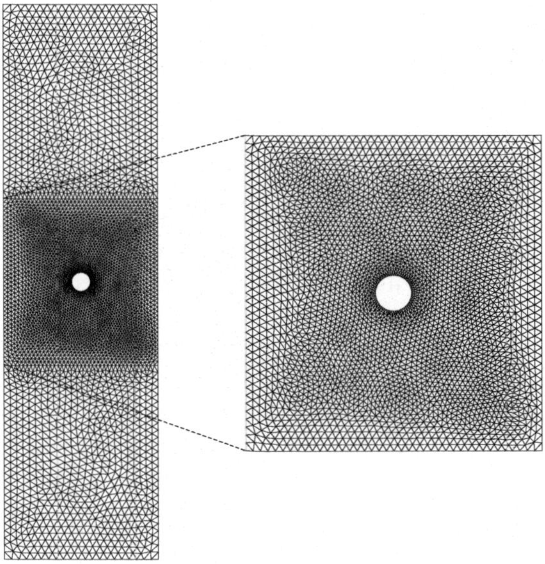

Fig. 4.12 Finite element mesh with 4,624 nodes and 8,958 constant strain triangular elements

$E = 14$ GPa, Poisson's ratio $\nu = 0.3$, cohesion $c = 10$ MPa, friction angle $\phi = 30°$, dilatancy angle $\psi = 25°$, and $H = 0$ (perfect plasticity). The loading and geometric configuration mimic a test on Berea sandstone conducted by Labuz et al. (2006).

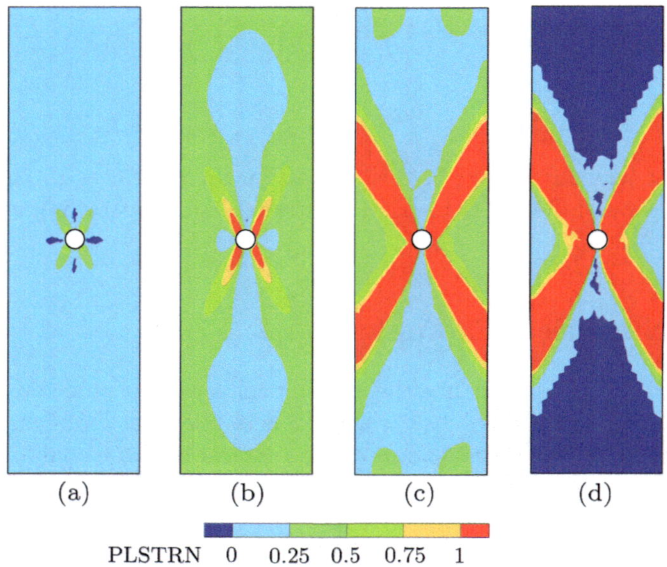

Fig. 4.13 Current plastic strain $\|\boldsymbol{\epsilon}^{\mathrm{P}}\|$ for a rectangular block with a circular hole at 0.58 mm vertical compression: (a) Drucker-Prager compression; (b) average; (c) tension; (d) Mohr-Coulomb. Color bar is plastic strain (PLSTRN) in percent

Figure 4.13 compares the yield zones in the block at a vertical compression of 0.58 mm predicted by the Mohr-Coulomb and Drucker-Prager plasticity models. Because the Drucker-Prager yield surface does not provide a unique approximation of the Mohr-Coulomb yield surface, we consider circular approximations passing through: (a) the compression corners, (b) halfway between the compression and tension corners, and (c) the tension corners of the Mohr-Coulomb yield surface. Similar approximations are made of the smoothed plastic potential function in terms of the angle of dilatancy. In general, the yield zone produced by the Mohr-Coulomb theory is more localized along narrow bands, as compared to the yield zones produced by the Drucker-Prager theory that are more diffuse. Figure 4.14 depicts the variation of axial stress with vertical compression, where we observe that the Mohr-Coulomb and Drucker-Prager-tension simulations generate softer responses compared to the responses generated by the Drucker-Prager compression and average simulations.

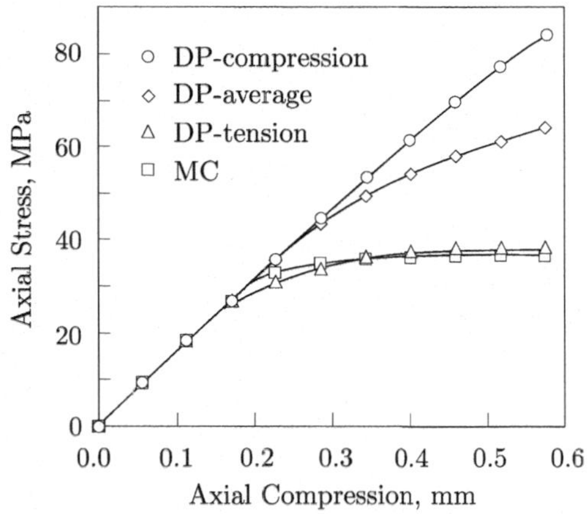

Fig. 4.14 Load-displacement responses of a rectangular block with a circular hole. MC = Mohr-Coulomb, DP = Drucker-Prager.

4.15 Summary

The three-invariant Mohr-Coulomb yield criterion is a classic plasticity model for cohesive-frictional materials. However, because the yield surface is not smooth, this model is not suitable for three-dimensional problems. Smooth approximations for the Mohr-Coulomb theory are available. They can be implemented easily using spectral decomposition of stresses and strains, along with a return-mapping algorithm in principal stress space. A closed-form expression for the algorithmic stress-strain tensor is also available for these three-invariant models.

5 Finite Deformation

5.1 Introduction

Many applications of plasticity theory involve large deformation for which the infinitesimal theory developed in the previous chapters may not be appropriate. Even if the stretching or straining of a material is small, the infinitesimal formulation is insufficient if the domain of interest experiences large rotation. This chapter focuses on the theoretical formulation and finite element implementation of a finite deformation theory of elastoplasticity. Whereas numerous finite deformation theories abound in the literature, we shall focus mainly on a formulation based on a multiplicative decomposition of the deformation gradient, which gives rise to so-called multiplicative plasticity theory. The initial development of multiplicative plasticity theory is due to Lee (1969). Since then, it has become a standard platform of many computer codes for nonlinear analysis of boundary-value problems in the finite deformation range. The theory is based on the notion of reference and current configurations, as well as on the existence of an intermediate stress-free configuration to which a body will return when it is unloaded. Nonlinear continuum mechanics is a key ingredient of the finite deformation theory of elastoplasticity, so this chapter begins with a brief review of this subject. We present continuum and discrete formulations of elastoplasticity similar to those used in the infinitesimal theory, and conclude the chapter by developing the standard weak and matrix forms of the finite element equations for solving elastoplastic boundary-value problem in the finite deformation range.

5.2 Basic Kinematics

Consider a material point X whose initial placement in the reference configuration $\mathcal{B}$ is given by the position vector $\boldsymbol{X}$, as shown in Fig. 5.1. The motion of $\mathcal{B}$ is denoted by $\boldsymbol{\varphi} : \mathcal{B} \to \mathbb{R}^{n_{\mathrm{sd}}}$, where n_{sd} is the number of spatial dimensions (typically taken as 3). We denote the current position of X by the vector $\boldsymbol{x}$, and assume that there is a mapping between $\boldsymbol{x}$ and $\boldsymbol{X}$ of the form

$$\boldsymbol{x} = \boldsymbol{\varphi}(\boldsymbol{X}, t) . \tag{5.1}$$

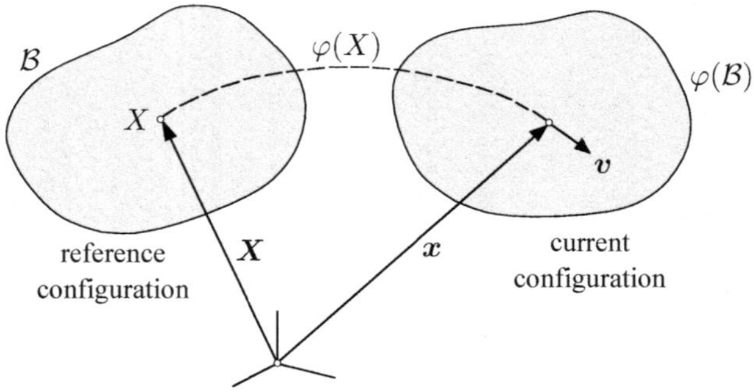

Fig. 5.1 Reference and current configurations, motion φ, and trajectory (dashed curve) of material point X

If (5.1) is invertible, then for each current placement $\boldsymbol{x}$ we can identify a material point X with reference position $\boldsymbol{X}$ from the inverse relation

$$\boldsymbol{X} = \varphi^{-1}(\boldsymbol{x}, t) . \tag{5.2}$$

The material velocity of X is tangent to its trajectory, and is given by

$$\boldsymbol{V}(\boldsymbol{X}, t) = \frac{\partial \varphi(\boldsymbol{X}, t)}{\partial t} . \tag{5.3}$$

The material acceleration of X is given by

$$\boldsymbol{A}(\boldsymbol{X}, t) = \frac{\partial \boldsymbol{V}(\boldsymbol{X}, t)}{\partial t} = \frac{\partial^2 \varphi(\boldsymbol{X}, t)}{\partial t^2} . \tag{5.4}$$

From the inverse mapping (5.2), the velocity vector (5.3) can be written in the spatial form

$$\boldsymbol{v}(\boldsymbol{x}, t) = \frac{\partial \varphi(\varphi^{-1}(\boldsymbol{x}, t), t)}{\partial t} , \tag{5.5}$$

where $\boldsymbol{v}(\boldsymbol{x}, t)$ is a spatial description of the changing velocity field at any point $\boldsymbol{x}$.

Consider a given spatial velocity field $\boldsymbol{v}(\boldsymbol{x}, t)$. The material acceleration at $\boldsymbol{x}$ can be obtained from the chain rule as

$$\boldsymbol{a}(\boldsymbol{x}, t) = \frac{\partial \boldsymbol{v}}{\partial t}\bigg|_{\boldsymbol{X}} = \frac{\partial \boldsymbol{v}}{\partial t} + \frac{\partial \boldsymbol{v}}{\partial \boldsymbol{x}} \cdot \frac{\partial \varphi(\boldsymbol{X}, t)}{\partial t} \equiv \frac{\partial \boldsymbol{v}}{\partial t} + \boldsymbol{v} \cdot \boldsymbol{\nabla} \boldsymbol{v} , \tag{5.6}$$

where $(\boldsymbol{v} \cdot \boldsymbol{\nabla} \boldsymbol{v})_i = v_j \partial_j v_i$, and $\partial(\cdot)/\partial t \equiv \partial(\cdot)/\partial t|_{\boldsymbol{x}}$ is the time derivative with $\boldsymbol{x}$ held fixed. The derivative $\partial \boldsymbol{v}/\partial t$ is called the spatial acceleration. In general, the material time derivative of any scalar, vector, or tensor-valued function $\phi = \phi(\boldsymbol{x}, t)$ is

$$\frac{\mathrm{d}\phi}{\mathrm{d}t} = \frac{\partial \phi}{\partial t}\bigg|_{\boldsymbol{X}} = \Big(\frac{\partial}{\partial t} + \boldsymbol{v} \cdot \boldsymbol{\nabla}\Big) \phi . \tag{5.7}$$

Depending on the independent variables, we can define the velocity and acceleration of a material point X as

$$\boldsymbol{V}(\boldsymbol{X},t) = \boldsymbol{v}(\varphi(\boldsymbol{X},t),t) \qquad \text{and} \qquad \boldsymbol{A}(\boldsymbol{X},t) = \boldsymbol{a}(\varphi(\boldsymbol{X},t),t)\,. \tag{5.8}$$

If the mapping is invertible, we can write

$$\boldsymbol{v} = \boldsymbol{V} \circ \varphi_t^{-1} \qquad \text{and} \qquad \boldsymbol{a} = \boldsymbol{A} \circ \varphi_t^{-1}\,, \tag{5.9}$$

where $\circ$ denotes a composition.

The deformation gradient is the derivative of the deformation with respect to $\boldsymbol{X}$,

$$\boldsymbol{F}(\boldsymbol{X},t) = \frac{\partial\varphi(\boldsymbol{X},t)}{\partial\boldsymbol{X}}\,. \tag{5.10}$$

From the chain rule, $\mathrm{d}\boldsymbol{x} = \boldsymbol{F}\cdot\mathrm{d}\boldsymbol{X}$, so the deformation gradient maps a differential segment $\mathrm{d}\boldsymbol{X}$ in the reference configuration to a differential segment $\mathrm{d}\boldsymbol{x}$ in the current configuration. The local Jacobian is

$$J(\boldsymbol{X},t) = \det[\boldsymbol{F}(\boldsymbol{X},t)] > 0\,, \tag{5.11}$$

where a positive value is true provided that the mapping is invertible. The Jacobian maps a differential volume $\mathrm{d}V$ in the reference configuration to a differential volume $\mathrm{d}v$ in the current configuration, or the local mass densities ρ_0 and ρ in the reference and current configurations, respectively. The mapping takes the form

$$J = \mathrm{d}v/\mathrm{d}V = \rho_0/\rho\,. \tag{5.12}$$

We can also map a differential area $\mathrm{d}A$ with unit normal $\boldsymbol{N}$ in the reference configuration to a differential area $\mathrm{d}a$ with unit normal $\boldsymbol{n}$ in the current configuration. This mapping is provided by Nanson's formula (see Fig. 5.2),

$$\boldsymbol{n}\mathrm{d}a = J\boldsymbol{F}^{-\mathsf{T}}\cdot\boldsymbol{N}\mathrm{d}A = J\boldsymbol{N}\cdot\boldsymbol{F}^{-1}\mathrm{d}A\,. \tag{5.13}$$

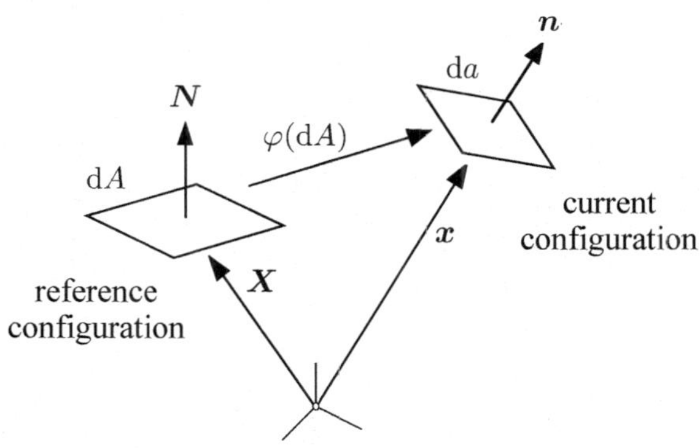

Fig. 5.2 Differential areas in the reference and current configurations

The time derivative of the deformation gradient is

$$\dot{\boldsymbol{F}} = \frac{\partial \dot{\boldsymbol{\varphi}}(\boldsymbol{X},t)}{\partial \boldsymbol{X}} = \frac{\partial \boldsymbol{V}}{\partial \boldsymbol{X}} = \boldsymbol{l} \cdot \boldsymbol{F}, \qquad \boldsymbol{l} = \boldsymbol{\nabla v}, \tag{5.14}$$

where $\boldsymbol{l}$ is the velocity gradient. The symmetric and skew-symmetric parts of $\boldsymbol{l}$ are

$$\boldsymbol{d} = \frac{1}{2}(\boldsymbol{\nabla v} + \boldsymbol{v\nabla}), \qquad \boldsymbol{w} = \frac{1}{2}(\boldsymbol{\nabla v} - \boldsymbol{v\nabla}), \tag{5.15}$$

where $\boldsymbol{d}$ is the rate of deformation tensor and $\boldsymbol{w}$ is the vorticity tensor.

Let $\boldsymbol{R}$ denote a proper orthogonal tensor. The deformation gradient admits the polar decompositions

$$\boldsymbol{F} = \boldsymbol{R} \cdot \boldsymbol{U} = \boldsymbol{V} \cdot \boldsymbol{R}, \tag{5.16}$$

where $\boldsymbol{U}$ and $\boldsymbol{V}$ are the symmetric right and left stretch tensors, respectively. As additional measures of deformation we define the right and left Cauchy-Green deformation tensors, $\boldsymbol{C}$ and $\boldsymbol{b}$, respectively, as

$$\boldsymbol{C} = \boldsymbol{F}^{\mathsf{T}} \cdot \boldsymbol{F}, \qquad \boldsymbol{b} = \boldsymbol{F} \cdot \boldsymbol{F}^{\mathsf{T}}. \tag{5.17}$$

Since $\boldsymbol{R} \cdot \boldsymbol{R}^{\mathsf{T}} = \boldsymbol{R}^{\mathsf{T}} \cdot \boldsymbol{R} = \boldsymbol{1}$, we have

$$\boldsymbol{C} = \boldsymbol{U}^2, \qquad \boldsymbol{b} = \boldsymbol{V}^2. \tag{5.18}$$

5.3 Spectral Representation

The tensors $\boldsymbol{U}$, $\boldsymbol{V}$, $\boldsymbol{C}$, and $\boldsymbol{b}$ are all symmetric, so they are amenable to spectral representation as follows,

$$\boldsymbol{U} = \sum_{A=1}^{3} \lambda_A \boldsymbol{N}^{(A)} \otimes \boldsymbol{N}^{(A)}, \qquad \boldsymbol{V} = \sum_{A=1}^{3} \lambda_A \boldsymbol{n}^{(A)} \otimes \boldsymbol{n}^{(A)}, \tag{5.19}$$

and

$$\boldsymbol{C} = \sum_{A=1}^{3} \lambda_A^2 \boldsymbol{N}^{(A)} \otimes \boldsymbol{N}^{(A)}, \qquad \boldsymbol{b} = \sum_{A=1}^{3} \lambda_A^2 \boldsymbol{n}^{(A)} \otimes \boldsymbol{n}^{(A)}. \tag{5.20}$$

The eigenvalues λ_A^2 and associated unit eigenvectors $\boldsymbol{N}^{(A)}$ and $\boldsymbol{n}^{(A)}$ can be obtained from the eigenvalue problem

$$\boldsymbol{C} \cdot \boldsymbol{N}^{(A)} = \lambda_A^2 \boldsymbol{N}^{(A)}, \qquad \boldsymbol{b} \cdot \boldsymbol{n}^{(A)} = \lambda_A^2 \boldsymbol{n}^{(A)}. \tag{5.21}$$

The eigenvalues λ_A for $A = 1, 2, 3$ are the principal stretches, and $\boldsymbol{N}^{(A)}$ and $\boldsymbol{n}^{(A)}$ are the corresponding Lagrangian and Eulerian principal axes. Both the

Lagrangian and Eulerian principal triads rotate with deformation. They are rotated by the proper orthogonal tensor $\boldsymbol{R}$, which is given by

$$\boldsymbol{R} = \sum_{A=1}^{3} \boldsymbol{n}^{(A)} \otimes \boldsymbol{N}^{(A)} . \tag{5.22}$$

From the polar decomposition of $\boldsymbol{F}$, it follows that

$$\boldsymbol{F} = \sum_{A=1}^{3} \lambda_A \boldsymbol{n}^{(A)} \otimes \boldsymbol{N}^{(A)} . \tag{5.23}$$

Next, we consider the time derivatives in spectral form. We recall from Sec. 4.3 that

$$\dot{\boldsymbol{N}}^{(A)} = \sum_{B \neq A} \Omega_{BA} \boldsymbol{N}^{(B)} , \qquad \dot{\boldsymbol{n}}^{(A)} = \sum_{B \neq A} \omega_{BA} \boldsymbol{n}^{(B)} , \tag{5.24}$$

where Ω_{BA} and ω_{BA} are components of the spins of the Lagrangian and Eulerian principal axes, respectively. Accordingly, we write

$$\begin{aligned} \dot{\boldsymbol{U}} &= \sum_{A=1}^{3} \dot{\lambda}_A \boldsymbol{N}^{(A)} \otimes \boldsymbol{N}^{(A)} + \sum_{A=1}^{3} \sum_{B \neq A} \Omega_{AB} (\lambda_B - \lambda_A) \boldsymbol{N}^{(A)} \otimes \boldsymbol{N}^{(B)} \\ \dot{\boldsymbol{C}} &= \sum_{A=1}^{3} 2\lambda_A \dot{\lambda}_A \boldsymbol{N}^{(A)} \otimes \boldsymbol{N}^{(A)} + \sum_{A=1}^{3} \sum_{B \neq A} \Omega_{AB} (\lambda_B^2 - \lambda_A^2) \boldsymbol{N}^{(A)} \otimes \boldsymbol{N}^{(B)} , \end{aligned} \tag{5.25}$$

and

$$\begin{aligned} \dot{\boldsymbol{V}} &= \sum_{A=1}^{3} \dot{\lambda}_A \boldsymbol{n}^{(A)} \otimes \boldsymbol{n}^{(A)} + \sum_{A=1}^{3} \sum_{B \neq A} \omega_{AB} (\lambda_B - \lambda_A) \boldsymbol{n}^{(A)} \otimes \boldsymbol{n}^{(B)} \\ \dot{\boldsymbol{b}} &= \sum_{A=1}^{3} 2\lambda_A \dot{\lambda}_A \boldsymbol{n}^{(A)} \otimes \boldsymbol{n}^{(A)} + \sum_{A=1}^{3} \sum_{B \neq A} \omega_{AB} (\lambda_B^2 - \lambda_A^2) \boldsymbol{n}^{(A)} \otimes \boldsymbol{n}^{(B)} . \end{aligned} \tag{5.26}$$

Since $\boldsymbol{R}$ defines the rotational separation between the Lagrangian and Eulerian principal axes, then $\dot{\boldsymbol{R}}$ should depend on both spins,

$$\dot{\boldsymbol{R}} = \sum_{A=1}^{3} \sum_{B \neq A} (\omega_{AB} - \Omega_{AB}) \boldsymbol{n}^{(A)} \otimes \boldsymbol{N}^{(B)} . \tag{5.27}$$

For future use we shall introduce $\widehat{\boldsymbol{\Omega}}$ as the relative spin between the Eulerian and Lagrangian triads. With respect to the current placement, we have

$$\widehat{\boldsymbol{\Omega}} = \dot{\boldsymbol{R}} \cdot \boldsymbol{R}^{\mathsf{T}} = \sum_{A=1}^{3} \sum_{B \neq A} (\omega_{AB} - \Omega_{AB}) \boldsymbol{n}^{(A)} \otimes \boldsymbol{n}^{(B)} . \tag{5.28}$$

Note that $\widehat{\boldsymbol{\Omega}}$ is skew-symmetric. For completeness, we also write the time derivative of $\boldsymbol{F}$ in component form,

$$\dot{\boldsymbol{F}} = \sum_{A=1}^{3} \dot{\lambda}_A \boldsymbol{n}^{(A)} \otimes \boldsymbol{N}^{(A)} + \sum_{A=1}^{3} \sum_{B \neq A} (\omega_{AB}\lambda_B - \Omega_{AB}\lambda_A) \boldsymbol{n}^{(A)} \otimes \boldsymbol{N}^{(B)} \,. \tag{5.29}$$

The above equation reflects the rate of stretching and spins of the two triads.

The velocity gradient $\boldsymbol{l}$ can be written in component form as well. We recall that

$$\boldsymbol{l} = \dot{\boldsymbol{F}} \cdot \boldsymbol{F}^{-1} = \dot{\boldsymbol{V}} \cdot \boldsymbol{V}^{-1} + \boldsymbol{V} \cdot \widehat{\boldsymbol{\Omega}} \cdot \boldsymbol{V}^{-1} \,, \tag{5.30}$$

Accordingly, we can write

$$\boldsymbol{l} = \sum_{A=1}^{3} \dot{\epsilon}_A \boldsymbol{n}^{(A)} \otimes \boldsymbol{n}^{(A)} + \sum_{A=1}^{3} \sum_{B \neq A} \left(\omega_{AB} - \Omega_{AB} \frac{\lambda_A}{\lambda_B} \right) \boldsymbol{n}^{(A)} \otimes \boldsymbol{n}^{(B)} \tag{5.31}$$

where

$$\epsilon_A = \log(\lambda_A) \,, \qquad A = 1, 2, 3 \tag{5.32}$$

are the logarithmic principal stretches (note: log = ln = natural logarithm). The first summation in (5.31) represents the rate of deformation tensor,

$$\boldsymbol{d} = \sum_{A=1}^{3} \dot{\epsilon}_A \boldsymbol{n}^{(A)} \otimes \boldsymbol{n}^{(A)} \,, \tag{5.33}$$

whereas the second summation is the vorticity tensor

$$\boldsymbol{w} = \sum_{A=1}^{3} \sum_{B \neq A} \left(\omega_{AB} - \Omega_{AB} \frac{\lambda_A}{\lambda_B} \right) \boldsymbol{n}^{(A)} \otimes \boldsymbol{n}^{(B)} \,. \tag{5.34}$$

Note that the vorticity tensor depends not only on the spins of the Eulerian and Lagrangian triads but also on the principal stretches.

Exercise 5.1. A thin film is subjected to a deformation gradient $\boldsymbol{F}(\boldsymbol{X}, t)$. A differential area $\mathrm{d}A$ on the film with unit normal $\boldsymbol{N}$ in the reference configuration is mapped to a differential area $\mathrm{d}a$ with unit normal $\boldsymbol{n}$ in the current configuration. Find the relation between the thickness h_0 of the film in the reference configuration to the thickness h in the current configuration.

Exercise 5.2. Let ω_{AB} denote the AB-components of the spin of the Eulerian principal axes of $\boldsymbol{V}$, and let $\widehat{\omega}_{AB} = \boldsymbol{n}^{(A)} \cdot \widehat{\boldsymbol{\Omega}} \cdot \boldsymbol{n}^{(B)}$, where $\widehat{\boldsymbol{\Omega}} = \dot{\boldsymbol{R}} \cdot \boldsymbol{R}^{\mathrm{T}}$ is the relative spin as defined previously. Given ω_{AB}, $\widehat{\omega}_{AB}$, and the principal stretches λ_A and λ_B, express the vorticity tensor $\boldsymbol{w}$ in spectral form using the Eulerian triad as a point of reference.

5.4 Stress Tensors

So far, we have only used the symmetric Cauchy stress tensor $\boldsymbol{\sigma}$ for the infinitesimal theory of plasticity. In the finite deformation theory, however, it is convenient to define other stress measures associated with either the reference configuration, or the current configuration, or both.

We first define the symmetric Kirchhoff stress tensor $\boldsymbol{\tau}$ as

$$\boldsymbol{\tau} = J\boldsymbol{\sigma} \qquad \Longleftrightarrow \qquad \tau_{ij} = J\sigma_{ij} \,. \tag{5.35}$$

We note that both $\boldsymbol{\sigma}$ and $\boldsymbol{\tau}$ are spatial tensors in the current configuration. The first Piola-Kirchhoff stress tensor $\boldsymbol{P}$ is defined as

$$\boldsymbol{P} = \boldsymbol{\tau} \cdot \boldsymbol{F}^{-\mathsf{T}} \qquad \Longleftrightarrow \qquad P_{iA} = \tau_{ij} F_{jA}^{-\mathsf{T}} \,. \tag{5.36}$$

It is a two-point stress tensor that measures the resultant force acting on area $\mathrm{d}a$ in the current configuration per unit area $\mathrm{d}A$ in the reference configuration. To elaborate the meaning of the stress tensor $\boldsymbol{P}$, consider the resultant force $\mathrm{d}\boldsymbol{\mathcal{F}} = \boldsymbol{\sigma} \cdot \boldsymbol{n} \, \mathrm{d}a$ acting on an area $\mathrm{d}a$ with unit normal $\boldsymbol{n}$ in the current configuration. Then, $\boldsymbol{P}$ is defined so that the same resultant force is obtained from $\mathrm{d}\boldsymbol{\mathcal{F}} = \boldsymbol{P} \cdot \boldsymbol{N} \mathrm{d}A$. In fact, if we use Nanson's formula (5.13), then $\mathrm{d}\boldsymbol{\mathcal{F}} = \boldsymbol{\sigma} \cdot \boldsymbol{n} \, \mathrm{d}a = \boldsymbol{\sigma} \cdot (J\boldsymbol{F}^{-\mathsf{T}} \cdot \boldsymbol{N})\mathrm{d}A$, and we recover precisely the definition for $\boldsymbol{P}$. The transpose $\boldsymbol{P}^{\mathsf{T}}$ is often called the nominal stress tensor (Ogden 1984), which relates $\mathrm{d}\boldsymbol{\mathcal{F}} = \boldsymbol{n} \cdot \boldsymbol{\sigma} \, \mathrm{d}a = \boldsymbol{N} \cdot \boldsymbol{P}^{\mathsf{T}} \mathrm{d}A$.

Next, we define the second Piola-Kirchhoff stress $\boldsymbol{S}$ as

$$\boldsymbol{S} = \boldsymbol{F}^{-1} \cdot \boldsymbol{P} = \boldsymbol{F}^{-1} \cdot \boldsymbol{\tau} \cdot \boldsymbol{F}^{-\mathsf{T}} \qquad \Longleftrightarrow \qquad S_{AB} = F_{Ai}^{-1} \tau_{ij} F_{jB}^{-\mathsf{T}} \,, \tag{5.37}$$

which relates the force in the reference configuration to an area in the reference configuration. The force in the reference configuration is obtained by applying the pull-back operator $\boldsymbol{F}^{-1}$ on the force in the current configuration. The nature of the stress tensors $\boldsymbol{\tau}$, $\boldsymbol{P}$ and $\boldsymbol{S}$ can be inferred from their subscripts: lower-case subscripts denote spatial indices, whereas upper-case subscripts denote referential indices, see Malvern (1969) and Marsden and Hughes (1983).

5.5 Objectivity and Isotropy

Objectivity (frame invariance, frame indifference) pertains to the satisfaction of certain rules of tensor transformation under superposed rigid body motions. Let $\boldsymbol{x} = \boldsymbol{\varphi}(\boldsymbol{X}, t)$ denote a mapping of a material point X from the reference position $\boldsymbol{X}$ to a current position $\boldsymbol{x}$. A rigid body motion on the current configuration is defined by a uniform translation $\boldsymbol{c}(t)$ and a rotation $\boldsymbol{Q}(t)$,

$$\boldsymbol{x}^{+} = \boldsymbol{c}(t) + \boldsymbol{Q}(t) \cdot \boldsymbol{x} \,. \tag{5.38}$$

Note that both the vector $\boldsymbol{c}(t)$ and the proper orthogonal motion $\boldsymbol{Q}(t)$ depend only on time, and therefore the above motion does not induce stretching or contraction of any segment in the body.

Under the above rigid body motion, the corresponding deformation gradient is

$$\boldsymbol{F}^{+}(\boldsymbol{X},t) = \frac{\partial \boldsymbol{x}^{+}}{\partial \boldsymbol{X}} = \boldsymbol{Q}(t) \cdot \boldsymbol{F}(\boldsymbol{X},t) . \tag{5.39}$$

Accordingly, the velocity gradient is

$$\boldsymbol{l}^{+} = \dot{\boldsymbol{F}}^{+} \cdot (\boldsymbol{F}^{+})^{-1} = \boldsymbol{Q} \cdot \boldsymbol{l} \cdot \boldsymbol{Q}^{\mathsf{T}} + \dot{\boldsymbol{Q}} \cdot \boldsymbol{Q}^{\mathsf{T}} . \tag{5.40}$$

We see that the velocity gradient does not transform objectively because of the additional term $\dot{\boldsymbol{Q}} \cdot \boldsymbol{Q}^{\mathsf{T}}$. The rate of deformation tensor transforms objectively,

$$\boldsymbol{d}^{+} = \mathrm{sym}(\boldsymbol{l}^{+}) = \boldsymbol{Q} \cdot \boldsymbol{d} \cdot \boldsymbol{Q}^{\mathsf{T}} , \tag{5.41}$$

but the vorticity tensor does not,

$$\boldsymbol{w}^{+} = \mathrm{skw}(\boldsymbol{l}^{+}) = \boldsymbol{Q} \cdot \boldsymbol{w} \cdot \boldsymbol{Q}^{\mathsf{T}} + \dot{\boldsymbol{Q}} \cdot \boldsymbol{Q}^{\mathsf{T}} . \tag{5.42}$$

The Cauchy stress tensor $\boldsymbol{\sigma}$ is objective in the sense

$$\boldsymbol{\sigma}^{+} = \boldsymbol{Q} \cdot \boldsymbol{\sigma} \cdot \boldsymbol{Q}^{\mathsf{T}} . \tag{5.43}$$

However, the Cauchy stress rate tensor $\dot{\boldsymbol{\sigma}}$ is not objective, which can be demonstrated as follows. Take $\boldsymbol{\sigma} = \boldsymbol{\sigma}(\boldsymbol{x}, t)$; then, the material time derivative is given by

$$\dot{\boldsymbol{\sigma}} = \frac{\partial \boldsymbol{\sigma}}{\partial t} + \boldsymbol{v} \cdot \boldsymbol{\nabla} \boldsymbol{\sigma} . \tag{5.44}$$

From (5.43) and (5.44), we write

$$\dot{\boldsymbol{\sigma}}^{+} = \boldsymbol{Q} \cdot \dot{\boldsymbol{\sigma}} \cdot \boldsymbol{Q}^{\mathsf{T}} + (\dot{\boldsymbol{Q}} \cdot \boldsymbol{Q}^{\mathsf{T}}) \cdot \boldsymbol{\sigma}^{+} - \boldsymbol{\sigma}^{+} \cdot (\dot{\boldsymbol{Q}} \cdot \boldsymbol{Q}^{\mathsf{T}}) , \tag{5.45}$$

which is not objective due to the presence of the last two terms.

To preserve objectivity of the stress rate, modified time derivatives are often used for the rate-constitutive equation. However, because $\dot{\boldsymbol{Q}} \cdot \boldsymbol{Q}^{\mathsf{T}}$ is not uniquely defined, there is an infinite number of possible objective stress rates. The Jaumann-Zaremba stress rate of the Kirchhoff stress is a co-rotated derivative relative to the spatial axes, and replaces $\dot{\boldsymbol{Q}} \cdot \boldsymbol{Q}^{\mathsf{T}}$ with the skew-symmetric vorticity tensor $\boldsymbol{w}$, yielding

$$\overset{\triangledown}{\boldsymbol{\tau}} = \dot{\boldsymbol{\tau}} - \boldsymbol{w} \cdot \boldsymbol{\tau} + \boldsymbol{\tau} \cdot \boldsymbol{w} . \tag{5.46}$$

The Green-McInnis-Naghdi stress rate of the Kirchhoff stress tensor replaces $\dot{\boldsymbol{Q}} \cdot \boldsymbol{Q}^{\mathsf{T}}$ with $\widehat{\boldsymbol{\Omega}} = \dot{\boldsymbol{R}} \cdot \boldsymbol{R}$, yielding

$$\overset{\circ}{\boldsymbol{\tau}} = \dot{\boldsymbol{\tau}} - \widehat{\boldsymbol{\Omega}} \cdot \boldsymbol{\tau} + \boldsymbol{\tau} \cdot \widehat{\boldsymbol{\Omega}} . \tag{5.47}$$

Finally, the Lie derivative of the Kirchhoff stress tensor, also known as the Truesdell stress rate, utilizes the full velocity gradient, and takes the form

$$\mathcal{L}_v \boldsymbol{\tau} = \dot{\boldsymbol{\tau}} - \boldsymbol{l} \cdot \boldsymbol{\tau} - \boldsymbol{\tau} \cdot \boldsymbol{l}^{\mathsf{T}} . \tag{5.48}$$

The notion of objectivity plays a central role in the formulation of the constitutive equation, since the material response may not depend on the choice of reference frame. A class of material response that automatically satisfies the requirement of objectivity is provided by a hyperelastic material. In a hyperelastic material, the stress is obtained from a stored energy function that depends on the local deformation. Now, consider a stored energy function of the form $\Psi^{\mathrm{e}}(\boldsymbol{X}, \boldsymbol{F})$, and write the first Piola-Kirchhoff stress tensor for any material point X as

$$\boldsymbol{P}(\boldsymbol{X}, t) = \frac{\partial \Psi^{\mathrm{e}}(\boldsymbol{X}, \boldsymbol{F})}{\partial \boldsymbol{F}} . \tag{5.49}$$

The stored energy function $\Psi^{\mathrm{e}}(\boldsymbol{X}, \boldsymbol{F})$ is frame invariant if

$$\Psi^{\mathrm{e}}(\boldsymbol{X}, \boldsymbol{F}) = \Psi^{\mathrm{e}}(\boldsymbol{X}, \boldsymbol{F}^{+}) , \qquad \boldsymbol{F}^{+} = \boldsymbol{Q} \cdot \boldsymbol{F} , \tag{5.50}$$

for all possible orthogonal transformations induced by $\boldsymbol{Q}$.

The stored energy function is left rotationally invariant if it varies with $\boldsymbol{F}$ through the right Cauchy-Green deformation tensor $\boldsymbol{C}$, for then,

$$\boldsymbol{C}^{+} = (\boldsymbol{F}^{+})^{\mathsf{T}} \cdot \boldsymbol{F}^{+} = \boldsymbol{F}^{\mathsf{T}} \cdot \boldsymbol{Q}^{\mathsf{T}} \cdot \boldsymbol{Q} \cdot \boldsymbol{F} = \boldsymbol{F}^{\mathsf{T}} \cdot \boldsymbol{F} \equiv \boldsymbol{C} . \tag{5.51}$$

Hence, the restriction of frame invariance allows the replacement of $\boldsymbol{F}$ with $\boldsymbol{C}$,

$$\Psi^{\mathrm{e}}(\boldsymbol{X}, \boldsymbol{F}) = \widetilde{\Psi}^{\mathrm{e}}(\boldsymbol{X}, \boldsymbol{C}) . \tag{5.52}$$

By making the stored energy function left rotationally invariant the number of independent deformation variables in three dimensions is reduced from nine to six (i.e., the number of independent components of the symmetric tensor $\boldsymbol{C}$).

With frame invariance, the hyperelastic constitutive equation for the first Piola-Kirchhoff stress tensor becomes

$$\boldsymbol{P} = \frac{\partial \widetilde{\Psi}^{\mathrm{e}}(\boldsymbol{X}, \boldsymbol{C})}{\partial \boldsymbol{C}} \cdot \frac{\partial \boldsymbol{C}}{\partial \boldsymbol{F}} = 2\boldsymbol{F} \cdot \frac{\partial \widetilde{\Psi}^{\mathrm{e}}(\boldsymbol{X}, \boldsymbol{C})}{\partial \boldsymbol{C}} . \tag{5.53}$$

Since $\boldsymbol{S} = \boldsymbol{F}^{-1} \cdot \boldsymbol{P}$, we get the hyperelastic constitutive equation for $\boldsymbol{S}$,

$$\boldsymbol{S} = 2 \frac{\partial \widetilde{\Psi}^{\mathrm{e}}(\boldsymbol{X}, \boldsymbol{C})}{\partial \boldsymbol{C}} . \tag{5.54}$$

Noting that $\boldsymbol{\tau} = \boldsymbol{F} \cdot \boldsymbol{S} \cdot \boldsymbol{F}^{\mathsf{T}}$, we get the hyperelastic constitutive equation for $\boldsymbol{\tau}$,

$$\boldsymbol{\tau} = 2\boldsymbol{F} \cdot \frac{\partial \widetilde{\Psi}^{\mathrm{e}}(\boldsymbol{X}, \boldsymbol{C})}{\partial \boldsymbol{C}} \cdot \boldsymbol{F}^{\mathsf{T}} . \tag{5.55}$$

Next, we consider a smaller subset of functional relationships satisfying the property of isotropy. By isotropy at a point $\boldsymbol{X}$, we mean that the material is insensitive to superposed rigid body motions on the reference configuration. Note that isotropy is not the same as frame indifference: the former pertains to some particular property of material response, whereas the latter is a fundamental principle of mechanics that must hold for all possible response functions (Simo 1998).

Let $\boldsymbol{X} \in \mathcal{B}$ be a point in the reference configuration and consider a superposed rigid body motion defined by

$$\boldsymbol{X}^+ = \boldsymbol{r} + \boldsymbol{Q} \cdot \boldsymbol{X} . \tag{5.56}$$

In this case the deformation gradient transforms according to

$$\boldsymbol{F}^+ = \frac{\partial \boldsymbol{x}}{\partial \boldsymbol{X}^+} = \boldsymbol{F} \cdot \boldsymbol{Q}^{\mathsf{T}} . \tag{5.57}$$

The right and left Cauchy-Green deformation tensors transform according to

$$\boldsymbol{C}^+ = \boldsymbol{Q} \cdot \boldsymbol{C} \cdot \boldsymbol{Q}^{\mathsf{T}} , \qquad \boldsymbol{b}^+ = \boldsymbol{b} , \tag{5.58}$$

respectively. We see that the left Cauchy-Green deformation tensor $\boldsymbol{b}$ satisfies the definition of isotropy but not necessarily that of objectivity. Frame invariance, along with isotropy, thus requires that

$$\widetilde{\Psi}^{\mathrm{e}}(\boldsymbol{X}, \boldsymbol{C}) = \widetilde{\Psi}^{\mathrm{e}}(\boldsymbol{X}, \boldsymbol{Q} \cdot \boldsymbol{C} \cdot \boldsymbol{Q}^{\mathsf{T}}) . \tag{5.59}$$

This requirement is met by a stored energy function that depends solely on the invariants of $\boldsymbol{C}$, or on the eigenvalues of $\boldsymbol{C}$ (or on the eigenvalues of $\boldsymbol{b}$). Thus, with frame invariance and isotropy, the number of independent deformation variables is reduced further to three.

Let $\overline{\lambda}_A = \lambda_A^2$ for $A = 1, 2, 3$ be the eigenvalues of $\boldsymbol{C}$ (or $\boldsymbol{b}$). Frame invariance along with isotropy allows the replacement

$$\Psi^{\mathrm{e}}(\boldsymbol{X}, \boldsymbol{F}) = \widehat{\Psi}^{\mathrm{e}}(\boldsymbol{\varphi}^{-1}(\boldsymbol{x}), \overline{\lambda}_1, \overline{\lambda}_2, \overline{\lambda}_3) = \psi^{\mathrm{e}}(\boldsymbol{X}, \boldsymbol{b}) , \tag{5.60}$$

where it is assumed that ψ^{e} depends on $\boldsymbol{b}$ through its eigenvalues. In this case, the first Piola-Kirchhoff stress tensor is determined from the hyperelastic constitutive equation

$$\boldsymbol{P} = \frac{\partial \psi^{\mathrm{e}}(\boldsymbol{X}, \boldsymbol{b})}{\partial \boldsymbol{b}} \cdot \frac{\partial \boldsymbol{b}}{\partial \boldsymbol{F}} = 2 \frac{\partial \psi^{\mathrm{e}}(\boldsymbol{X}, \boldsymbol{b})}{\partial \boldsymbol{b}} \cdot \boldsymbol{F} . \tag{5.61}$$

The constitutive equation for the second Piola-Kirchhoff stress tensor becomes

$$\boldsymbol{S} = 2\boldsymbol{F}^{-1} \cdot \frac{\partial \psi^{\mathrm{e}}(\boldsymbol{X}, \boldsymbol{b})}{\partial \boldsymbol{b}} \cdot \boldsymbol{F} . \tag{5.62}$$

Finally, the symmetric Kirchhoff stress tensor is determined from the hyperelastic constitutive equation

$$\boldsymbol{\tau} = 2\frac{\partial \psi^{\mathrm{e}}}{\partial \boldsymbol{b}} \cdot \boldsymbol{b} = 2\boldsymbol{b} \cdot \frac{\partial \psi^{\mathrm{e}}}{\partial \boldsymbol{b}} . \tag{5.63}$$

Note that $\partial\psi^{\mathrm{e}}/\partial\boldsymbol{b}$ and $\boldsymbol{b}$ are now coaxial and therefore commute by the assumption of isotropy.

Exercise 5.3. Show that $\widehat{\boldsymbol{\Omega}} = \boldsymbol{w}$ when $\dot{\boldsymbol{U}} = \boldsymbol{0}$. In this case, the Jaumann-Zaremba stress rate coincides with the Green-McInnis stress rate.

5.6 Multiplicative Plasticity: Kinematics

We assume that the reference configuration for a simple body $\mathcal{B}$ at some uniform base temperature θ_0 is undeformed and stress-free. Let $X \in \mathcal{B}$ denote any material point in this body, and denote its position vector by $\boldsymbol{X}$. Further, let $\varphi(\mathcal{B})$ denote the final deformed configuration after following some complex elastoplastic deformation, and assume a mapping $\boldsymbol{x} = \varphi(\boldsymbol{X})$. We define the intermediate configuration as one that results if each point in the body is unstressed and reduced to the uniform base temperature θ_0.

Let $\boldsymbol{\chi}$ denote the position vector of X in the intermediate configuration, and consider the sequential mapping (see Fig. 5.3)

$$\boldsymbol{\chi} = \varphi^{\mathrm{p}}(\boldsymbol{X},t)\,, \qquad \boldsymbol{x} = \varphi^{\mathrm{e}}(\boldsymbol{\chi}) = \varphi^{\mathrm{e}}(\varphi^{\mathrm{p}}(\boldsymbol{X},t)) \equiv \varphi(\boldsymbol{X},t)\,. \tag{5.64}$$

Then, by the chain rule we have

$$\boldsymbol{F}(\boldsymbol{X},t) = \frac{\partial \varphi}{\partial \boldsymbol{X}} = \frac{\partial \varphi^{\mathrm{e}}}{\partial \boldsymbol{\chi}} \cdot \frac{\partial \varphi^{\mathrm{p}}}{\partial \boldsymbol{X}} = \boldsymbol{F}^{\mathrm{e}}(\boldsymbol{X},t) \cdot \boldsymbol{F}^{\mathrm{p}}(\boldsymbol{X},t)\,, \tag{5.65}$$

where

$$\boldsymbol{F}^{\mathrm{e}}(\boldsymbol{X},t) = \frac{\partial \varphi^{\mathrm{e}}}{\partial \boldsymbol{\chi}} \circ \varphi^{\mathrm{p}}\,, \qquad \boldsymbol{F}^{\mathrm{p}}(\boldsymbol{X},t) = \frac{\partial \varphi^{\mathrm{p}}}{\partial \boldsymbol{X}}\,. \tag{5.66}$$

We define $\boldsymbol{F}^{\mathrm{e}}$ and $\boldsymbol{F}^{\mathrm{p}}$ as the elastic and plastic components of deformation gradient, respectively, and equation (5.65) as the multiplicative decomposition of $\boldsymbol{F}$.

The concept of multiplicative split was motivated by crystal plasticity since $\boldsymbol{F}^{\mathrm{p}}$ may be viewed as the deformation induced when single crystals slip via dislocation motion, while $\boldsymbol{F}^{\mathrm{e}}$ rotates and distorts the crystal lattice (Lee 1969). As for the existence of an intermediate configuration, we can argue that if a body is reduced to the uniform base temperature θ_0 and all the external applied loads are removed, a residual stress distribution may in fact remain due to prior plastic flow. In order to totally remove all the residual stresses the body will have to be dissected to smaller elements, resulting in a discontinuous mapping φ^{e}. Fortunately, in the multiplicative plasticity theory, we do not need to know the intermediate configuration, so the deformation is not essentially modified by any discontinuous mapping.

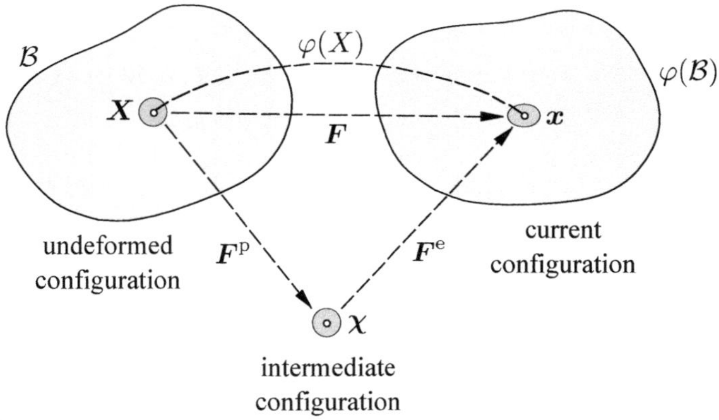

Fig. 5.3 Intermediate configuration for material point X following a multiplicative decomposition of the deformation gradient

Along with the multiplicative decomposition (5.65) we define elastic right and left Cauchy-Green deformation tensors

$$\overline{\boldsymbol{C}}^{\mathrm{e}} = \boldsymbol{F}^{\mathrm{e}\mathsf{T}} \cdot \boldsymbol{F}^{\mathrm{e}}, \qquad \boldsymbol{b}^{\mathrm{e}} = \boldsymbol{F}^{\mathrm{e}} \cdot \boldsymbol{F}^{\mathrm{e}\mathsf{T}}. \tag{5.67}$$

Note that $\overline{\boldsymbol{C}}^{\mathrm{e}}$ is a covariant tensor field defined with respect to the intermediate configuration (implied by the superposed line), whereas $\boldsymbol{b}^{\mathrm{e}}$ is a contravariant tensor field defined with respect to the current configuration. From the multiplicative decomposition of deformation gradient, we determine $\boldsymbol{b}^{\mathrm{e}}$ as

$$\boldsymbol{b}^{\mathrm{e}} = \boldsymbol{F} \cdot \overline{\boldsymbol{C}}^{\mathrm{p}\,-1} \cdot \boldsymbol{F}^{\mathsf{T}}, \qquad \overline{\boldsymbol{C}}^{\mathrm{p}} = \boldsymbol{F}^{\mathrm{p}\mathsf{T}} \cdot \boldsymbol{F}^{\mathrm{p}}. \tag{5.68}$$

Time differentiation of $\boldsymbol{b}^{\mathrm{e}}$ yields the expression

$$\dot{\boldsymbol{b}}^{\mathrm{e}} = \boldsymbol{l} \cdot \boldsymbol{b}^{\mathrm{e}} + \boldsymbol{b}^{\mathrm{e}} \cdot \boldsymbol{l}^{\mathsf{T}} + \mathcal{L}_v \boldsymbol{b}^{\mathrm{e}}, \qquad \mathcal{L}_v \boldsymbol{b}^{\mathrm{e}} = \boldsymbol{F} \cdot \frac{\partial \overline{\boldsymbol{C}}^{\mathrm{p}\,-1}}{\partial t} \cdot \boldsymbol{F}^{\mathsf{T}}, \tag{5.69}$$

where $\mathcal{L}_v \boldsymbol{b}^{\mathrm{e}}$ is the Lie derivative of $\boldsymbol{b}^{\mathrm{e}}$ (see Section 5.5).

Exercise 5.4. A differential area $\mathrm{d}a$ with unit normal $\boldsymbol{n}$ in the current configuration maps with a differential area $\mathrm{d}A^*$ with unit normal $\boldsymbol{N}^*$ in the intermediate configuration. Determine the two-point stress tensor $\boldsymbol{P}^*$ in the intermediate configuration so that $\mathrm{d}\boldsymbol{\mathcal{F}} = \boldsymbol{\sigma} \cdot \boldsymbol{n}\,\mathrm{d}a = \boldsymbol{P}^* \cdot \boldsymbol{N}^* \mathrm{d}A^*$.

5.7 Free Energy, Yield Function, and Plastic Flow Evolution

We restrict this section to a rotationally invariant, isotropic stored energy function. Motivated by the result shown in equation (5.60), we write the free energy function in the form

$$\psi = \psi(\boldsymbol{b}^{\mathrm{e}}, \boldsymbol{\xi}), \qquad \boldsymbol{b}^{\mathrm{e}} = \boldsymbol{F}^{\mathrm{e}} \cdot \boldsymbol{F}^{\mathrm{eT}}, \tag{5.70}$$

where $\boldsymbol{\xi}$ is a vector of strain-like plastic internal variable analogous to the cumulative plastic strain λ used in the infinitesimal theory, see the developments in Sec. 3.9. We assume that ψ can be decomposed additively into an elastic stored energy function $\psi^{\mathrm{e}}(\boldsymbol{b}^{\mathrm{e}})$ and a part $\psi^{\mathrm{p}}(\boldsymbol{\xi})$ that depends solely on the plastic internal variable $\boldsymbol{\xi}$. This results in complete uncoupling of the elastic and plastic responses. The dissipation inequality of thermodynamics then takes the following form analogous to (3.82),

$$\mathcal{D} = \boldsymbol{\tau} : \boldsymbol{d} - \frac{\mathrm{d}\psi}{\mathrm{d}t} \geq 0, \tag{5.71}$$

where

$$\frac{\mathrm{d}\psi}{\mathrm{d}t} = \frac{\partial\psi}{\partial\boldsymbol{b}^{\mathrm{e}}} : \dot{\boldsymbol{b}}^{\mathrm{e}} + \frac{\partial\psi}{\partial\boldsymbol{\xi}} \cdot \dot{\boldsymbol{\xi}}. \tag{5.72}$$

Inserting the expression for the time derivative of $\boldsymbol{b}^{\mathrm{e}}$ from (5.69) yields

$$\frac{\mathrm{d}\psi}{\mathrm{d}t} = \left(2\frac{\partial\psi}{\partial\boldsymbol{b}^{\mathrm{e}}} \cdot \boldsymbol{b}^{\mathrm{e}}\right) : \left(\boldsymbol{l} + \frac{1}{2}(\mathcal{L}_v\boldsymbol{b}^{\mathrm{e}}) \cdot \boldsymbol{b}^{\mathrm{e}\,-1}\right) + \frac{\partial\psi}{\partial\boldsymbol{\xi}} \cdot \dot{\boldsymbol{\xi}}. \tag{5.73}$$

By restriction to isotropy $\partial\psi/\partial\boldsymbol{b}^{\mathrm{e}}$ and $\boldsymbol{b}^{\mathrm{e}}$ commute, and their tensor product is symmetric, so the skew-symmetric term of $\boldsymbol{l}$ drops out. Thus, (5.71) becomes

$$\mathcal{D} = \left(\boldsymbol{\tau} - 2\frac{\partial\psi}{\partial\boldsymbol{b}^{\mathrm{e}}} \cdot \boldsymbol{b}^{\mathrm{e}}\right) : \boldsymbol{d} + \left(2\frac{\partial\psi}{\partial\boldsymbol{b}^{\mathrm{e}}} \cdot \boldsymbol{b}^{\mathrm{e}}\right) : \left(-\frac{1}{2}(\mathcal{L}_v\boldsymbol{b}^{\mathrm{e}}) \cdot \boldsymbol{b}^{\mathrm{e}\,-1}\right) - \frac{\partial\psi}{\partial\boldsymbol{\xi}} \cdot \dot{\boldsymbol{\xi}} \geq 0. \tag{5.74}$$

Since (5.74) must hold for all admissible processes, a standard argument leads to the following hyperelastic constitutive equation

$$\boldsymbol{\tau} = 2\frac{\partial\psi}{\partial\boldsymbol{b}^{\mathrm{e}}} \cdot \boldsymbol{b}^{\mathrm{e}} = 2\boldsymbol{b}^{\mathrm{e}} \cdot \frac{\partial\psi}{\partial\boldsymbol{b}^{\mathrm{e}}}, \tag{5.75}$$

and the following reduced dissipation inequality

$$\mathcal{D} = \boldsymbol{\tau} : \left(-\frac{1}{2}(\mathcal{L}_v\boldsymbol{b}^{\mathrm{e}}) \cdot \boldsymbol{b}^{\mathrm{e}\,-1}\right) + \boldsymbol{\kappa} \cdot \dot{\boldsymbol{\xi}} \geq 0, \tag{5.76}$$

where $\boldsymbol{\kappa} = -\partial\psi/\partial\boldsymbol{\xi}$ is a vector of stress-like plastic internal variables conjugate to $\boldsymbol{\xi}$.

Consider a yield function of the form

$$f(\boldsymbol{\tau}, \boldsymbol{\kappa}) \leq 0. \tag{5.77}$$

From the reduced dissipation inequality (5.76), we identify the expression inside the big parentheses as the plastic flow direction, i.e.,

$$-\frac{1}{2}(\mathcal{L}_v\boldsymbol{b}^{\mathrm{e}}) \cdot \boldsymbol{b}^{\mathrm{e}\,-1} = \dot{\gamma}\frac{\partial g}{\partial\boldsymbol{\tau}}, \tag{5.78}$$

where g is the plastic potential function and $\dot{\gamma}$ is a nonnegative plastic multiplier (Note a slight change in the notation for the plastic multiplier from the usual symbol $\dot{\lambda}$. This is made so as not to confuse the plastic multiplier with the principal stretches). Maximum plastic dissipation is ensured by assuming an associative flow rule, i.e., $\partial g/\partial \boldsymbol{\tau} \parallel \partial f/\partial \boldsymbol{\tau}$, along with associative hardening,

$$\dot{\boldsymbol{\xi}} = \dot{\gamma}\frac{\partial f}{\partial \boldsymbol{\kappa}} . \tag{5.79}$$

In this case, the reduced dissipation inequality becomes

$$\mathcal{D} = \dot{\gamma}\Big(\boldsymbol{\tau} : \frac{\partial f}{\partial \boldsymbol{\tau}} + \boldsymbol{\kappa} \cdot \frac{\partial f}{\partial \boldsymbol{\kappa}}\Big) \geq 0 . \tag{5.80}$$

Similar to the infinitesimal formulation, maximum plastic dissipation implies: (a) convexity of the yield function f; (b) associative flow rule; and (c) the Kuhn-Tucker conditions

$$\dot{\gamma} \geq 0 , \qquad f(\boldsymbol{\tau}, \boldsymbol{\kappa}) \leq 0 , \qquad \dot{\gamma} f(\boldsymbol{\tau}, \boldsymbol{\kappa}) = 0 . \tag{5.81}$$

5.8 Elastoplastic Tangent Operator

To further highlight the interesting similarity between the infinitesimal formulations presented in previous chapters and the finite deformation formulation, we derive below the continuum elastoplastic tangent operator in the finite deformation range. In view of the requirement of objectivity, we shall note at the outset that some of the expressions will contain so-called initial stress terms, in addition to the moduli terms. We shall also invoke the property of isotropy so that we can use all of the nice results we developed from the preceding section.

Consider a rate constitutive equation of the form

$$\dot{\boldsymbol{\tau}} = \boldsymbol{\alpha}^{\mathrm{e}} : \Big(\boldsymbol{l} - \dot{\gamma}\frac{\partial g}{\partial \boldsymbol{\tau}}\Big) , \qquad \dot{\gamma} \geq 0 . \tag{5.82}$$

In the above equation, $\boldsymbol{\alpha}^{\mathrm{e}}$ is a rank-four elastic tangent tensor analogous to the elastic constitutive operator $\boldsymbol{c}^{\mathrm{e}}$ of the infinitesimal theory. As usual, the plastic flow direction is obtained from the flow rule, which we rewrite below as

$$\mathcal{L}_v \boldsymbol{b}^{\mathrm{e}} = -2\dot{\gamma}\frac{\partial g}{\partial \boldsymbol{\tau}} \cdot \boldsymbol{b}^{\mathrm{e}} = -2\dot{\gamma}\boldsymbol{b}^{\mathrm{e}} \cdot \frac{\partial g}{\partial \boldsymbol{\tau}} . \tag{5.83}$$

Since $\boldsymbol{\tau}$ is a symmetric tensor, the flow rule indicated in (5.83) necessarily implies that $\partial g/\partial \boldsymbol{\tau}$ is a symmetric tensor, thereby eliminating any plastic spin. It also suggests coaxiality of $\mathcal{L}_v \boldsymbol{b}^{\mathrm{e}}$ with $\boldsymbol{b}^{\mathrm{e}}$ and $\partial g/\partial \boldsymbol{\tau}$, allowing their product to commute. Plastic spin can be accommodated by a more general flow rule, see Dafalias (1998).

To develop an expression for $\boldsymbol{\alpha}^{\mathrm{e}}$, we first write $\boldsymbol{\tau}$ and $\boldsymbol{b}^{\mathrm{e}}$ in spectral form,

$$\boldsymbol{\tau} = \sum_{A=1}^{3} \tau_A \boldsymbol{n}^{(A)} \otimes \boldsymbol{n}^{(A)}, \qquad \boldsymbol{b}^{\mathrm{e}} = \sum_{A=1}^{3} \overline{\lambda}^{\mathrm{e}}_A \boldsymbol{n}^{(A)} \otimes \boldsymbol{n}^{(A)}, \tag{5.84}$$

where the principal values are

$$\tau_A = 2\left(\frac{\partial \widehat{\psi}^{\mathrm{e}}}{\partial \overline{\lambda}^{\mathrm{e}}_A}\right)\overline{\lambda}^{\mathrm{e}}_A \equiv \frac{\partial \widetilde{\psi}^{\mathrm{e}}}{\partial \epsilon^{\mathrm{e}}_A}, \qquad \epsilon^{\mathrm{e}}_A = \log(\lambda^{\mathrm{e}}_A) = \frac{1}{2}\log(\overline{\lambda}^{\mathrm{e}}_A), \tag{5.85}$$

for $A = 1, 2, 3$, and $\widetilde{\psi}^{\mathrm{e}}$ is the stored energy function, a component of the free energy, which depends only on $\boldsymbol{b}^{\mathrm{e}}$ through the logarithmic elastic principal stretches. Both $\dot{\boldsymbol{\tau}}$ and $\dot{\boldsymbol{b}}^{\mathrm{e}}$ are symmetric second order tensors with the same spins (by isotropy), so the equivalent tangent relation to (5.75) is given by

$$\dot{\boldsymbol{\tau}} = \boldsymbol{\phi}^{\mathrm{e}} : \dot{\boldsymbol{b}}^{\mathrm{e}}, \tag{5.86}$$

where $\boldsymbol{\phi}^{\mathrm{e}}$ is a rank-four stress-deformation tensor with minor symmetries on its first and last two indices,

$$\begin{aligned}\boldsymbol{\phi}^{\mathrm{e}} &= \sum_{A=1}^{3}\sum_{B=1}^{3}\left(\frac{\partial \tau_A}{\partial \overline{\lambda}^{\mathrm{e}}_B}\right)\boldsymbol{m}^{(A)} \otimes \boldsymbol{m}^{(B)} \\ &+ \frac{1}{2}\sum_{A=1}^{3}\sum_{B\neq A}\left(\frac{\tau_B - \tau_A}{\overline{\lambda}^{\mathrm{e}}_B - \overline{\lambda}^{\mathrm{e}}_A}\right)\left(\boldsymbol{m}^{(AB)} \otimes \boldsymbol{m}^{(AB)} + \boldsymbol{m}^{(AB)} \otimes \boldsymbol{m}^{(BA)}\right),\end{aligned} \tag{5.87}$$

and where $\boldsymbol{m}^{(A)} = \boldsymbol{n}^{(A)} \otimes \boldsymbol{n}^{(A)}$, $\boldsymbol{m}^{(AB)} = \boldsymbol{n}^{(A)} \otimes \boldsymbol{n}^{(B)}$, etc. Inserting expression (5.69) for $\dot{\boldsymbol{b}}^{\mathrm{e}}$ gives

$$\begin{aligned}\boldsymbol{\alpha}^{\mathrm{e}} &= \sum_{A=1}^{3}\sum_{B=1}^{3} a^{\mathrm{e}}_{AB}\boldsymbol{m}^{(A)} \otimes \boldsymbol{m}^{(B)} \\ &+ \sum_{A=1}^{3}\sum_{B\neq A}\left(\frac{\tau_B - \tau_A}{\overline{\lambda}^{\mathrm{e}}_B - \overline{\lambda}^{\mathrm{e}}_A}\right)\left(\overline{\lambda}^{\mathrm{e}}_B\boldsymbol{m}^{(AB)} \otimes \boldsymbol{m}^{(AB)} + \overline{\lambda}^{\mathrm{e}}_A\boldsymbol{m}^{(AB)} \otimes \boldsymbol{m}^{(BA)}\right),\end{aligned} \tag{5.88}$$

where

$$a^{\mathrm{e}}_{AB} = 2\left(\frac{\partial \tau_A}{\partial \overline{\lambda}^{\mathrm{e}}_B}\right)\overline{\lambda}^{\mathrm{e}}_B \equiv \frac{\partial \tau_A}{\partial \epsilon^{\mathrm{e}}_B}, \qquad \epsilon^{\mathrm{e}}_B = \log(\lambda^{\mathrm{e}}_B) = \frac{1}{2}\log(\overline{\lambda}^{\mathrm{e}}_B). \tag{5.89}$$

We now use the elastic tangent operator $\boldsymbol{\alpha}^{\mathrm{e}}$ to develop an expression for the elastoplastic tangent operator, see Borja (2002a) for a more detailed development. First, we impose the consistency condition

$$\frac{\partial f}{\partial \boldsymbol{\tau}} : \dot{\boldsymbol{\tau}} + \dot{\gamma} H = 0, \tag{5.90}$$

where H is the generalized plastic modulus in the Kirchhoff stress space. Solving for the plastic multiplier, we get

$$\dot{\gamma} = \frac{1}{\chi}\frac{\partial f}{\partial \boldsymbol{\tau}} : \boldsymbol{\alpha}^{\mathrm{e}} : \boldsymbol{l}, \tag{5.91}$$

where

$$\chi = \frac{\partial f}{\partial \boldsymbol{\tau}} : \boldsymbol{\alpha}^{\mathrm{e}} : \frac{\partial g}{\partial \boldsymbol{\tau}} + H = \sum_{A=1}^{3}\sum_{B=1}^{3} \frac{\partial f}{\partial \tau_A} a^{\mathrm{e}}_{AB} \frac{\partial g}{\partial \tau_B} + H\,. \tag{5.92}$$

Substituting the result into (5.82) gives

$$\dot{\boldsymbol{\tau}} = \boldsymbol{\alpha}^{\mathrm{ep}} : \boldsymbol{l} \equiv \boldsymbol{\alpha}^{\mathrm{ep}} : \boldsymbol{d}, \tag{5.93}$$

where

$$\begin{aligned} \boldsymbol{\alpha}^{\mathrm{ep}} &= \sum_{A=1}^{3}\sum_{B=1}^{3} a^{\mathrm{ep}}_{AB} \boldsymbol{m}^{(A)} \otimes \boldsymbol{m}^{(B)} \\ &+ \sum_{A=1}^{3}\sum_{B\neq A} \left(\frac{\tau_B - \tau_A}{\overline{\lambda}^{\mathrm{e}}_B - \overline{\lambda}^{\mathrm{e}}_A}\right)(\overline{\lambda}^{\mathrm{e}}_B \boldsymbol{m}^{(AB)} \otimes \boldsymbol{m}^{(AB)} + \overline{\lambda}^{\mathrm{e}}_A \boldsymbol{m}^{(AB)} \otimes \boldsymbol{m}^{(BA)})\,, \end{aligned} \tag{5.94}$$

and $a^{\mathrm{ep}}_{AB} = \partial \tau_A / \partial \epsilon_B$ are the elastoplastic moduli in principal axes. Note that $\boldsymbol{d}$ can replace $\boldsymbol{l}$ in (5.93) due to the minor symmetry of $\boldsymbol{\alpha}^{\mathrm{ep}}$ with respect to its last two indices, see the exercise below.

Exercise 5.5. A hyperelastic material is subjected to combined tension and shear in plane strain. The deformations are given by: $x_1 = X_1 + k_1 X_2$, $x_2 = k_2 X_2$, and $x_3 = X_3$, where k_1 and k_2 are time-like variables. Assuming an elastic stored energy function that is quadratic in the principal logarithmic stretches, and of the form

$$\widetilde{\psi}^{\mathrm{e}} = \frac{1}{2}\lambda(\epsilon_1 + \epsilon_2 + \epsilon_3)^2 + \mu(\epsilon_1^2 + \epsilon_2^2 + \epsilon_3^2)\,,$$

where λ and μ are the Lamé parameters, calculate: (a) the left Cauchy-Green deformation tensor $\boldsymbol{b}$; (b) the Kirchhoff stress tensor $\boldsymbol{\tau}$; (c) the Cauchy stress tensor $\boldsymbol{\sigma}$; (d) the first Piola-Kirchhoff stress tensor $\boldsymbol{P}$; and (e) the second Piola-Kirchhoff stress tensor $\boldsymbol{S}$. Assume $\lambda = 100$ GPa and $\mu = 45$ GPa (typical of copper), and evaluate your results at the instant $k_1 = 2k_2 = 0.10$.

Exercise 5.6. Show that the tangent operators $\boldsymbol{\alpha}^{\mathrm{e}}$ and $\boldsymbol{\alpha}^{\mathrm{ep}}$ given in (5.88) and (5.94) have minor symmetries with respect to their first two and last two indices. Hint: without expanding, work on the four indices of α_{ijkl} and recall that A and B are dummy indices.

5.9 Momentum Balance and Weak Form

We refer again to the body $\mathcal{B}$ shown in Fig. 5.1, and denote the surface boundary of this body by $\partial\mathcal{B}$. Assume surface decompositions of the form

$$\partial\mathcal{B} = \overline{\partial\mathcal{B}_{\boldsymbol{u}} \cup \partial\mathcal{B}_{\boldsymbol{t}}}\,, \qquad \varnothing = \partial\mathcal{B}_{\boldsymbol{u}} \cap \partial\mathcal{B}_{\boldsymbol{t}}\,, \tag{5.95}$$

where $\partial\mathcal{B}_{\boldsymbol{u}}$ and $\partial\mathcal{B}_{\boldsymbol{t}}$ are non-overlapping portions of the boundary where displacements and surface tractions, respectively, are prescribed, and where the long overbar denotes a closure. Note that $\mathcal{B}$ is fixed in the reference configuration, but $\varphi(\mathcal{B})$ moves with deformation. This makes it more convenient to impose relevant conservation laws on the reference configuration than on the current configuration. Throughout this section, we shall thus adopt a Lagrangian description for convenience.

To develop a point-wise expression for balance of linear momentum, consider any arbitrary volume $\mathcal{V} \in \mathcal{B}$ with surface boundary $\partial\mathcal{V}$ over which the nominal surface traction vector $\boldsymbol{t}_0 = \boldsymbol{P}\cdot\boldsymbol{N}$ acts. Letting $\boldsymbol{G}$ denote the gravity acceleration vector, balance of linear momentum over the volume $\mathcal{V}$ takes the form

$$\int_{\partial\mathcal{V}} \boldsymbol{t}_0 \,\mathrm{d}A + \int_{\mathcal{V}} \rho_0 \boldsymbol{G}\,\mathrm{d}V = \frac{\mathrm{d}}{\mathrm{d}t}\int_{\mathcal{V}} \frac{1}{2}\rho_0 \boldsymbol{V}\cdot\boldsymbol{V}\,\mathrm{d}V\,, \tag{5.96}$$

where ρ_0 is the reference mass density of the body, and $\mathrm{d}(\cdot)/\mathrm{d}t$ denotes a material time derivative that operates directly on the terms inside the volume integral. Using the divergence theorem on the surface integral, and noting that $\mathcal{V}$ is an arbitrary volume, we obtain the following point-wise expression for balance of linear momentum

$$\mathrm{DIV}\,\boldsymbol{P} + \rho_0\boldsymbol{G} = \rho_0\boldsymbol{A}(\boldsymbol{X},t)\,, \qquad \forall\boldsymbol{X}\in\mathcal{B}\,, \tag{5.97}$$

where $(\mathrm{DIV}\,\boldsymbol{P})_i = \partial P_{iA}/\partial X_A$. The above equation must be augmented with boundary conditions

$$\boldsymbol{u} = \widehat{\boldsymbol{u}} \quad \text{on} \quad \partial\mathcal{B}_{\boldsymbol{u}} \qquad \text{and} \qquad \boldsymbol{P}\cdot\boldsymbol{N} = \widehat{\boldsymbol{t}}_0 \quad \text{on} \quad \partial\mathcal{B}_{\boldsymbol{t}}\,, \tag{5.98}$$

and initial condition

$$\boldsymbol{u}(\boldsymbol{X},0) = \widehat{\boldsymbol{u}}_0(\boldsymbol{X})\,, \qquad \forall\boldsymbol{X}\in\mathcal{B}\,. \tag{5.99}$$

Developing the weak form of the boundary-value problem requires trial and weighting functions

$$\mathcal{U} := \{\boldsymbol{u} : \mathcal{B}\rightarrow\mathbb{R}^{n_{\mathrm{dim}}} \mid u_i \in H^1, u_i = \widehat{u}_i \text{ on } \partial\mathcal{B}_{u_i}\}\,, \tag{5.100}$$

and

$$\mathcal{V} := \{\boldsymbol{\eta} : \mathcal{B}\rightarrow\mathbb{R}^{n_{\mathrm{dim}}} \mid \eta_i \in H^1, \eta_i = 0 \text{ on } \partial\mathcal{B}_{u_i}\}\,, \tag{5.101}$$

where H^1 is the first Sobolev space and n_{dim} is the number of spatial dimensions. The variational equation then develops following standard lines (i.e., integration by parts and divergence theorem):

$$\int_{\mathcal{B}} \text{GRAD}\,\boldsymbol{\eta} : \boldsymbol{P}\,\text{d}V = \int_{\mathcal{B}} \boldsymbol{\eta} \cdot \rho_0(\boldsymbol{G} - \boldsymbol{A})\,\text{d}V + \int_{\partial\mathcal{B}_t} \boldsymbol{\eta} \cdot \widehat{\boldsymbol{t}}_0\,\text{d}A\,, \tag{5.102}$$

where $(\text{GRAD}\,\boldsymbol{\eta})_{iA} = \partial\eta_i/\partial X_A$.

The volume integral on the left of (5.102) typically appears in different forms and requires further elaboration. From the definition of the first Piola-Kirchhoff stress tensor $\boldsymbol{P}$ in (5.36), we write

$$\text{GRAD}\,\boldsymbol{\eta} : \boldsymbol{P} = \frac{\partial\eta_i}{\partial X_A} P_{iA} = \frac{\partial\eta_i}{\partial X_A} \tau_{ij} F_{jA}^{-\mathsf{T}} = \frac{\partial\eta_i}{\partial X_A}\frac{\partial X_A}{\partial x_j}\tau_{ij} = \boldsymbol{\nabla}^{\text{s}}\boldsymbol{\eta} : \boldsymbol{\tau}\,, \tag{5.103}$$

where $\boldsymbol{\nabla}^{\text{s}}$ is the usual symmetric gradient operator

$$\boldsymbol{\nabla}^{\text{s}}\boldsymbol{\eta} = \frac{1}{2}(\boldsymbol{\nabla}\boldsymbol{\eta} + \boldsymbol{\eta}\boldsymbol{\nabla})\,, \tag{5.104}$$

which replaces the spatial gradient operator $\boldsymbol{\nabla}$ due to the symmetry of $\boldsymbol{\tau}$. Thus, the volume integral on the left of (5.102) takes the following equivalent forms

$$\int_{\mathcal{B}} \text{GRAD}\,\boldsymbol{\eta} : \boldsymbol{P}\,\text{d}V = \int_{\mathcal{B}} \boldsymbol{\nabla}^{\text{s}}\boldsymbol{\eta} : \boldsymbol{\tau}\,\text{d}V = \int_{\varphi(\mathcal{B})} \boldsymbol{\nabla}^{\text{s}}\boldsymbol{\eta} : \boldsymbol{\sigma}\,\text{d}v\,, \tag{5.105}$$

where $\text{d}v = J\text{d}V$ following the usual transformation induced by the Jacobian determinant. As noted earlier, integration over $\varphi(\mathcal{B})$ is undesirable because the volume is moving. However, we can still integrate the spatial tensor $\boldsymbol{\tau}$ over the reference volume $\mathcal{B}$ without having to deal with a moving domain. The form of the variational equation of interest is

$$\int_{\mathcal{B}} \boldsymbol{\nabla}^{\text{s}}\boldsymbol{\eta} : \boldsymbol{\tau}\,\text{d}V = \int_{\mathcal{B}} \boldsymbol{\eta} \cdot \rho_0(\boldsymbol{G} - \boldsymbol{A})\,\text{d}V + \int_{\partial\mathcal{B}_t} \boldsymbol{\eta} \cdot \widehat{\boldsymbol{t}}_0\,\text{d}A\,. \tag{5.106}$$

More specifically, we want to impose the above variational equation at discrete time stations. In the context of conventional nonlinear finite element analysis, this requires a stress-point integration algorithm for the Kirchhoff stress $\boldsymbol{\tau}$ at the numerical integration points.

Exercise 5.7. From the definitions of the stress tensors $\boldsymbol{\tau}$, $\boldsymbol{P}$, and $\boldsymbol{S}$, and from the relationships among the kinematical variables, show that $\boldsymbol{\tau} : \boldsymbol{d} = \boldsymbol{P} : \dot{\boldsymbol{F}} = \boldsymbol{S} : \dot{\boldsymbol{C}}/2$. Suggestion: Use the indicial notation.

Exercise 5.8. Let $\boldsymbol{\eta} \in \mathcal{V}$ denote a weighting function satisfying the homogeneous boundary condition on $\partial\mathcal{B}_{\boldsymbol{u}}$. Show that $\text{GRAD}\,\boldsymbol{\eta} : \boldsymbol{P} = \boldsymbol{D\eta} : \boldsymbol{S}$, where $\boldsymbol{D\eta} = \text{sym}(\text{GRAD}\,\boldsymbol{\eta} \cdot \boldsymbol{F}^{\mathsf{T}})$.

5.10 Stress-Point Integration

In the actual solution of boundary-value problems, it is essential that the rate form of the constitutive equation be integrated so as to quantify the local stresses and deformations at discrete time stations, or time instants, where the equilibrium conditions are then enforced. Consider, for example, a time interval $[t_n, t_{n+1}]$ for an arbitrary material point $X \in \mathcal{B}$. We assume that the variables

$$\boldsymbol{F}_n = \frac{\partial \boldsymbol{\varphi}_n}{\partial \boldsymbol{X}}, \qquad \boldsymbol{b}^{\mathrm{e}}_n = \boldsymbol{F}^{\mathrm{e}}_n \cdot \boldsymbol{F}^{\mathrm{e}\,\mathsf{T}}_n, \qquad \boldsymbol{\xi}_n = \boldsymbol{\xi}|_{t_n} \tag{5.107}$$

have already been calculated at time t_n. The hyperelastic constitutive equation (5.75) can be used to evaluate the Kirchhoff stress tensor $\boldsymbol{\tau}_n$ from the given value of $\boldsymbol{b}^{\mathrm{e}}_n$. We assume that the calculated solution at time t_n is statically admissible in the sense that $\boldsymbol{\tau}_n$ and $\boldsymbol{\xi}_n$ satisfy the local yield condition.

Following the standard deformation-driven format, we then prescribe a local displacement field $\boldsymbol{u}_t : \varphi(\mathcal{B}) \times [t_n, t_{n+1}] \to \mathbb{R}^{n_{\mathrm{dim}}}$ over the time interval in question. This is equivalent to prescribing the relative deformation gradient over the same time interval,

$$\boldsymbol{f}_t = \frac{\partial \boldsymbol{\varphi}_t}{\partial \boldsymbol{x}_n} = \boldsymbol{F}_t \cdot \boldsymbol{F}_n^{-1} = \mathbf{1} + \frac{\partial \boldsymbol{u}_t}{\partial \boldsymbol{x}_n}, \qquad \forall t \in [t_n, t_{n+1}]\,. \tag{5.108}$$

Even though the equation above defines a continuous evolution over the time interval in question, we are actually interested in the calculated values at discrete time stations t_{n+1}, t_{n+2}, and so on. For brevity in presentation, we shall then drop the subscript t and take the unsubscripted variables to mean either a time evolution or the desired (as yet calculated) values at t_{n+1}. It may be worthwhile to mention that the relative deformation gradient $\boldsymbol{f}$ as defined above is the analog of the incremental strain tensor $\Delta\boldsymbol{\epsilon}$ used in the infinitesimal formulation.

In the finite deformation range, we want to determine the statically admissible values of $\boldsymbol{b}^{\mathrm{e}}$ and $\boldsymbol{\xi}$ given the relative deformation gradient $\boldsymbol{f}$. The relevant rate equation to integrate is (5.69) for the time derivative of $\boldsymbol{b}^{\mathrm{e}}$, which we rewrite below in the predictor-corrector format as follows

$$\dot{\boldsymbol{b}}^{\mathrm{e}} = \underbrace{\boldsymbol{l} \cdot \boldsymbol{b}^{\mathrm{e}} + \boldsymbol{b}^{\mathrm{e}} \cdot \boldsymbol{l}^{\mathsf{T}}}_{\text{predictor}} - \underbrace{(-\mathcal{L}_v \boldsymbol{b}^{\mathrm{e}})}_{\text{corrector}}\,. \tag{5.109}$$

The predictor value is what would arise if there was no plastic flow, while the corrector value takes care of the 'overshoot' in the presence of plastic flow. Note that the plastic corrector is nothing else but the negative of the Lie derivative of $\boldsymbol{b}^{\mathrm{e}}$.

Unlike in the infinitesimal formulation where the incremental predictor-corrector format remains additive, the incremental version of the predictor-corrector algorithm in the finite deformation range now takes a multiplicative form. More specifically, the predictor-corrector format is facilitated by a so-called product-formula algorithm (Simo 1998, Simo and Hughes 1998). In a nutshell, the algorithm consists of two steps. The first step is the incremental counterpart of the predictor step, which is given by the product formula

$$\boldsymbol{b}_{n+1}^{\mathrm{e\,tr}} \equiv \boldsymbol{b}^{\mathrm{e\,tr}} = \boldsymbol{f} \cdot \boldsymbol{b}_n^{\mathrm{e}} \cdot \boldsymbol{f}^{\mathsf{T}} . \tag{5.110}$$

Note that $\boldsymbol{b}^{\mathrm{e}} = \boldsymbol{b}^{\mathrm{e\,tr}}$ if there is no plastic flow. The second step is the incremental plastic corrector in the presence of plastic flow, which is realized by by integrating the Lie derivative (5.83). In spectral form, we write

$$\mathcal{L}_v \boldsymbol{b}^{\mathrm{e}} = \sum_{A=1}^{3} \mathcal{L}_v \overline{\lambda}_A^{\mathrm{e}} \boldsymbol{m}^{(A)} = -2\dot{\gamma} \sum_{A=1}^{3} \frac{\partial g}{\partial \tau_A} \overline{\lambda}_A^{\mathrm{e}} \boldsymbol{m}^{(A)} , \tag{5.111}$$

where $\overline{\lambda}_A^{\mathrm{e}} = \lambda_A^{\mathrm{e}\,2}$ are the principal values of $\boldsymbol{b}^{\mathrm{e}}$, and $\mathcal{L}_v \overline{\lambda}_A^{\mathrm{e}}$ are the principal values of $\mathcal{L}_v \boldsymbol{b}^{\mathrm{e}}$. Note in this case that the Lie derivative of $\boldsymbol{b}^{\mathrm{e}}$ has no spin, which means that the plastic correction takes place at fixed principal directions.

From the coaxiality of the relevant tensors, the principal values of the left- and right-hand sides of (5.111) must be equal. Therefore, we can simply work on the principal values, which are scalar quantities. Integrating both sides, and enforcing the limits of integration, which are: (a) $\overline{\lambda}_A^{\mathrm{e\,tr}}$ at $t = t_n$, and $\overline{\lambda}_A^{\mathrm{e}}$ at $t = t_{n+1}$, we get

$$\log\left(\frac{\overline{\lambda}_A^{\mathrm{e}}}{\overline{\lambda}_A^{\mathrm{e\,tr}}}\right) = -2\Delta\gamma \frac{\partial g}{\partial \tau_A}\bigg|_{n+1} , \qquad \text{no sum on } A , \tag{5.112}$$

where $\overline{\lambda}_A^{\mathrm{e\,tr}}$, $A = 1, 2, 3$, are the principal values of $\boldsymbol{b}^{\mathrm{e\,tr}}$. Note that a simple backward integration has been employed on the plastic flow direction $\partial g / \partial \tau_A$, which could also vary within the load interval. Expanding the left-hand side of the above equation and noting that $\log(\overline{\lambda}_A) = 2\log(\lambda_A) = 2\epsilon_A$, we obtain the desired incremental update equation in the space of principal elastic logarithmic stretches as follows,

$$\epsilon_A^{\mathrm{e}} = \epsilon_A^{\mathrm{e\,tr}} - \Delta\gamma \frac{\partial g}{\partial \tau_A} , \qquad A = 1, 2, 3 , \tag{5.113}$$

where $\epsilon_A^{\mathrm{e}} = \log(\lambda_A^{\mathrm{e}})$ and $\epsilon_A^{\mathrm{e\,tr}} = \log(\lambda_A^{\mathrm{e\,tr}})$. The above predictor-corrector equation, first presented by Simo (1992), preserves the additive return-mapping format of the infinitesimal theory.

The plastic internal variable $\boldsymbol{\xi}$ could also change with deformation, although in general the update equation for this variable depends on the specific constitutive model used for the material. Without getting into complex plastic hardening responses (see Chapter 6 for more realistic examples), we shall assume for now that a simple linear update of the following form is valid for the material

$$\boldsymbol{\xi} = \boldsymbol{\xi}_n + \Delta\gamma \boldsymbol{h} , \tag{5.114}$$

where $\boldsymbol{h}$ defines the direction of hardening. The plastic multiplier $\Delta\gamma$ can be determined by imposing the discrete consistency condition $f(\boldsymbol{\tau}, \boldsymbol{\xi}) = 0$.

Box 5.1 summarizes the implementation of the algorithm. With the relative deformation gradient $\boldsymbol{f}$ driving the solution, the predictor values $\boldsymbol{b}^{\mathrm{e\,tr}}$ and $\boldsymbol{\tau}^{\mathrm{tr}}$ are calculated explicitly (Step 1) to check for yielding (Step 2). The spectral form for $\boldsymbol{b}^{\mathrm{e\,tr}}$ (Step 3) is given by

$$\boldsymbol{b}^{\mathrm{e\,tr}} = \sum_{A=1}^{3} \overline{\lambda}_A^{\mathrm{e\,tr}} \boldsymbol{m}^{(A)}, \tag{5.115}$$

where the spectral directions are

$$\boldsymbol{m}^{(A)} = \left(\frac{\boldsymbol{b}^{\mathrm{e\,tr}} - \overline{\lambda}_B^{\mathrm{e\,tr}}\boldsymbol{1}}{\overline{\lambda}_B^{\mathrm{e\,tr}} - \overline{\lambda}_A^{\mathrm{e\,tr}}}\right) \cdot \left(\frac{\boldsymbol{b}^{\mathrm{e\,tr}} - \overline{\lambda}_C^{\mathrm{e\,tr}}\boldsymbol{1}}{\overline{\lambda}_C^{\mathrm{e\,tr}} - \overline{\lambda}_A^{\mathrm{e\,tr}}}\right), \tag{5.116}$$

for $A = 1, 2, 3$, with $B = 1 + \mathrm{mod}(3, A)$ and $C = 1 + \mathrm{mod}(3, B)$, see Simo (1998). Alternatively, $\boldsymbol{m}^{(A)}$ can be constructed from available computer packages (Borja et al. 2003).

Step 1. Compute $\boldsymbol{b}^{\mathrm{e\,tr}} = \boldsymbol{f} \cdot \boldsymbol{b}_n^{\mathrm{e}} \cdot \boldsymbol{f}^{\mathsf{T}}, \quad \boldsymbol{\tau}^{\mathrm{tr}} = 2\boldsymbol{b}^{\mathrm{e\,tr}} \cdot \partial\psi^{\mathrm{e}}/\partial\boldsymbol{b}^{\mathrm{e\,tr}}$
Step 2. Check $f(\boldsymbol{\tau}^{\mathrm{tr}}, \boldsymbol{\xi}_n) > 0$?
No, set $\boldsymbol{b}^{\mathrm{e}} = \boldsymbol{b}^{\mathrm{e\,tr}}$ and exit.
Step 3. Yes, spectrally decompose $\boldsymbol{b}^{\mathrm{e\,tr}} = \sum_A^3 \overline{\lambda}_A^{\mathrm{e\,tr}} \boldsymbol{m}^{(A)}$
Step 4. Plastic corrector: iterate for $\Delta\gamma$ satisfying

$$\begin{aligned} \epsilon_A^{\mathrm{e}} &= \epsilon_A^{\mathrm{e\,tr}} - \Delta\gamma \partial g/\partial\tau_A \rightarrow \boldsymbol{b}^{\mathrm{e}} \rightarrow \boldsymbol{\tau} \\ \boldsymbol{\xi} &= \boldsymbol{\xi}_n + \Delta\gamma\boldsymbol{h} \\ f(\boldsymbol{\tau}, \boldsymbol{\xi}) &= 0 \end{aligned}$$

Step 5. Exit with the corrected values of $\boldsymbol{b}^{\mathrm{e}}$ and $\boldsymbol{\xi}$.

Box 5.1 Predictor-corrector algorithm for multiplicative plasticity: isotropic case

The plastic corrector phase (Step 4) typically entails a local iteration for the plastic multiplier $\Delta\gamma$ due to the nonlinearity introduced by the plastic flow direction $\partial g/\partial\tau_A$. The symbol "$\rightarrow \boldsymbol{b}^{\mathrm{e}} \rightarrow \boldsymbol{\tau}$" in Box 5.1 suggests sequential calculations leading to the Kirchhoff stress, i.e.,

$$\boldsymbol{b}^{\mathrm{e}} = \sum_{A=1}^{3} \overline{\lambda}_A^{\mathrm{e}} \boldsymbol{m}^{(A)}, \qquad \boldsymbol{\tau} = \sum_{A=1}^{4} \tau_A \boldsymbol{m}^{(A)}, \tag{5.117}$$

where $\overline{\lambda}_A^{\mathrm{e}} = (\lambda_A^{\mathrm{e}})^2$ and $\epsilon_A^{\mathrm{e}} = \log(\lambda_A^{\mathrm{e}})$. Note that the spectral directions of $\boldsymbol{b}^{\mathrm{e\,tr}}$ and $\boldsymbol{b}^{\mathrm{e}}$ are the same.

5.11 Algorithmic Tangent Operator

Consider the expression

$$W_{INT} = \int_{\mathcal{B}} \mathrm{GRAD}\,\boldsymbol{\eta} : \boldsymbol{P}\,\mathrm{d}V = \int_{\mathcal{B}} \boldsymbol{\nabla}^{\mathrm{s}}\boldsymbol{\eta} : \boldsymbol{\tau}\,\mathrm{d}V\,. \tag{5.118}$$

These volume integrals appear in (5.102) and (5.106) and have the physical significance of being the virtual work done by the internal stresses acting within $\mathcal{B}$. We are interested in taking the first variation of this volume integral with respect to the displacement field $\boldsymbol{u}$ for purposes of developing a consistent algorithmic tangent operator for nonlinear finite element calculations.

We need two preliminary results. The first is the variation of the deformation gradient $\boldsymbol{F}$ from the directional derivative

$$\delta\boldsymbol{F} = \frac{\mathrm{d}}{\mathrm{d}\epsilon}\bigg|_{\epsilon=0} \boldsymbol{F}(\boldsymbol{\varphi} + \epsilon\delta\boldsymbol{u}) = \boldsymbol{\nabla}\delta\boldsymbol{u}\cdot\boldsymbol{F}\,. \tag{5.119}$$

The second is the variation of the inverse of $\boldsymbol{F}$ from the identity $\delta(\boldsymbol{F}\cdot\boldsymbol{F}^{-1}) = \delta\boldsymbol{F}\cdot\boldsymbol{F}^{-1} + \boldsymbol{F}\cdot\delta\boldsymbol{F}^{-1} = \boldsymbol{0}$, which gives

$$\delta\boldsymbol{F}^{-1} = -\boldsymbol{F}^{-1}\cdot\boldsymbol{\nabla}\delta\boldsymbol{u}\,. \tag{5.120}$$

From these, we can evaluate the first variation of W_{INT} as

$$\begin{aligned} \delta W_{INT} &= \int_{\mathcal{B}} \mathrm{GRAD}\,\boldsymbol{\eta} : \delta\boldsymbol{P}\,\mathrm{d}V \\ &= \int_{\mathcal{B}} \mathrm{GRAD}\,\boldsymbol{\eta} : (\delta\boldsymbol{\tau}\cdot\boldsymbol{F}^{-\mathsf{T}} + \boldsymbol{\tau}\cdot\delta\boldsymbol{F}^{-\mathsf{T}})\,\mathrm{d}V \\ &= \int_{\mathcal{B}} \boldsymbol{\nabla}^{\mathrm{s}}\boldsymbol{\eta} : \delta\boldsymbol{\tau}\,\mathrm{d}V - \int_{\mathcal{B}} \boldsymbol{\nabla}\boldsymbol{\eta} : (\boldsymbol{\tau}\ominus\boldsymbol{1}) : \boldsymbol{\nabla}\delta\boldsymbol{u}\,\mathrm{d}V\,, \end{aligned} \tag{5.121}$$

where $(\boldsymbol{\tau}\ominus\boldsymbol{1})_{ijkl} = \tau_{il}\delta_{jk}$ is a fourth-order tensor representing the so-called initial stress term. From the tangent constitutive relation $\delta\boldsymbol{\tau} = \boldsymbol{\alpha} : \boldsymbol{\nabla}\delta\boldsymbol{u}$, the first variation of W_{INT} simplifies to

$$\delta W_{INT} = \int_{\mathcal{B}} \boldsymbol{\nabla}\boldsymbol{\eta} : \mathsf{a} : \boldsymbol{\nabla}\delta\boldsymbol{u}\,\mathrm{d}V\,, \tag{5.122}$$

where

$$\mathsf{a} = \boldsymbol{\alpha} - \boldsymbol{\tau}\ominus\boldsymbol{1}\,. \tag{5.123}$$

Note that the full gradients of $\boldsymbol{\eta}$ and $\delta\boldsymbol{u}$ are required for the first variation, whereas only the symmetric components of the gradient appear in the infinitesimal formulation.

To obtain the algorithmic tangent tensor $\boldsymbol{\alpha}$, we first write the variation of the Kirchhoff stress tensor from the spectral form $(5.117)_2$ as

$$\delta\boldsymbol{\tau} = \sum_{A=1}^{3} \delta\tau_A \boldsymbol{n}^{(A)} \otimes \boldsymbol{n}^{(A)} + \sum_{A=1}^{3} \sum_{B\neq A} \omega_{AB}(\tau_B - \tau_A)\boldsymbol{n}^{(A)} \otimes \boldsymbol{n}^{(B)} , \tag{5.124}$$

where the ω_{AB}'s are the relevant spins. The variation of $\boldsymbol{b}^{\mathrm{e\,tr}}$ takes a similar form,

$$\delta\boldsymbol{b}^{\mathrm{e\,tr}} = \sum_{A=1}^{3} \delta\overline{\lambda}_A^{\mathrm{e\,tr}} \boldsymbol{n}^{(A)} \otimes \boldsymbol{n}^{(A)} + \sum_{A=1}^{3} \sum_{B\neq A} \omega_{AB}(\overline{\lambda}_B^{\mathrm{e\,tr}} - \overline{\lambda}_A^{\mathrm{e\,tr}})\boldsymbol{n}^{(A)} \otimes \boldsymbol{n}^{(B)} , \tag{5.125}$$

where we note once again that the spins of the two tensors are the same.

Following the developments of the previous sections, we write

$$\delta\boldsymbol{\tau} = \boldsymbol{\phi} : \delta\boldsymbol{b}^{\mathrm{e\,tr}} , \tag{5.126}$$

where $\delta\boldsymbol{b}^{\mathrm{e\,tr}}$ is obtained from the first variation of (5.110) as

$$\delta\boldsymbol{b}^{\mathrm{e\,tr}} = \delta\boldsymbol{f} \cdot \boldsymbol{b}_n^{\mathrm{e}} \cdot \boldsymbol{f}^{\mathsf{T}} + \boldsymbol{f} \cdot \boldsymbol{b}_n^{\mathrm{e}} \cdot \delta\boldsymbol{f}^{\mathsf{T}} = (\boldsymbol{\nabla}\delta\boldsymbol{u}) \cdot \boldsymbol{b}^{\mathrm{e\,tr}} + \boldsymbol{b}^{\mathrm{e\,tr}} \cdot (\delta\boldsymbol{u}\boldsymbol{\nabla}) . \tag{5.127}$$

It is useful to compare (5.126) with (5.86), for then, the form for $\boldsymbol{\phi}$ can be readily determined as

$$\begin{aligned} \boldsymbol{\phi} &= \sum_{A=1}^{3} \sum_{B=1}^{3} \left(\frac{\partial\tau_A}{\partial\overline{\lambda}_B^{\mathrm{e\,tr}}}\right) \boldsymbol{m}^{(A)} \otimes \boldsymbol{m}^{(B)} \\ &+ \frac{1}{2} \sum_{A=1}^{3} \sum_{B\neq A} \left(\frac{\tau_B - \tau_A}{\overline{\lambda}_B^{\mathrm{e\,tr}} - \overline{\lambda}_A^{\mathrm{e\,tr}}}\right) (\boldsymbol{m}^{(AB)} \otimes \boldsymbol{m}^{(AB)} + \boldsymbol{m}^{(AB)} \otimes \boldsymbol{m}^{(BA)}) , \end{aligned} \tag{5.128}$$

where $\boldsymbol{m}^{(A)} = \boldsymbol{n}^{(A)} \otimes \boldsymbol{n}^{(A)}$, $\boldsymbol{m}^{(AB)} = \boldsymbol{n}^{(A)} \otimes \boldsymbol{n}^{(B)}$, etc. Combining the above equation with (5.127) yields the desired tangent operator

$$\begin{aligned} \boldsymbol{\alpha} &= \sum_{A=1}^{3} \sum_{B=1}^{3} a_{AB} \boldsymbol{m}^{(A)} \otimes \boldsymbol{m}^{(B)} \\ &+ \sum_{A=1}^{3} \sum_{B\neq A} \left(\frac{\tau_B - \tau_A}{\overline{\lambda}_B^{\mathrm{e\,tr}} - \overline{\lambda}_A^{\mathrm{e\,tr}}}\right) (\overline{\lambda}_B^{\mathrm{e\,tr}} \boldsymbol{m}^{(AB)} \otimes \boldsymbol{m}^{(AB)} + \overline{\lambda}_A^{\mathrm{e\,tr}} \boldsymbol{m}^{(AB)} \otimes \boldsymbol{m}^{(BA)}) , \end{aligned} \tag{5.129}$$

where

$$a_{AB} = 2\left(\frac{\partial\tau_A}{\partial\overline{\lambda}_B^{\mathrm{e\,tr}}}\right)\overline{\lambda}_B^{\mathrm{e\,tr}} \equiv \frac{\partial\tau_A}{\partial\epsilon_B^{\mathrm{e\,tr}}} , \qquad \epsilon_B^{\mathrm{e\,tr}} = \log(\lambda_B^{\mathrm{e\,tr}}) = \frac{1}{2}\log(\overline{\lambda}_B^{\mathrm{e\,tr}}) \tag{5.130}$$

is the algorithmic stress-strain matrix in principal axes. The calculations of $a_{AB} = \partial\tau_A/\partial\epsilon_B^{\mathrm{e\,tr}}$ depend on the algorithm for determining τ_A. For further details, we refer the reader to Chapter 4, Sections 11 and 12, where many striking similarities with the infinitesimal formulation can be noted.

Before closing this chapter, we note the role played by the first and second Piola-Kirchhoff stress tensors in the constitutive model formulation of finite deformation elastoplasticity. Recalling that $\boldsymbol{\tau} = \boldsymbol{P} \cdot \boldsymbol{F}^{\mathsf{T}}$, where $\boldsymbol{P}$ is the first Piola-Kirchhoff stress tensor, we obtain the first variation

$$\delta\boldsymbol{\tau} = (\mathbf{A} : \delta\boldsymbol{F}) \cdot \boldsymbol{F}^{\mathsf{T}} + \boldsymbol{P} \cdot \delta\boldsymbol{F}^{\mathsf{T}} , \tag{5.131}$$

where $\mathbf{A} = \partial\boldsymbol{P}/\partial\boldsymbol{F}$ is the first tangential tensor with components $\mathsf{A}_{iAjB} = \partial P_{iA}/\partial F_{jB}$. Note that this tensor is 'two-point' in the sense that it has upper-case and lower-case subscripts corresponding to the reference and current configurations, respectively. The first term on the right-hand side of (5.131) is equal to $\mathbf{a} : \boldsymbol{\nabla}\delta\boldsymbol{u}$, whereas the second terms is equal to $(\boldsymbol{\tau} \ominus \mathbf{1}) : \boldsymbol{\nabla}\delta\boldsymbol{u}$. Since $\boldsymbol{\alpha} = \mathbf{a} + (\boldsymbol{\tau} \ominus \mathbf{1})$, we see that the components of $\mathbf{a}$ can be obtained from the push-forward of the large indices of the first tangential tensor, i.e.,

$$\mathsf{a}_{ikjl} = F_{kA} F_{lB} \mathsf{A}_{iAjB} . \tag{5.132}$$

We also recall that $\boldsymbol{\tau} = \boldsymbol{F} \cdot \boldsymbol{S} \cdot \boldsymbol{F}^{\mathsf{T}}$, where $\boldsymbol{S}$ is the second Piola-Kirchhoff stress tensor. Taking the first variation gives

$$\delta\boldsymbol{\tau} = \boldsymbol{F} \cdot (\mathbf{C} : \delta\boldsymbol{C}) \cdot \boldsymbol{F}^{\mathsf{T}} + \boldsymbol{F} \cdot \boldsymbol{S} \cdot \delta\boldsymbol{F}^{\mathsf{T}} + \delta\boldsymbol{F} \cdot \boldsymbol{S} \cdot \boldsymbol{F}^{\mathsf{T}} , \tag{5.133}$$

where $\mathbf{C} = \partial\boldsymbol{S}/\partial\boldsymbol{C}$ is the second tangential tensor with components $\mathsf{C}_{ABCD} = \partial S_{AB}/\partial C_{DC}$ and $\boldsymbol{C}$ is the right Cauchy-Green deformation tensor (note the difference in the fonts, see Marsen and Hughes (1983) for the same notations). The first term on the right-hand side of (5.133) is equal to $\mathbf{c} : \boldsymbol{d}$, where $\boldsymbol{d}$ is the rate of deformation, and $\mathbf{c}$ is a spatial tangential tensor obtained from the push-forward of all four large indices of $\mathbf{C}$, i.e.,

$$\mathsf{c}_{ijkl} = 2 F_{iA} F_{jB} F_{kC} F_{lD} \mathsf{C}_{ABCD} . \tag{5.134}$$

The second term is equal to $(\boldsymbol{\tau} \ominus \mathbf{1}) : \boldsymbol{\nabla}\delta\boldsymbol{u}$, whereas the third term is equal to $(\boldsymbol{\tau} \oplus \mathbf{1}) : \boldsymbol{\nabla}\delta\boldsymbol{u}$, where $(\boldsymbol{\tau} \oplus \mathbf{1})_{ijkl} = \tau_{jl}\delta_{ik}$.

In summary, the relations among the three spatial tangential tensors are

$$\boldsymbol{\alpha} = \mathbf{c} + \boldsymbol{\tau} \oplus \mathbf{1} + \boldsymbol{\tau} \ominus \mathbf{1} = \mathbf{a} + \boldsymbol{\tau} \ominus \mathbf{1} . \tag{5.135}$$

It may be helpful to remember that $\mathbf{a}$ emanates from the first Piola-Kirchhoff stress tensor $\boldsymbol{P}$, $\mathbf{c}$ is derived from the second Piola-Kirchhoff stress tensor $\boldsymbol{S}$, and $\boldsymbol{\alpha}$ is associated with the symmetric Kirchhoff stress tensor $\boldsymbol{\tau}$.

Exercise 5.9. Use the indicial notation to show that $(\mathbf{A} : \delta\boldsymbol{F}) \cdot \boldsymbol{F}^{\mathsf{T}} = \mathbf{a} : \boldsymbol{\nabla}\delta\boldsymbol{u}$ and $\boldsymbol{F} \cdot (\mathbf{C} : \delta\boldsymbol{C}) \cdot \boldsymbol{F}^{\mathsf{T}} = \mathbf{c} : \boldsymbol{\nabla}\delta\boldsymbol{u}$.

Exercise 5.10. Show that $\mathbf{c}$ has minor symmetries with respect to its first two and last two indices. Then, show that $\dot{\boldsymbol{\tau}} = \mathbf{c} : \boldsymbol{d} + \boldsymbol{l} \cdot \boldsymbol{\tau} + \boldsymbol{\tau} \cdot \boldsymbol{l}^{\mathsf{T}}$, where $\boldsymbol{d}$ is the rate of deformation tensor. This means that $\mathbf{c} : \boldsymbol{d} \equiv \mathcal{L}_v \boldsymbol{\tau}$ is the Lie derivative of $\boldsymbol{\tau}$, cf. (5.48).

5.12 Summary

A formulation based on multiplicative decomposition of deformation gradient naturally gives rise to multiplicative theory of elastoplasticity. It exploits numerous elements of nonlinear continuum mechanics, including various definitions of stresses and deformation. For the multiplicative plasticity formulation, the Lie derivative of the Kirchhoff stress tensor has been identified as compatible with the formulation and is the objective stress rate chosen in this work. Under the assumption of isotropy, the spectral decomposition technique is once again exploited in both the theoretical developments and numerical implementation of the theory.

6 Cap Models

6.1 Introduction

Elastic materials often exhibit a linear stress-strain response, but a nonlinear response does not necessarily imply inelastic behavior. The definition of an elastic material is quite broad: if it stores but does not dissipate energy, and if it returns to its undeformed shape when the loads are removed, then the material is elastic. Nonlinear elastic behavior could be due to the elastic parameters being intrinsically dependent on the state of stress (material nonlinearity), or to the large deformation that developed in the specimen during testing (geometric nonlinearity). The elastic properties of geomaterials, such as soil, rock, and concrete, are known to depend on the confining pressure and density, but these materials are also known to have a very small elastic range. On the other hand, rubber can sustain a very large deformation and still behave elastically.

Figure 6.1 shows the variation of Young's modulus E and Poisson's ratio ν with confining pressure for two rocks. Figure 6.2 shows a similar variation of shear wave velocities with density and confining pressure for two sands. In both figures, the strains experienced by the material during testing are too small to alter the geometry of the specimen. Therefore, they reflect material nonlinearity in the elastic response characterized by state-dependent elastic properties.

Variable elastic moduli typically are incorporated into the constitutive formulation using the framework of hypoelasticity (Borja and Lee 1990). This approach has a serious drawback in that on a closed elastic loop, the incremental constitutive equation could either generate or dissipate energy (Zytynski et al. 1978), thus violating the most important axiom of an elastic response. An overarching rule on any elastic model is that it should be conservative in the sense that no energy may be generated or lost during a closed elastic cycle. This is fulfilled by a hyperelastic formulation, a constitutive framework that relies on the existence of a stored energy function.

Nonlinear elasticity is used in the formulation of critical state plasticity models. The term 'critical state' pertains to a condition in which a material can be sheared without plastic volume change (Roscoe and Burland 1968, Schofield and Wroth 1968). Depending on the stress state relative to a so-called critical state line, shearing can induce plastic compaction or plastic

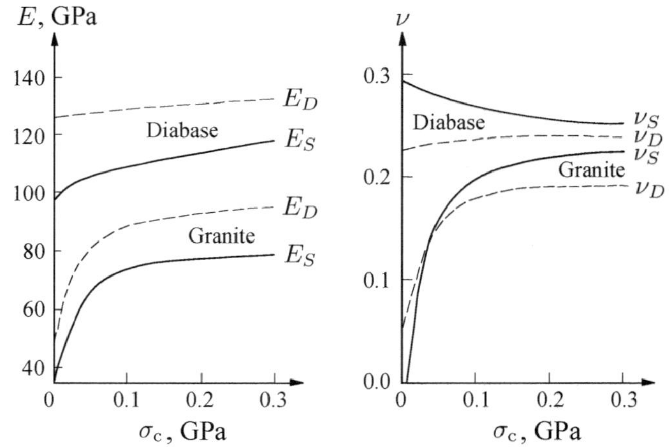

Fig. 6.1 Static and dynamic values of Young's modulus E and Poisson's ratio ν for two rocks as functions of confining pressure σ_c. Dynamic values are from wave velocities. Data from Brace (1965) and Simmons and Brace (1965). Figure reproduced from Jaeger and Cook (1976).

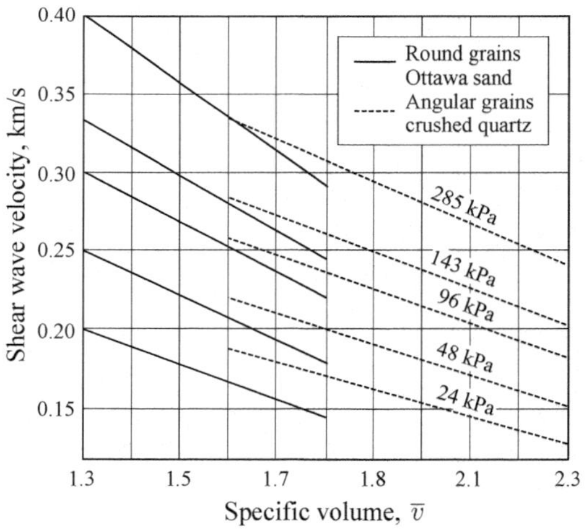

Fig. 6.2 Shear wave velocities for two sands as functions of specific volume $\overline{v}$ and confining pressure σ_c. Data from Hardin and Richart (1963).

dilation. To describe yielding in compression, as well as the accompanying inelastic compaction, the yield surface is provided with a compression cap that can expand or shrink depending on the mechanism of deformation.

In this chapter we present a sampling of plasticity models that employ a compression cap to capture plastic compaction in geomaterials. Different mechanisms can trigger the movement of the compression cap, and here we present three contrasting styles of hardening and softening: a hardening/softening mechanism for clay driven by the volumetric component of plastic strain, a hardening/softening law for sand driven by the deviatoric component of plastic strain, and a third mechanism for concrete driven by the full plastic strain tensor, including its volumetric and deviatoric components. We combine these cap plasticity models with hyperelasticity and present an elegant numerical stress-point integration technique based on return mapping in strain space that accommodates the intricate coupling of the elastic and plastic deformation responses.

6.2 Infinitesimal Strain Hyperelasticity

A material is hyperelastic if there exists a stored energy function $\psi^{\mathrm{e}}(\boldsymbol{\epsilon}^{\mathrm{e}})$ such that

$$\boldsymbol{\sigma} = \frac{\partial \psi^{\mathrm{e}}(\boldsymbol{\epsilon}^{\mathrm{e}})}{\partial \boldsymbol{\epsilon}^{\mathrm{e}}} . \tag{6.1}$$

The tangential elasticity tensor is obtained from the second derivative

$$\boldsymbol{c}^{\mathrm{e}} = \frac{\partial \boldsymbol{\sigma}}{\partial \boldsymbol{\epsilon}^{\mathrm{e}}} = \frac{\partial^2 \psi^{\mathrm{e}}(\boldsymbol{\epsilon}^{\mathrm{e}})}{\partial \boldsymbol{\epsilon}^{\mathrm{e}} \otimes \partial \boldsymbol{\epsilon}^{\mathrm{e}}} . \tag{6.2}$$

The tensor $\boldsymbol{c}^{\mathrm{e}}$ is positive definite if the stored energy is a convex function of the elastic strain. It possesses the major symmetry and two minor symmetries with respect to the first two and last two indices. If ψ^{e} is a scalar quadratic function of $\boldsymbol{\epsilon}^{\mathrm{e}}$, then the tangential tensor $\boldsymbol{c}^{\mathrm{e}}$ is constant. In this chapter, we accommodate a possible nonlinear variation of $\boldsymbol{c}^{\mathrm{e}}$ with elastic strain tensor $\boldsymbol{\epsilon}^{\mathrm{e}}$, which implies that the elastic stored energy function is not necessarily a quadratic function of $\boldsymbol{\epsilon}^{\mathrm{e}}$.

Consider a hyperelastic material for which the elastic stored energy is an isotropic function of $\boldsymbol{\epsilon}^{\mathrm{e}}$. Thus,

$$\psi^{\mathrm{e}}(\boldsymbol{\epsilon}^{\mathrm{e}}) = \psi^{\mathrm{e}}(\boldsymbol{Q} \cdot \boldsymbol{\epsilon}^{\mathrm{e}} \cdot \boldsymbol{Q}^{\mathsf{T}}) \tag{6.3}$$

for any rotation tensor $\boldsymbol{Q}$ belonging to the proper orthogonal group. As in the previous chapters, we claim that the elastic stored energy may be expressed as a function of the elastic strain invariants,

$$\psi^{\mathrm{e}}(\boldsymbol{\epsilon}^{\mathrm{e}}) = \widetilde{\psi}^{\mathrm{e}}(\mathcal{I}_1(\boldsymbol{\epsilon}^{\mathrm{e}}), \mathcal{I}_2(\boldsymbol{\epsilon}^{\mathrm{e}}), \mathcal{I}_3(\boldsymbol{\epsilon}^{\mathrm{e}})) . \tag{6.4}$$

Equivalently, ψ^{e} may be expressed as a scalar function of the principal elastic strains,

$$\psi^{\mathrm{e}}(\boldsymbol{\epsilon}^{\mathrm{e}}) = \widehat{\psi}^{\mathrm{e}}(\epsilon_1^{\mathrm{e}}, \epsilon_2^{\mathrm{e}}, \epsilon_3^{\mathrm{e}}) = \widehat{\psi}^{\mathrm{e}}(\epsilon_1^{\mathrm{e}}, \epsilon_3^{\mathrm{e}}, \epsilon_2^{\mathrm{e}}) = \ldots \tag{6.5}$$

Suppose that the stored energy function is expressed in terms of the principal elastic strains. Then the Cauchy stress tensor admits the spectral representation (see Chapter 4)

$$\boldsymbol{\sigma} = \sum_{A=1}^{3} \sigma_A \boldsymbol{n}^{(A)} \otimes \boldsymbol{n}^{(A)}, \qquad \sigma_A = \frac{\partial \widehat{\psi}^{\mathrm{e}}}{\partial \epsilon_A^{\mathrm{e}}}. \tag{6.6}$$

The corresponding tangential elasticity tensor is

$$\begin{aligned} \boldsymbol{c}^{\mathrm{e}} &= \sum_{A=1}^{3}\sum_{B=1}^{3} \left(\frac{\partial^2 \widehat{\psi}^{\mathrm{e}}}{\partial \epsilon_A^{\mathrm{e}} \partial \epsilon_B^{\mathrm{e}}}\right) \boldsymbol{m}^{(A)} \otimes \boldsymbol{m}^{(B)} \\ &+ \frac{1}{2}\sum_{A=1}^{3}\sum_{B\neq A} \left(\frac{\sigma_B - \sigma_A}{\epsilon_B^{\mathrm{e}} - \epsilon_A^{\mathrm{e}}}\right)(\boldsymbol{m}^{(AB)} \otimes \boldsymbol{m}^{(AB)} + \boldsymbol{m}^{(AB)} \otimes \boldsymbol{m}^{(BA)}). \end{aligned} \tag{6.7}$$

Because there is a unique stress for every elastic strain, the elasticity tensor $\boldsymbol{c}^{\mathrm{e}}$ must have an inverse, which we denote by the compliance tensor $\boldsymbol{d}^{\mathrm{e}} = \boldsymbol{c}^{\mathrm{e}-1}$, where

$$\begin{aligned} \boldsymbol{d}^{\mathrm{e}} &= \sum_{A=1}^{3}\sum_{B=1}^{3} \left(\frac{\partial^2 \widehat{\psi}^{\mathrm{e}}}{\partial \epsilon_A^{\mathrm{e}} \partial \epsilon_B^{\mathrm{e}}}\right)^{-1} \boldsymbol{m}^{(A)} \otimes \boldsymbol{m}^{(B)} \\ &+ \frac{1}{2}\sum_{A=1}^{3}\sum_{B\neq A} \left(\frac{\epsilon_B^{\mathrm{e}} - \epsilon_A^{\mathrm{e}}}{\sigma_B - \sigma_A}\right)(\boldsymbol{m}^{(AB)} \otimes \boldsymbol{m}^{(AB)} + \boldsymbol{m}^{(AB)} \otimes \boldsymbol{m}^{(BA)}). \end{aligned} \tag{6.8}$$

Consider the stored energy function $\widetilde{\psi}^{\mathrm{e}}$ of the invariants and assume that this function depends only on the first two invariants of the elastic strain. Specifically, we consider the volumetric and deviatoric elastic strain invariants

$$\epsilon_v^{\mathrm{e}} = \operatorname{tr}(\boldsymbol{\epsilon}^{\mathrm{e}}), \qquad \epsilon_s^{\mathrm{e}} = \sqrt{\frac{2}{3}}\,\|\boldsymbol{e}^{\mathrm{e}}\|, \qquad \boldsymbol{e}^{\mathrm{e}} = \boldsymbol{\epsilon}^{\mathrm{e}} - \frac{1}{3}\epsilon_v^{\mathrm{e}}\mathbf{1}. \tag{6.9}$$

Taking $\psi^{\mathrm{e}} = \widetilde{\psi}^{\mathrm{e}}(\epsilon_v^{\mathrm{e}}, \epsilon_s^{\mathrm{e}})$, the chain rule gives

$$\boldsymbol{\sigma} = \frac{\partial \widetilde{\psi}^{\mathrm{e}}}{\partial \epsilon_v^{\mathrm{e}}}\frac{\partial \epsilon_v^{\mathrm{e}}}{\partial \boldsymbol{\epsilon}^{\mathrm{e}}} + \frac{\partial \widetilde{\psi}^{\mathrm{e}}}{\partial \epsilon_s^{\mathrm{e}}}\frac{\partial \epsilon_s^{\mathrm{e}}}{\partial \boldsymbol{\epsilon}^{\mathrm{e}}}. \tag{6.10}$$

Setting the mean normal stress p and the von Mises stress q as

$$p = \frac{\partial \widetilde{\psi}^{\mathrm{e}}}{\partial \epsilon_v^{\mathrm{e}}} \equiv \frac{1}{3}\operatorname{tr}(\boldsymbol{\sigma}), \qquad q = \frac{\partial \widetilde{\psi}^{\mathrm{e}}}{\partial \epsilon_s^{\mathrm{e}}} \equiv \sqrt{\frac{3}{2}}\,\|\boldsymbol{s}\|, \tag{6.11}$$

where $\boldsymbol{s}$ is the deviatoric component of $\boldsymbol{\sigma}$, then (6.6) becomes

$$\boldsymbol{\sigma} = p\mathbf{1} + \sqrt{\frac{2}{3}}\,q\widehat{\boldsymbol{n}}, \tag{6.12}$$

where $\widehat{\boldsymbol{n}} = \boldsymbol{e}^{\mathrm{e}}/\|\boldsymbol{e}^{\mathrm{e}}\| = \sqrt{2/3}\,\boldsymbol{e}^{\mathrm{e}}/\epsilon_s^{\mathrm{e}}$.

Taking the derivative of (6.11) with respect to the elastic strain invariants yields the Hessian matrix (Borja et al. 1997, Borja and Tamagnini 1998)

$$\boldsymbol{D}^{\mathrm{e}} = \begin{bmatrix} D^{\mathrm{e}}_{11} & D^{\mathrm{e}}_{12} \\ D^{\mathrm{e}}_{21} & D^{\mathrm{e}}_{22} \end{bmatrix} = \begin{bmatrix} \partial^2/\partial\epsilon^{\mathrm{e}}_v\partial\epsilon^{\mathrm{e}}_v & \partial^2/\partial\epsilon^{\mathrm{e}}_v\partial\epsilon^{\mathrm{e}}_s \\ \partial^2/\partial\epsilon^{\mathrm{e}}_s\partial\epsilon^{\mathrm{e}}_v & \partial^2/\partial\epsilon^{\mathrm{e}}_s\partial\epsilon^{\mathrm{e}}_s \end{bmatrix} \widetilde{\psi}^{\mathrm{e}} . \tag{6.13}$$

The above matrix relates the differential stress and elastic strain invariants according to the relation

$$\begin{Bmatrix} \delta p \\ \delta q \end{Bmatrix} = \begin{bmatrix} D^{\mathrm{e}}_{11} & D^{\mathrm{e}}_{12} \\ D^{\mathrm{e}}_{21} & D^{\mathrm{e}}_{22} \end{bmatrix} \begin{Bmatrix} \delta\epsilon^{\mathrm{e}}_v \\ \delta\epsilon^{\mathrm{e}}_s \end{Bmatrix} . \tag{6.14}$$

The Hessian matrix is symmetric provided that $\widetilde{\psi}^{\mathrm{e}}$ is twice-differentiable. If $D^{\mathrm{e}}_{12} \neq 0$, then (6.14) reflects a full coupling in the sense that a volumetric elastic strain increment produces a deviatoric stress increment, and vice versa. In terms of the elements of the Hessian matrix, the tangential elasticity tensor $\boldsymbol{c}^{\mathrm{e}}$ can be constructed as

$$\boldsymbol{c}^{\mathrm{e}} = \frac{2}{3} D^{\mathrm{e}}_{22} \boldsymbol{I} + \Big(D^{\mathrm{e}}_{11} - \frac{2}{9} D^{\mathrm{e}}_{22} \Big) \boldsymbol{1} \otimes \boldsymbol{1} + \sqrt{\frac{2}{3}}\, D^{\mathrm{e}}_{12} (\boldsymbol{1} \otimes \widehat{\boldsymbol{n}} + \widehat{\boldsymbol{n}} \otimes \boldsymbol{1}) , \tag{6.15}$$

where $\boldsymbol{I}$ is the rank-four symmetric identity operator defined in Section 3.2. Let $\boldsymbol{E}^{\mathrm{e}} = \boldsymbol{D}^{\mathrm{e}-1}$; the tangential compliance tensor takes the form

$$\begin{aligned} \boldsymbol{d}^{\mathrm{e}} &= \frac{3}{2} D^{\mathrm{e}-1}_{22} \boldsymbol{I} + \Big(\frac{1}{9} E^{\mathrm{e}}_{11} - \frac{1}{2} D^{\mathrm{e}-1}_{22} \Big) \boldsymbol{1} \otimes \boldsymbol{1} + \frac{1}{\sqrt{6}} E^{\mathrm{e}}_{12} (\boldsymbol{1} \otimes \widehat{\boldsymbol{n}} + \widehat{\boldsymbol{n}} \otimes \boldsymbol{1}) \\ &\quad + \frac{3}{2} (E^{\mathrm{e}}_{22} - D^{\mathrm{e}-1}_{22}) \widehat{\boldsymbol{n}} \otimes \widehat{\boldsymbol{n}} . \end{aligned} \tag{6.16}$$

Exercise 6.1. Construct stored energy functions $\widetilde{\psi}^{\mathrm{e}}$ and $\widehat{\psi}^{\mathrm{e}}$ for a linearly elastic Hookean material with constant elastic bulk and shear moduli, K and μ, respectively. Then, evaluate the Hessian matrix $\boldsymbol{D}^{\mathrm{e}}$ for the function $\widetilde{\psi}^{\mathrm{e}}$ and state whether this matrix is diagonal or fully populated.

6.3 Critical State Theory

Critical state theory provides a general framework for describing the compactive and dilative inelastic behavior of granular materials such as soil and rock. Inelastic volume change, whether by compaction or dilation, is interpreted as shear-induced and not a result of simple isotropic compression or extension. At sufficiently low confining pressures, granular materials typically exhibit shear-induced dilation, whereas at sufficiently high pressures they exhibit shear-induced compaction. However, a material yielding at critical state neither compacts nor dilates when sheared. Therefore, the condition at critical state is

$$\mathrm{tr}(\dot{\boldsymbol{\epsilon}}^{\mathrm{p}}) = 0 \qquad \text{at} \qquad \boldsymbol{\sigma} = \boldsymbol{\sigma}_{\mathrm{cs}} , \tag{6.17}$$

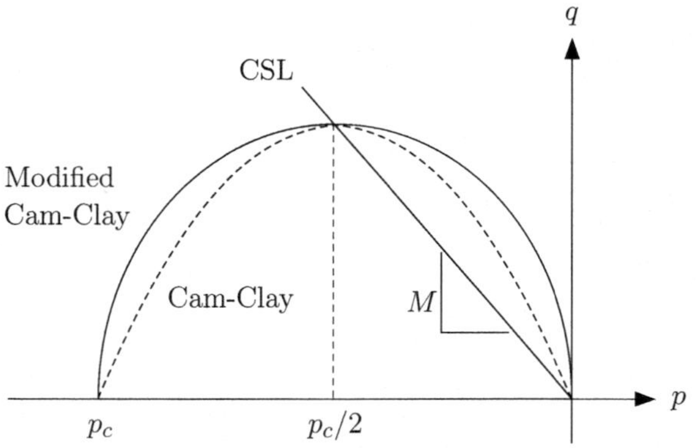

Fig. 6.3 Cam-Clay and modified Cam-Clay yield surfaces on p-q plane

where $\boldsymbol{\sigma}_{\text{cs}}$ is the stress tensor at any point on the critical state line (CSL). The CSL separates the compression cap from the dilation side of the yield surface.

The modified Cam-Clay (MCC) plasticity theory introduced by Roscoe and Burland (1968) is the most widely used critical state model. The model, named after the river Cam, evolved from the Cam-Clay (CC) model of Schofield and Wroth (1968) and incorporates pressure sensitivity and hardening/softening responses. The yield function for the MCC model is given by

$$\widetilde{f} = \frac{q^2}{M^2} + p(p - p_c) \leq 0\,, \qquad p = \frac{1}{3}\text{tr}(\boldsymbol{\sigma})\,, \qquad q = \sqrt{\frac{3}{2}}\|\boldsymbol{s}\|\,, \tag{6.18}$$

where M is a positive constant and $p_c < 0$ is a stress-like plastic internal variable called the preconsolidation stress. The yield surface defined by $\widetilde{f} = 0$ is a function of the first two stress invariants and plots as a semi-ellipse on the p-q plane (Fig. 6.3). By comparison, the Cam-Clay yield surface has the shape of a bullet with vertices at the intersection points with the p-axis.

Experimental evidence suggests that an associative flow rule is valid on the compression side of the CC and MCC yield surfaces. This results in the following expression for the plastic strain increment

$$\dot{\boldsymbol{\epsilon}}^{\text{p}} = \dot{\lambda}\frac{\partial \widetilde{f}}{\partial \boldsymbol{\sigma}} = \dot{\lambda}\Big(\frac{1}{3}v\mathbf{1} + \sqrt{\frac{3}{2}}e\widehat{\boldsymbol{n}}\Big)\,, \qquad v = 2p - p_c\,, \qquad e = \frac{2q}{M^2}\,. \tag{6.19}$$

Normality of the plastic flow ensures a nonnegative plastic work, as can be seen from

$$\mathcal{D}^{\text{p}} = \boldsymbol{\sigma} : \dot{\boldsymbol{\epsilon}}^{\text{p}} = \dot{\lambda}(pv + qe) = \dot{\lambda}\Big(pv + \frac{2q^2}{M^2}\Big) = \dot{\lambda}pp_c \geq 0\,, \tag{6.20}$$

since p and p_c are both negative numbers. At the critical state, $v = 0$ and $p_c = 2p$. Thus, the CSL takes the form

$$q = -Mp\,. \tag{6.21}$$

As noted previously, the material can be sheared without volume change at any point on the CSL. Therefore, the CSL may be considered as a limiting failure envelope for frictional materials. The slope M of the CSL can be related to the critical state friction angle ϕ_{cs} through the equation

$$M = \frac{6\sin\phi_{\text{cs}}}{3 - \sin\phi_{\text{cs}}}\,, \tag{6.22}$$

where ϕ_{cs} is determined from the shear strength of a material deforming plastically at a constant volume. The yield surface expands or contracts depending on whether $-p_c$ increases or decreases. For a material yielding in isotropic compression, the stress path is defined by $p = p_c$ and $q = 0$. Thus, we can develop a hardening law for p_c by performing loading-unloading tests in isotropic compression. For geomaterials such as clay, which consist of a mixture of solid and void, a hardening law is typically expressed in terms of the specific volume $\overline{v} > 0$, defined as

$$\overline{v} = 1 + \overline{e}\,, \qquad \overline{e} = V_v/V_s\,, \tag{6.23}$$

where V_v is the volume of void, V_s is the volume of solid, and $\overline{e}$ is the void ratio.

Based on experimental tests on clays, Hashiguchi and Ueno (1977) and Butterfield (1979) proposed a bilogarithmic compressibility law of the form

$$\dot{\epsilon}_v = \frac{\dot{\overline{v}}}{\overline{v}} = -c_c \frac{\dot{p}_c}{p_c}\,, \tag{6.24}$$

where $c_c > 0$ is a compressibility index. When integrated over a finite load increment, the above compressibility law produces a linear plot of $\log(\overline{v})$ with $\log(-p_c)$ as shown in Fig. 6.4a (note: $\log = \ln$). A nominal measure of volumetric strain rate is obtained by substituting $\overline{v}_0$ in lieu of $\overline{v}$, yielding

$$\dot{\epsilon}_v \approx \frac{\dot{\overline{v}}}{\overline{v}_0} = -c_c \frac{\dot{p}_c}{p_c}\,, \tag{6.25}$$

and we obtain a linear plot between $\overline{v}/\overline{v}_0$ and $\log(-p_c)$ as shown in Fig. 6.4b (Roscoe and Burland 1968). These two hardening laws are accurate up to moderately high pressures. At great depths (i.e., several kilometers) beneath the ground surface, where the pressures can be very high, the two compressibility laws should not be used because they give negative values of $\overline{v}$, which is not meaningful.

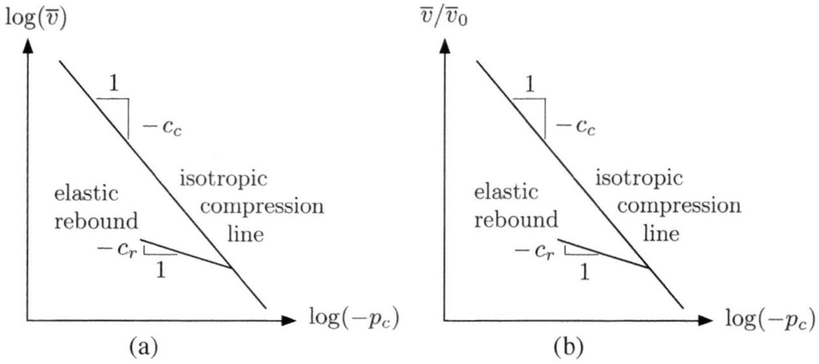

Fig. 6.4 Hardening laws for clay derived from isotropic compression tests: (a) bilogarithmic; (b) semilogarithmic

When a granular material is released from initial compression, a small amount of swelling takes place. This is the elastic rebound, and the corresponding elastic volumetric strain increment is given by

$$\dot{\epsilon}_v^{\mathrm{e}} = -c_r \frac{\dot{p}_c}{p_c}\,, \tag{6.26}$$

where $c_r > 0$ is the elastic compressibility index, see Fig. 6.4. The elastic rebound cannot exceed the initial volumetric compression, so we must have $0 < c_r < c_c$. The plastic component of the volumetric strain rate is given by

$$\dot{\epsilon}_v^{\mathrm{p}} = \dot{\epsilon}_v - \dot{\epsilon}_v^{\mathrm{e}} = -(c_c - c_r)\frac{\dot{p}_c}{p_c} = \mathrm{tr}(\dot{\boldsymbol{\epsilon}}^{\mathrm{p}})\,. \tag{6.27}$$

Therefore,

$$\dot{p}_c = -\frac{\mathrm{tr}(\dot{\boldsymbol{\epsilon}}^{\mathrm{p}})}{c_c - c_r} p_c\,. \tag{6.28}$$

This is the hardening law for the CC and MCC models.

The mechanism for expansion and contraction of the yield surface is shown in Fig. 6.5. On the compression side the yield surface expands because $\mathrm{tr}(\dot{\boldsymbol{\epsilon}}^{\mathrm{p}}) < 0$ (hardening), whereas on the dilation side it contracts (softening). At the critical state the size of the yield surface is fixed (perfect plasticity). It may be worthwhile to compare the above hardening law with the one we used for J_2 plasticity: The hardening law for the CC and MCC models is driven by the volumetric component of plastic strain, whereas the hardening law for J_2 plasticity is determined by the deviatoric component of plastic strain.

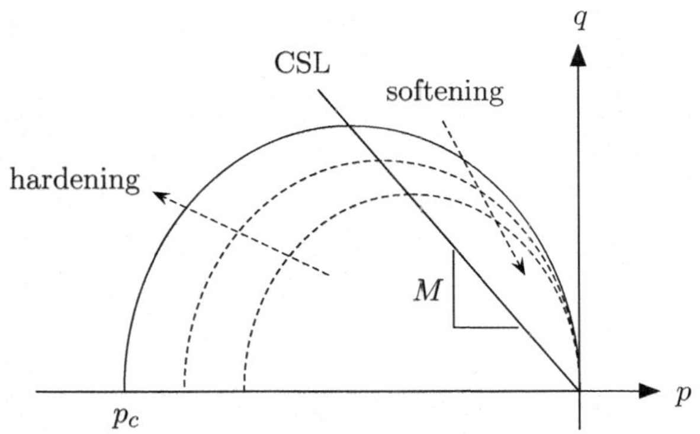

Fig. 6.5 Hardening and softening for the modified Cam-Clay plasticity model

6.4 Hyperelastic Law for MCC Model

A straight-line bilogarithmic elastic compression law (see Fig. 6.4a) produces the following elastic rate constitutive equation

$$\dot{p} = -\left(\frac{p}{c_r}\right)\dot{\epsilon}_v^{\mathrm{e}} \,. \tag{6.29}$$

The quantity inside the parentheses, $K := -p/c_r$, may be viewed as the tangential elastic bulk modulus in compression and is seen to vary linearly with $-p$. This suggests an exponential relationship between p and $\dot{\epsilon}_v^{\mathrm{e}}$. The compressibility law does not say anything about the elastic shear modulus μ, so we postulate two cases.

Case 1: μ = constant. In this case, the stress invariants p and q can be evaluated independently as

$$p = p_0 \exp\left(\frac{\epsilon_{v0} - \epsilon_v^{\mathrm{e}}}{c_r}\right), \qquad q = 3\mu\epsilon_s^{\mathrm{e}} \,, \tag{6.30}$$

where $p = p_0$ when $\epsilon_v^{\mathrm{e}} = \epsilon_{v0}^{\mathrm{e}}$. This results in an elastic stored energy function of the form

$$\widetilde{\psi}^{\mathrm{e}}(\epsilon_v^{\mathrm{e}}, \epsilon_s^{\mathrm{e}}) = \int p \, \mathrm{d}\epsilon_v^{\mathrm{e}} + \int q \, \mathrm{d}\epsilon_s^{\mathrm{e}} = -p_0 c_r \exp\left(\frac{\epsilon_{v0} - \epsilon_v^{\mathrm{e}}}{c_r}\right) + \frac{3}{2}\mu(\epsilon_s^{\mathrm{e}})^2 + \widetilde{\psi}_0 \,, \tag{6.31}$$

where $\widetilde{\psi}_0$ is a constant of integration.

Case 2: μ = variable. This is motivated by Figs. 6.1 and 6.2, which suggest that for granular materials the elastic shear modulus may increase with the confining pressure. However, according to our hyperelastic formulation, there should be full coupling between the elastic volumetric and deviatoric responses if both μ and K are allowed to vary with $-p$ (Houlsby 1985). Consider then an elastic stored energy function of the form

$$\widetilde{\psi}^{\rm e}(\epsilon_v^{\rm e},\epsilon_s^{\rm e}) = -p_0 c_r \exp\left(\frac{\epsilon_{v0}-\epsilon_v^{\rm e}}{c_r}\right) - \frac{3}{2}c_\mu p_0 \exp\left(\frac{\epsilon_{v0}-\epsilon_v^{\rm e}}{c_r}\right)(\epsilon_s^{\rm e})^2 + \widetilde{\psi}_0\,, \quad (6.32)$$

where $c_\mu > 0$. We obtain the first two stress invariants from the derivatives

$$\left.\begin{aligned} p &= \frac{\partial\widetilde{\psi}^{\rm e}}{\partial\epsilon_v^{\rm e}} = p_0\Big(1+\frac{3c_\mu}{2c_r}(\epsilon_s^{\rm e})^2\Big)\exp\left(\frac{\epsilon_{v0}-\epsilon_v^{\rm e}}{c_r}\right)\\ q &= \frac{\partial\widetilde{\psi}^{\rm e}}{\partial\epsilon_s^{\rm e}} = -3c_\mu p_0\exp\left(\frac{\epsilon_{v0}-\epsilon_v^{\rm e}}{c_r}\right)\epsilon_s^{\rm e} \end{aligned}\right\}. \quad (6.33)$$

Accordingly, the elements of the Hessian matrix $\boldsymbol{D}^{\rm e}$ are

$$\left.\begin{aligned} D_{11}^{\rm e} &= -\frac{p_0}{c_r}\Big(1+\frac{3c_\mu}{2c_r}(\epsilon_s^{\rm e})^2\Big)\exp\left(\frac{\epsilon_{v0}-\epsilon_v^{\rm e}}{c_r}\right)\\ D_{22}^{\rm e} &= -3c_\mu p_0\exp\left(\frac{\epsilon_{v0}-\epsilon_v^{\rm e}}{c_r}\right)\\ D_{12}^{\rm e} &= D_{21}^{\rm e} = \frac{3p_0c_\mu\epsilon_s^{\rm e}}{c_r}\exp\left(\frac{\epsilon_{v0}-\epsilon_v^{\rm e}}{c_r}\right) \end{aligned}\right\}. \quad (6.34)$$

Note that $\boldsymbol{D}^{\rm e}$ is fully populated, suggesting that indeed there is coupling between the elastic volumetric and deviatoric responses. The equivalent elastic bulk modulus K is $D_{11}^{\rm e} = -p/c_r$, which varies linearly with the compressive stress. The equivalent elastic shear modulus μ is $D_{22}^{\rm e}$, which also varies linearly with $-p$ if we set $\epsilon_s^{\rm e} = 0$ on the first line of (6.33). The coupling term is represented by $D_{12}^{\rm e}$, which is not zero, resulting in coupled volumetric and deviatoric responses.

We should emphasize at the outset that the presented exponential form for the stored energy function was motivated by the experimentally observed dependence of the elastic moduli on the confinement for some materials, and is valid only over a certain range of elastic strains. It does not accommodate all stress conditions, including the trivial stress-free condition ($p = 0$).

Exercise 6.2. Assume an elastic stored energy function of the form

$$\widetilde{\psi}^{\rm e}(\epsilon_v^{\rm e},\epsilon_s^{\rm e}) = -p_0 c_r \exp\left(\frac{\epsilon_{v0}-\epsilon_v^{\rm e}}{c_r}\right) + \frac{3}{2}\mu(\epsilon_s^{\rm e})^2 + \widetilde{\psi}_0\,,$$

where

$$\mu = \mu_0 - c_\mu p_0 \exp\left(\frac{\epsilon_{v0}-\epsilon_v^{\rm e}}{c_r}\right),$$

and $\mu_0 > 0$ is constant. If $c_\mu = 0$ and $\mu_0 > 0$, we obtain a constant elastic shear modulus; if $\mu_0 = 0$ and $c_\mu > 0$, we obtain a variable elastic shear modulus. Now, consider the more general case where neither c_μ nor μ_0 is zero. Evaluate the elements of the Hessian matrix $\boldsymbol{D}^{\rm e}$ and state whether this matrix is diagonal or fully populated.

6.5 Three-Invariant Formulation

Yield strengths in granular materials are known to be higher in compression than in extension. To capture this feature, two-invariant models such as the MCC must be enhanced with the third stress invariant. In this section, we employ a simple enhancement utilizing Lode's angle θ presented in Chapter 4. To facilitate a geometric interpretation, it is convenient to introduce the (p, q, θ) representation analogous to the cylindrical Haigh-Westergaard coordinates (Chen and Han 1988), where the hydrostatic axis serves as the pole and any of the three positive (extensional) principal stress axes serves as the polar axis. The polar radius extends to all polar directions $0 \leq \theta \leq 2\pi$ using the three-fold symmetry shown in Fig. 6.6.

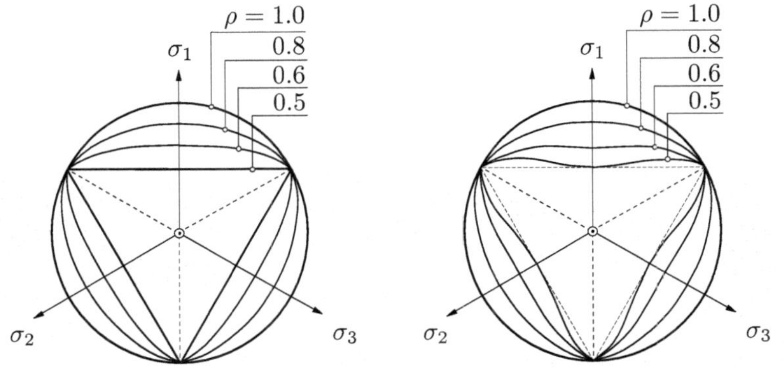

Fig. 6.6 Convex and non-convex yield functions obtained from the scaling functions of Willam and Warnke (1974) (left); and Gudehus (1973) and Argyris et al. (1974) (right).

We define the ellipticity

$$\rho = \|\boldsymbol{s}\|_{\text{ext}} / \|\boldsymbol{s}\|_{\text{com}} = q_{\text{ext}} / q_{\text{com}}\,, \qquad 1/2 \leq \rho \leq 1\,, \tag{6.35}$$

where $\|\boldsymbol{s}\|_{\text{com}}$ (alternatively, $\|\boldsymbol{s}\|_{\text{ext}}$) is the radius on the compressive (alternatively, extensional) principal stress axis. The ellipticity ρ describes the deviation from roundness of the cross section of the yield surface on the deviatoric plane. The upper bound $\rho = 1$ is for a perfectly circular cross section on the deviatoric plane, whereas the lower bound $\rho = 1/2$ is for a limiting triangular cross section (note: the yield surface loses convexity when $\rho < 1/2$, see Fig. 6.6).

For illustration purposes, we enhance the two-invariant MCC yield function to include the third stress invariant. The technique is general enough to accommodate any two-invariant plasticity models having a similar form of the yield function. The enhanced yield function takes the form

$$\widetilde{f} = \zeta^2 \frac{q^2}{M^2} + p(p - p_c) \leq 0\,, \tag{6.36}$$

where $\zeta = \zeta(\rho, \theta)$ is a scaling function designed to reproduce a higher yield stress in compression. The function ζ must satisfy two boundary conditions: (a) $\zeta = 1/\rho$ when $\theta = 0$; and (b) $\zeta = 1$ when $\theta = \pi/3$. Note that since ρ does not depend on the first stress invariant, all cross sections of the yield surface undergo similar scaling. Some possible forms for the scaling function ζ are presented below, with proper account for conditions under which either the convexity or smoothness of the yield function is lost.

Willam and Warnke (1974) proposed a scaling function of the form

$$\zeta(\rho, \theta) = \frac{4(1-\rho^2)\cos^2\theta + (2\rho - 1)^2}{2(1-\rho^2)\cos\theta + (2\rho-1)[4(1-\rho^2)\cos^2\theta + 5\rho^2 - 4\rho]^{1/2}}\,. \tag{6.37}$$

This function is smooth and convex in the range $1/2 < \rho \leq 1$. Perić and Ayari (2002) utilized this function to enhance the MCC model; Menétrey and Willam (1995) employed this function to develop a failure criterion for concrete.

Argyris et al. (1974) and Gudehus (1973) proposed a scaling function of the form

$$\zeta(\rho, \theta) = \frac{(1+\rho) + (1-\rho)\cos 3\theta}{2\rho}\,. \tag{6.38}$$

This function is smooth, but it is convex only for $7/9 \leq \rho \leq 1$ (Jiang and Pietruszczak 1988). For $1/2 \leq \rho < 7/9 \approx 0.778$ the yield function is not convex, see Fig. 6.6.

Boswell and Chen (1987) proposed a scaling function of the form

$$\zeta(\rho, \theta) = \frac{[3 - 4(1-\rho^2)\sin^2\theta]^{1/2}}{\sqrt{3}\rho}\,. \tag{6.39}$$

This function is convex within the region $1/2 \leq \rho \leq 1$; however, it is not smooth since $\partial\zeta/\partial\theta \neq 0$ at $\theta = \pi/3$.

6.6 Derivatives of Lode's Angle

Because Lode's angle θ plays a central role in the three-invariant formulation, we will evaluate its derivatives once and for all. Let $\phi(p, q, \theta)$ be any convex scalar function of the stress invariants p, q, and θ, for which we want to obtain the derivatives with respect to the Cauchy stress tensor $\boldsymbol{\sigma}$. If ϕ is the yield function, the stress derivative $\partial\phi/\partial\boldsymbol{\sigma}$ is used to impose the consistency condition. If ϕ is the plastic potential function, it would be necessary to obtain the second derivative $\partial^2\phi/(\partial\boldsymbol{\sigma} \otimes \partial\boldsymbol{\sigma})$ as well, since this is needed for the local Jacobian matrix. Extracting the derivatives by brute force is not an attractive option since it entails an enormous amount of work (Willam 2002). Fortunately, through the spectral representation technique, we can simplify the task of obtaining the derivatives by working in principal stress space.

We assume that ϕ can be decomposed additively in the form

$$\phi(p,q,\theta) = \phi_1(p) + \phi_2(q,\theta)\,. \tag{6.40}$$

The above equation defines the shape of the function ϕ on any meridional plane. By the chain rule, we obtain the first derivative of ϕ as

$$\frac{\partial \phi}{\partial \sigma_A} = \frac{1}{3}\frac{\partial \phi}{\partial p}\delta_A + \frac{\partial \phi}{\partial q}\frac{\partial q}{\partial \sigma_A} + \frac{\partial \phi}{\partial \theta}\frac{\partial \theta}{\partial \sigma_A}\,, \tag{6.41}$$

where $\delta_A = 1$ for $A = 1,2,3$ (this is a notational convenience to keep track of the indices). We also obtain the second derivative as

$$\begin{aligned}\frac{\partial^2 \phi}{\partial \sigma_A \partial \sigma_B} &= \frac{1}{9}\left(\frac{\partial^2 \phi}{\partial p^2}\right)\delta_A\delta_B + \frac{\partial \phi}{\partial q}\frac{\partial^2 q}{\partial \sigma_A \partial \sigma_B} + \frac{\partial \phi}{\partial \theta}\frac{\partial^2 \theta}{\partial \sigma_A \partial \sigma_B}\\ &+ \frac{\partial^2 \phi}{\partial q^2}\frac{\partial q}{\partial \sigma_A}\frac{\partial q}{\partial \sigma_B} + \frac{\partial^2 \phi}{\partial \theta^2}\frac{\partial \theta}{\partial \sigma_A}\frac{\partial \theta}{\partial \sigma_B}\\ &+ \frac{\partial^2 \phi}{\partial q \partial \theta}\left(\frac{\partial q}{\partial \sigma_A}\frac{\partial \theta}{\partial \sigma_B} + \frac{\partial \theta}{\partial \sigma_A}\frac{\partial q}{\partial \sigma_B}\right).\end{aligned} \tag{6.42}$$

With respect to the principal axes the trace functions reduce nicely to the forms

$$\mathrm{tr}(\boldsymbol{s}^2) = s_1^2 + s_2^2 + s_3^2\,, \qquad \mathrm{tr}(\boldsymbol{s}^3) = s_1^3 + s_2^3 + s_3^3\,, \qquad \text{etc.} \tag{6.43}$$

So, we readily obtain the first stress derivative of q as

$$\frac{\partial q}{\partial \sigma_A} = \sqrt{\frac{3}{2}}\frac{s_A}{\chi}\,, \qquad \chi = \sqrt{\mathrm{tr}(\boldsymbol{s}^2)}\,, \tag{6.44}$$

and the second derivative as

$$\frac{\partial^2 q}{\partial \sigma_A \partial \sigma_B} = \frac{\sqrt{3/2}}{\chi}\left(\delta_{AB} - \frac{1}{3}\delta_A\delta_B - \frac{s_A s_B}{\chi^2}\right), \tag{6.45}$$

where δ_{AB} is the Kronecker delta.

To obtain the first derivative of θ, we write (4.40) in the form

$$\frac{1}{\sqrt{6}}\cos 3\theta = \frac{\mathrm{tr}(\boldsymbol{s}^3)}{\chi^3} \equiv y\,. \tag{6.46}$$

Differentiating implicitly gives

$$-\left(\frac{\sqrt{6}}{2}\sin 3\theta\right)\frac{\partial \theta}{\partial \sigma_A} = \frac{\partial y}{\partial \sigma_A}\,, \tag{6.47}$$

where

$$\frac{\partial y}{\partial \sigma_A} = 3\,\frac{s_A^2}{\chi^3} - 3\,\frac{\mathrm{tr}(\boldsymbol{s}^3)s_A}{\chi^5} - \frac{\delta_A}{\chi}\,. \tag{6.48}$$

Solving,

$$\frac{\partial \theta}{\partial \sigma_A} = -\Big(\frac{2}{\sqrt{6}}\csc 3\theta\Big)\frac{\partial y}{\partial \sigma_A}\,, \qquad 0 < \theta < \frac{\pi}{3}\,. \tag{6.49}$$

We have a case of non-distinct principal stresses if θ is either zero or $\pi/3$. In this case, a simple perturbation would be necessary to arrive at three distinct eigenvalues. Recall that in Chapter 4 a similar perturbation has been suggested in the spectral decomposition of the Cauchy stress tensor to ensure that the three principal stresses are distinct.

Differentiating (6.47) once more yields

$$-\Big(\frac{\sqrt{6}}{2}\sin 3\theta\Big)\frac{\partial^2\theta}{\partial\sigma_A\partial\sigma_B} - \Big(\frac{3\sqrt{6}}{2}\cos 3\theta\Big)\frac{\partial\theta}{\partial\sigma_A}\frac{\partial\theta}{\partial\sigma_B} = \frac{\partial^2 y}{\partial\sigma_A\partial\sigma_B}\,, \tag{6.50}$$

where

$$\begin{aligned}\frac{\partial^2 y}{\partial\sigma_A\partial\sigma_B} &= 6\,\frac{\widehat{s}_{AB}}{\chi^3} - 3\,\frac{\mathrm{tr}(\boldsymbol{s}^3)}{\chi^5}\Big(\delta_{AB} - \frac{1}{3}\delta_A\delta_B - 5\,\frac{s_A s_B}{\chi^2}\Big)\\ &\quad + \frac{1}{\chi^3}(\delta_A s_B + s_A\delta_B) - \frac{9}{\chi^5}(s_A s_B^2 + s_A^2 s_B)\,,\end{aligned} \tag{6.51}$$

and $\widehat{s}_{AB} = s_A\delta_{AB}$ (no sum on A). Solving,

$$\frac{\partial^2\theta}{\partial\sigma_A\partial\sigma_B} = -\Big(\frac{2}{\sqrt{6}}\csc 3\theta\Big)\frac{\partial^2 y}{\partial\sigma_A\partial\sigma_B} - (3\cot 3\theta)\frac{\partial\theta}{\partial\sigma_A}\frac{\partial\theta}{\partial\sigma_B}\,, \qquad 0 < \theta < \frac{\pi}{3}\,. \tag{6.52}$$

The remaining derivatives in (6.42) depend on the specific form of the invariant function $\phi(p, q, \theta)$.

Exercise 6.3. Show that $\boldsymbol{\sigma} : (\partial\theta/\partial\boldsymbol{\sigma}) = 0$. Suggestion: Express θ in terms of the principal values of $\boldsymbol{s}$ and use implicit differentiation.

Exercise 6.4. Use the result from Exercise 6.3 to show that $\mathcal{D}^{\mathrm{p}} = \boldsymbol{\sigma} : \dot{\boldsymbol{\epsilon}}^{\mathrm{p}} \geq 0$ for the three-invariant modified Cam-Clay model with associative flow rule.

6.7 A Plasticity Model for Sand

Many variations of critical state theory for sand can be found in the literature. One of the simplest versions is a model called Nor-Sand proposed by Jefferies (1993), the name derived from its East Anglian heritage. Nor-Sand was developed to capture the dilatant behavior of dense sand during shearing. Borja and Andrade (2006) and Andrade and Borja (2006) enhanced the Nor-Sand formulation to include a hyperelastic regularization, a non-associative

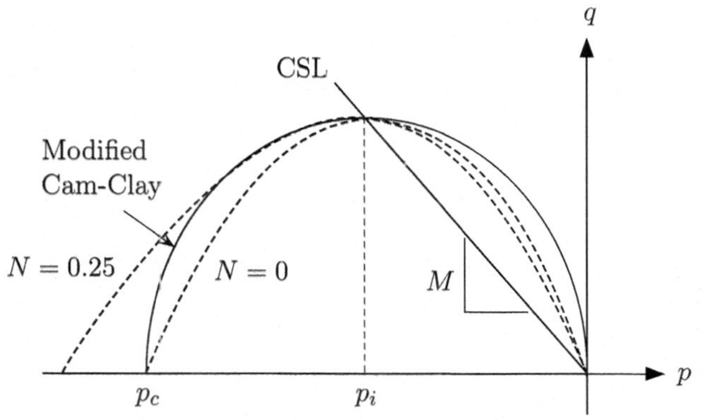

Fig. 6.7 Nor-Sand and modified Cam-Clay yield surfaces

plastic flow, and the contribution of the third stress invariant. In what follows we present the yield function for the enhanced version of this model.

The yield function is of the form

$$\widetilde{f} = \zeta q + \eta p \leq 0\,, \tag{6.53}$$

where $\zeta = \zeta(\rho, \theta)$ is a scaling function that varies with the ellipticity ρ and Lode's angle θ, and

$$\eta = \begin{cases} M[1 + \log(p_i/p)] & \text{if } N = 0; \\ (M/N)[1 - (1-N)(p/p_i)^{N/(1-N)}] & \text{if } N > 0. \end{cases} \tag{6.54}$$

The parameter M has the same meaning as in the MCC model, and $p_i < 0$ is the plastic internal variable, see Fig. 6.7. A closed-form expression for p_i is obtained by setting $\widetilde{f} = 0$,

$$\frac{p_i}{p} = \begin{cases} \exp(\eta/M - 1) & \text{if } N = 0; \\ [(1-N)/(1-\eta N/M)]^{(1-N)/N} & \text{if } N > 0. \end{cases} \tag{6.55}$$

The exponent parameter N determines the curvature of the yield surface on the hydrostatic axis, and typically has a value less than 0.4 (Jefferies 1993), see Fig. 6.7. If $\zeta = 1$ the yield function reduces to the CC yield function (Schofield and Wroth 1968) with vertices at $q = 0$, see Fig. 6.8.

We consider a three-invariant plastic potential function $\widetilde{g}$ of the form

$$\widetilde{g} = \overline{\zeta} q + \overline{\eta} p\,, \tag{6.56}$$

where

$$\overline{\eta} = \begin{cases} M[1 + \ln(\overline{p}_i/p)] & \text{if } \overline{N} = 0; \\ (M/\overline{N})[1 - (1-\overline{N})(p/\overline{p}_i)^{\overline{N}/(1-\overline{N})}] & \text{if } \overline{N} > 0. \end{cases} \tag{6.57}$$

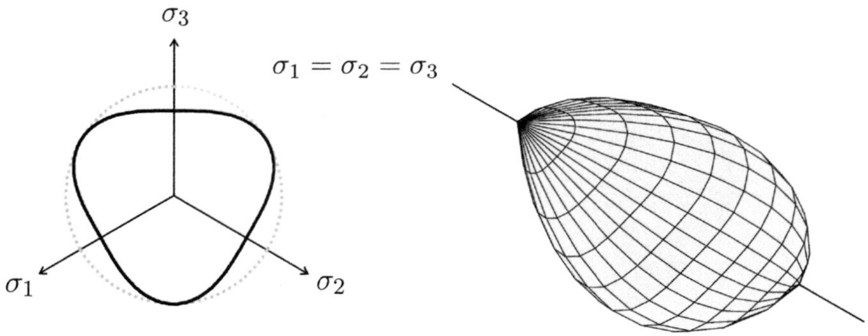

Fig. 6.8 Yield surface for sand enhanced with the Argyris et al. (1974) and Gudehus (1973) scaling function with $\rho = 0.78$. Cross section on deviatoric plane (left) and isometric view in principal stress space (right). Figure reproduced from Andrade and Borja (2006).

The parameter $\overline{N}$ controls the amount of volumetric nonassociativity of the plastic flow, whereas $\overline{\zeta} = \overline{\zeta}(\overline{\rho}, \theta)$ introduces deviatoric nonassociativity for $\overline{\rho} \neq \rho$. The variable $\overline{p}_i$ is a free parameter that can be chosen so that $\widetilde{g} = 0$ when $\widetilde{f} = 0$ through an equation that is very similar to (6.55). Plastic flow is associative if $\overline{N} = N$ and $\overline{p}_i = p_i$.

It is important to choose the parameters of the yield and plastic potential functions appropriately to ensure a nonnegative plastic dissipation. Using the identity $\boldsymbol{\sigma} : \partial\theta/\partial\boldsymbol{\sigma} = \sum_{A=1}^{3} \sigma_A (\partial\theta/\partial\sigma_A) = 0$ (see Exercise 6.3), we require that

$$\mathcal{D}^{\mathrm{p}} = \boldsymbol{\sigma} : \dot{\boldsymbol{\epsilon}}^{\mathrm{p}} = \dot{\lambda}\Big(p\frac{\partial\widetilde{g}}{\partial p} + q\frac{\partial\widetilde{g}}{\partial q}\Big) = \dot{\lambda}p\Big(\frac{M - \overline{\eta}}{1 - \overline{N}} + \overline{\eta}\Big) \geq 0\,, \tag{6.58}$$

since $\widetilde{f} = \widetilde{g} = 0$ when the material is yielding. Noting that $-\dot{\lambda}p \geq 0$ and $1 - \overline{N} \geq 0$ for physically meaningful values of $\overline{N}$, we require that $-\overline{N}\overline{\eta} + M \geq 0$. Now, for a stress point on the yield surface we have $\overline{\eta}/\overline{\zeta} = \eta/\zeta$. Hence, the condition $\mathcal{D}^{\mathrm{p}} \geq 0$ requires that

$$-M\frac{\overline{N}}{N}\frac{\overline{\zeta}}{\zeta}\Big[1 - (1 - N)\Big(\frac{p}{p_i}\Big)^{N/(1-N)}\Big] + M \geq 0\,. \tag{6.59}$$

The expression inside the square brackets is a monotonically decreasing function and is always less than unity for $p_i/(1 - N)^{(1-N)/N} \leq p \leq 0$. Therefore, the condition reduces to $\overline{N}\,\overline{\zeta} \leq N\zeta$, which is satisfied by the pair of conditions

$$\overline{N} \leq N \qquad \text{and} \qquad \overline{\zeta} \leq \zeta\,. \tag{6.60}$$

Both conditions are met by choosing a dilatancy angle ψ_c that is smaller than the critical state friction angle ϕ_c (Borja and Andrade 2006).

The hardening law for sand describes the evolution of p_i and takes the form

$$\dot{p}_i = -\sqrt{\frac{2}{3}} h(p_i - p_i^*) \|\dot{\boldsymbol{e}}^{\mathrm{p}}\| , \qquad \dot{\boldsymbol{e}}^{\mathrm{p}} = \dot{\boldsymbol{\epsilon}}^{\mathrm{p}} - \frac{1}{3}\mathrm{tr}(\dot{\boldsymbol{\epsilon}}^{\mathrm{p}})\mathbf{1} , \tag{6.61}$$

where $h > 0$ is a dimensionless constant and p_i^* is the size of the yield surface at the maximum possible plastic dilatancy (Jefferies 1993). The above hardening law for sand should be contrasted with the hardening law for MCC given in (6.28). In the MCC model, the size of the yield surface is driven by the volumetric component of plastic strain; in the sand model, the size is driven by the deviatoric component. The next section presents a plasticity model for concrete in which the movement of the compression cap is driven by the cumulative plastic strain as well as by the mean normal stress to capture increased ductility under increased confinement.

6.8 A Plasticity Model for Concrete

Concrete is a geomaterial that also exhibits complex load-deformation behavior. In this section, we consider a cap plasticity model proposed by Kang and Willam (1999) that fits nicely within the framework of the present three-invariant formulation. The discussion below focuses on the mechanism of yielding that includes hardening, failure, and softening.

Using the (p, q, θ) representation, the yield function f (called 'loading function' by Kang and Willam (1999)) is expressed as the sum of a failure function, a hardening function, and a softening function:

$$\widetilde{f} = F_1(p, q, \theta) + F_2(p, q, k) + F_3(p, q, c) \leq 0 , \tag{6.62}$$

where F_1 is a failure delimiter, F_2 accounts for isotropic hardening, and F_3 accounts for isotropic softening. We elucidate the individual components of the yield function below.

The surface defined by $F_1 = 0$ is the failure envelope similar to the three-invariant conical surfaces presented in Chapter 4. The form of the failure envelope proposed by Kang and Willam (1999) is

$$F_1(p, q, \theta) = \zeta \frac{q}{f_c'} - \frac{q_1}{f_c'} \left(\frac{p_0 - p}{p_0 - p_1} \right)^{\alpha} = 0 , \tag{6.63}$$

where f_c' is the ultimate compressive strength of concrete, which is used as a normalization parameter, and the exponent α is taken to be around 0.77. Since $1/2 < \alpha < 1$, the meridional sections are intermediate between a straight line and a parabola, suggesting a decrease in the effective friction angle with the mean confining pressure (see Fig. 6.9). The mean normal stress $p_0 = \sqrt{3} f_t'$ locates the vertex of the failure surface according to the tensile strength f_t' of concrete. The mean normal stress $p_1 = \sqrt{3} S f_c' k < 0$ defines the hydrostatic range in compression, defined by the dimensionless parameter

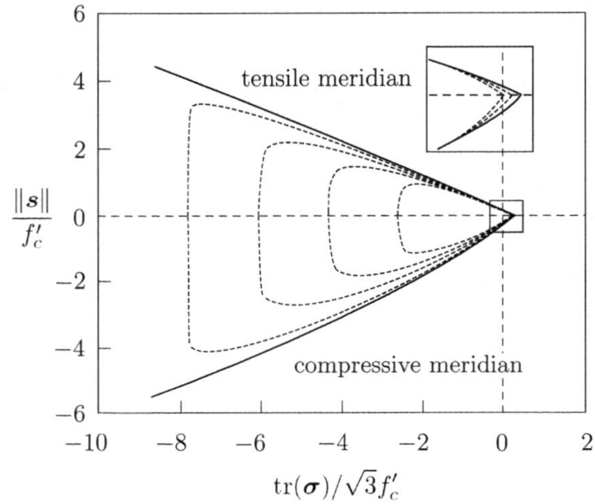

Fig. 6.9 Triaxial concrete plasticity model by Kang and Willam (1999): The yield surface $\widetilde{f} = 0$ is represented by the hardening caps (dashed curves), which collapse to the uncapped failure surface $F_1 = 0$ (solid curves) when the hardening parameter $k = 1$. Inset shows detail of the cohesion softening mechanism that reduces the concrete tensile strength to zero when the softening parameter $c = 0$. Note from the inset that the local cohesion softening is accompanied by a local friction hardening near the vertex.

S that locates the cap surface, and the dimensionless hardening parameter $k \in [0, 1]$.

The von Mises stress q_1 is selected so that the compressive meridian passes through the point of uniaxial compressive strength. The scaling function ζ follows the Willam-Warnke relation (6.37), but the ellipticity ρ now varies with the mean normal stress according to the relation

$$\rho = 1 - 0.5\left(\frac{p_0 - 5.5f'_c}{p - 5.5f'_c}\right). \tag{6.64}$$

This makes the ellipticity approach 1/2 near the region of tensile strength, and 1.0 as the confining pressure becomes very high. Hence, the cross-sections of the failure surface on the deviatoric plane are more triangular near the tensile strength region, and more circular at high pressures. These features are in agreement with the experimental data for plain concrete shown in Fig. 4.1.

In addition to the failure function F_1, the yield function $\widetilde{f}$ contains a compression cap given by the function

$$F_2(p, q, k) = -\frac{q_1}{f'_c}\left[\left(\frac{p - p_1}{p_0 - p_1}\right)^{\beta} - 1\right], \tag{6.65}$$

where the exponent β is given by the equation

$$\beta = \frac{1}{4}\left(\frac{1-k^2}{1-k_0^2}\right), \tag{6.66}$$

and where k_0 is a normalization parameter. The hardening parameter $k \in [0,1]$ increases monotonically with the cumulative plastic strain, as well as with the mean normal stress p to capture increased ductility under increased confinement, i.e.,

$$k = k(\varepsilon^{\mathrm{p}}, p), \qquad \varepsilon^{\mathrm{p}} = \int_t \|\dot{\boldsymbol{\epsilon}}^{\mathrm{p}}\| \, \mathrm{d}t, \tag{6.67}$$

where the plastic strain rate tensor $\dot{\boldsymbol{\epsilon}}^{\mathrm{p}}$ may contain both volumetric and deviatoric parts. It may be useful to compare the above hardening law with (6.28) and (6.61) for clay and sand, respectively. When the hardening parameter k reaches the limiting value of unity, the compression cap disappears, and the yield surface coincides with the failure surface.

The function F_3 is a mathematical description of softening induced by tensile decohesion on the failure surface, and takes the form

$$F_3(p,q,c) = \frac{q_1}{f_c'}(1-c)\left(\frac{p_0}{p_0 - p_1}\right)^{\alpha}\left(\frac{\vartheta - p}{\vartheta}\right)^2, \tag{6.68}$$

where $c \in [0,1]$ is a monotonically decreasing function of the equivalent tensile fracture strain, or crack opening displacement, and ϑ defines the transition point of brittle-ductile failure. The softening mechanism defines the movement of the vertex p_0 toward zero, signifying the loss of tensile strength with plastic deformation. The hardening surface is inactive during the softening process.

The Kang-Willam concrete model uses linear elasticity with constant elastic moduli. The flow rule is nonassociative with a plastic potential function that has a similar shape to the yield function, but is flatter on the p, q plane to avoid predicting excessive plastic dilatancy. The surfaces are all smooth except at the vertex.

6.9 Return Mapping in Strain Space

With a stress-dependent elasticity tensor one loses the advantage of an additive elastic-plastic operator split in stress space, since the predictor and corrector stresses are now related to the elastic and plastic strains in a nonlinear way. However, in the infinitesimal regime the additive split is still valid in the strain space. In the finite deformation regime, we have seen that the additive split is also valid in the space of principal logarithmic strains. In general, a return mapping algorithm in the strain space is more robust than the classical return mapping in the stress space because it can readily accommodate nonlinear elasticity.

Consider the strain tensor decomposition

$$\boldsymbol{\epsilon}^{\mathrm{e}}_{n+1} = \boldsymbol{\epsilon}^{\mathrm{e\,tr}}_{n+1} - \Delta\lambda \frac{\partial g}{\partial \boldsymbol{\sigma}_{n+1}}, \qquad \boldsymbol{\epsilon}^{\mathrm{e\,tr}}_{n+1} = \boldsymbol{\epsilon}^{\mathrm{e}}_{n} + \Delta\boldsymbol{\epsilon}, \tag{6.69}$$

where g is the plastic potential function and $\Delta\lambda$ is the discrete plastic multiplier. We readily see an additive split in the incremental strains irrespective of the elastic constitutive behavior. We view $\boldsymbol{\epsilon}^{\mathrm{e\,tr}}_{n+1}$ as an elastic strain predictor that would be the final elastic strain if the deformation was all elastic. Note that if a constant elasticity tensor $\boldsymbol{c}^{\mathrm{e}}$ operates on both sides of the above equation, the standard (additive) predictor-corrector split in stress space is recovered. Irrespective of $\boldsymbol{c}^{\mathrm{e}}$, the elastic predictor stress can be calculated from the hyperelastic equation

$$\boldsymbol{\sigma}^{\mathrm{tr}}_{n+1} = \left.\frac{\partial \psi^{\mathrm{e}}(\boldsymbol{\epsilon}^{\mathrm{e}})}{\partial \boldsymbol{\epsilon}^{\mathrm{e}}}\right|_{\boldsymbol{\epsilon}^{\mathrm{e\,tr}}_{n+1}}. \tag{6.70}$$

For isotropic hyperelasto-plasticity it suffices to work in principal axes. In this case, (6.69) becomes (see Section 4.4 for notation)

$$\sum_{A=1}^{3} \epsilon^{\mathrm{e}}_{A} \boldsymbol{n}^{(A)} \otimes \boldsymbol{n}^{(A)} = \sum_{A=1}^{3} \epsilon^{\mathrm{e\,tr}}_{A} \boldsymbol{n}^{\mathrm{tr}(A)} \otimes \boldsymbol{n}^{\mathrm{tr}(A)} - \Delta\lambda \sum_{A=1}^{3} \frac{\partial \widehat{g}}{\partial \sigma_A} \boldsymbol{n}^{(A)} \otimes \boldsymbol{n}^{(A)}, \tag{6.71}$$

after dropping the subscript '$n+1$' again for brevity. The above equation indicates that $\boldsymbol{n}^{(A)} \otimes \boldsymbol{n}^{(A)} = \boldsymbol{n}^{\mathrm{tr}(A)} \otimes \boldsymbol{n}^{\mathrm{tr}(A)}$. Therefore, we can simply work with principal values and write

$$\epsilon^{\mathrm{e}}_{A} = \epsilon^{\mathrm{e\,tr}}_{A} - \Delta\lambda \frac{\partial \widehat{g}}{\partial \sigma_A}, \qquad A = 1, 2, 3. \tag{6.72}$$

To complete the formulation of the local nonlinear problem, we need to impose the discrete hardening law, which we write symbolically as

$$\widehat{l}(\sigma_1, \sigma_2, \sigma_3, \kappa, \Delta\lambda) = \widehat{L}(\epsilon^{\mathrm{e}}_1, \epsilon^{\mathrm{e}}_2, \epsilon^{\mathrm{e}}_3, \kappa, \Delta\lambda) = 0. \tag{6.73}$$

We also need to impose the discrete consistency condition, which we write symbolically as

$$\widehat{f}(\sigma_1, \sigma_2, \sigma_3, \kappa) = \widehat{F}(\epsilon^{\mathrm{e}}_1, \epsilon^{\mathrm{e}}_2, \epsilon^{\mathrm{e}}_3, \kappa) = 0. \tag{6.74}$$

Note that these last two equations are specific to the constitutive model.

The next step is to solve the local nonlinear problem. This is done by first assembling the local residual vector $\boldsymbol{r}$ and the vector of unknowns $\boldsymbol{x}$ as

$$\boldsymbol{r}(\boldsymbol{x}) = \left\{ \begin{array}{c} \epsilon^{\mathrm{e}}_1 - \epsilon^{\mathrm{e\,tr}}_1 + \Delta\lambda g_1 \\ \epsilon^{\mathrm{e}}_2 - \epsilon^{\mathrm{e\,tr}}_2 + \Delta\lambda g_2 \\ \epsilon^{\mathrm{e}}_3 - \epsilon^{\mathrm{e\,tr}}_3 + \Delta\lambda g_3 \\ \widehat{L}(\epsilon^{\mathrm{e}}_1, \epsilon^{\mathrm{e}}_2, \epsilon^{\mathrm{e}}_3, \kappa, \Delta\lambda) \\ \widehat{F}(\epsilon^{\mathrm{e}}_1, \epsilon^{\mathrm{e}}_2, \epsilon^{\mathrm{e}}_3, \kappa) \end{array} \right\}, \qquad \boldsymbol{x} = \left\{ \begin{array}{c} \epsilon^{\mathrm{e}}_1 \\ \epsilon^{\mathrm{e}}_2 \\ \epsilon^{\mathrm{e}}_3 \\ \kappa \\ \Delta\lambda \end{array} \right\}, \tag{6.75}$$

where $g_A = \partial\widehat{g}/\partial\sigma_A$ for $A = 1, 2, 3$.

We have five nonlinear equations in five unknowns, for which we can use Newton's method to solve for the unknowns. To this end, we need to construct the local Jacobian

$$\boldsymbol{A}^k = \boldsymbol{r}'(\boldsymbol{x}^k) = \begin{bmatrix} c_{11} & c_{12} & c_{13} & \Delta\lambda\partial g_1/\partial\kappa & g_1 \\ c_{21} & c_{22} & c_{23} & \Delta\lambda\partial g_2/\partial\kappa & g_2 \\ c_{31} & c_{32} & c_{33} & \Delta\lambda\partial g_3/\partial\kappa & g_3 \\ \partial\widehat{L}/\partial\epsilon_1^{\mathrm{e}} & \partial\widehat{L}/\partial\epsilon_2^{\mathrm{e}} & \partial\widehat{L}/\partial\epsilon_3^{\mathrm{e}} & \partial\widehat{L}/\partial\kappa & \partial\widehat{L}/\partial\Delta\lambda \\ \partial\widehat{F}/\partial\epsilon_1^{\mathrm{e}} & \partial\widehat{F}/\partial\epsilon_2^{\mathrm{e}} & \partial\widehat{F}/\partial\epsilon_3^{\mathrm{e}} & \partial\widehat{F}/\partial\kappa & 0 \end{bmatrix},$$

$$c_{IJ} = \delta_{IJ} + \Delta\lambda\frac{\partial g_I}{\partial\epsilon_J^{\mathrm{e}}}, \qquad I, J = 1, 2, 3. \tag{6.76}$$

Recognizing that the yield and plastic potential functions are given in terms of the stresses and not of the elastic strains, we evaluate the strain gradients using the chain rule as

$$\frac{\partial\widehat{F}}{\partial\epsilon_I^{\mathrm{e}}} = \sum_{A=1}^{3}\frac{\partial\widehat{f}}{\partial\sigma_A}\frac{\partial^2\widehat{\psi}^{\mathrm{e}}}{\partial\epsilon_A^{\mathrm{e}}\partial\epsilon_I^{\mathrm{e}}}, \qquad \frac{\partial g_I}{\partial\epsilon_J^{\mathrm{e}}} = \sum_{A=1}^{3}\frac{\partial g_I}{\partial\sigma_A}\frac{\partial^2\widehat{\psi}^{\mathrm{e}}}{\partial\epsilon_A^{\mathrm{e}}\partial\epsilon_J^{\mathrm{e}}}. \tag{6.77}$$

The algorithm is summarized in Box 6.1. Note that a return map on the principal axes reduces the number of local unknowns, as well as ameliorates the treatment of the third stress invariant. The algorithm does not require that the Cauchy stresses be stored, rather, they are simply calculated from the elastic strains. From Chapter 5, we note that this approach is similar to the finite deformation algorithm where the Kirchhoff stress tensor $\boldsymbol{\tau}$ is simply calculated from the left Cauchy-Green deformation tensor $\boldsymbol{b}^{\mathrm{e}}$.

Step 1. Compute $\boldsymbol{\epsilon}_{n+1}^{\mathrm{e\,tr}} = \boldsymbol{\epsilon}_n^{\mathrm{e}} + \Delta\boldsymbol{\epsilon}$
Step 2. Spectrally decompose $\boldsymbol{\epsilon}_{n+1}^{\mathrm{e\,tr}} = \sum_{A=1}^{3}\epsilon_A^{\mathrm{e\,tr}}\boldsymbol{n}^{\mathrm{tr}(A)}\otimes\boldsymbol{n}^{\mathrm{tr}(A)}$
Step 3. Compute $\sigma_A^{\mathrm{tr}} = \partial\widehat{\psi}^{\mathrm{e}}/\partial\epsilon_A^{\mathrm{e}}$ at $\boldsymbol{\epsilon}_{n+1}^{\mathrm{e\,tr}}$
Step 4. Check $\widehat{f}(\sigma_1^{\mathrm{tr}}, \sigma_2^{\mathrm{tr}}, \sigma_3^{\mathrm{tr}}, \kappa_n) > 0$?
No, set $\boldsymbol{\sigma}_{n+1} = \sum_{A=1}^{3}\sigma_A^{\mathrm{tr}}\boldsymbol{n}^{\mathrm{tr}(A)}\otimes\boldsymbol{n}^{\mathrm{tr}(A)}$ and exit.
Step 5. Yes, solve for ϵ_1^{e}, ϵ_2^{e}, ϵ_3^{e}, and κ such that $\widehat{f}(\sigma_1, \sigma_2, \sigma_3, \kappa) = 0$
Step 6. Compute $\boldsymbol{\sigma}_{n+1} = \sum_{A=1}^{3}(\partial\widehat{\psi}^{\mathrm{e}}/\partial\epsilon_A^{\mathrm{e}})\boldsymbol{n}^{\mathrm{tr}(A)}\otimes\boldsymbol{n}^{\mathrm{tr}(A)}$ and exit.

Box 6.1 Strain-space return mapping in principal axes for isotropic hyperelastic-plastic model

The algorithm shown in Box 6.1 is useful for evaluating the consistent tangent operator $\boldsymbol{c}_{n+1}$ for global Newton iteration. The form for $\boldsymbol{c}_{n+1}$ is given in equation (4.119), where the algorithmic tangent matrix in principal axes is determined from the relation

$$a_{AB} := \frac{\partial \sigma_A}{\partial \epsilon_B^{\mathrm{e\,tr}}} = \sum_{C=1}^{3} \frac{\partial \sigma_A}{\partial \epsilon_C^{\mathrm{e}}} \frac{\partial \epsilon_C^{\mathrm{e}}}{\partial \epsilon_B^{\mathrm{e\,tr}}} = \sum_{C=1}^{3} \left(\frac{\partial^2 \widehat{\psi}^{\mathrm{e}}}{\partial \epsilon_A^{\mathrm{e}} \partial \epsilon_C^{\mathrm{e}}} \right) \frac{\partial \epsilon_C^{\mathrm{e}}}{\partial \epsilon_B^{\mathrm{e\,tr}}} \,. \tag{6.78}$$

Since $\boldsymbol{r}(\boldsymbol{x}^*) = \boldsymbol{0}$ at the locally converged state, it follows that

$$\frac{\partial r_I}{\partial \epsilon_B^{\mathrm{e\,tr}}} = \sum_{C=1}^{5} \underbrace{\frac{\partial r_I}{\partial x_C}\bigg|_{\epsilon_{n+1}^{\mathrm{e\,tr}}}}_{A_{IC}^*} \frac{\partial x_C}{\partial \epsilon_B^{\mathrm{e\,tr}}} + \underbrace{\frac{\partial r_I}{\partial \epsilon_B^{\mathrm{e\,tr}}}\bigg|_{\boldsymbol{x}}}_{-\delta_{IB}} = 0 \,, \tag{6.79}$$

for $I = 1, \ldots, 5$; $B = 1, 2, 3$. Inverting the above equation gives

$$\frac{\partial x_C}{\partial \epsilon_B^{\mathrm{e\,tr}}} = - \sum_{I=1}^{5} (A_{CI}^*)^{-1} \frac{\partial r_I}{\partial \epsilon_B^{\mathrm{e\,tr}}}\bigg|_{\boldsymbol{x}} \,. \tag{6.80}$$

The components of $\partial x_C / \partial \epsilon_B^{\mathrm{e\,tr}}$ provide the required terms $\partial \epsilon_C^{\mathrm{e}} / \partial \epsilon_B^{\mathrm{e\,tr}}$ appearing in expression (6.78) for the algorithmic tangent matrix a_{AB}.

Exercise 6.5. For a two-invariant plasticity model the return mapping algorithm in elastic strain invariant space is given by

$$\epsilon_v^{\mathrm{e}} = \epsilon_v^{\mathrm{e\,tr}} - \Delta\lambda v \,, \qquad \epsilon_s^{\mathrm{e}} = \epsilon_s^{\mathrm{e\,tr}} - \Delta\lambda e \,,$$

where $v = \partial \widetilde{g} / \partial p$ and $e = \partial \widetilde{g} / \partial q$ define the plastic flow directions in the two invariant spaces, which are both evaluated implicitly at the final converged state. Take the final Cauchy stress as $\boldsymbol{\sigma} = p\boldsymbol{1} + \sqrt{2/3}q\widehat{\boldsymbol{n}}$. Determine the final values of p, q, and $\widehat{\boldsymbol{n}}$ from the converged values of ϵ_v^{e} and ϵ_s^{e}.

6.10 Finite Deformation

In this section we develop a finite deformation version of the MCC model. Similar formulations can be used for other isotropic constitutive models such as those described in the previous sections.

We recall from Section 5.5 the various levels of representation for the stored energy function in the finite deformation range,

$$\Psi^{\mathrm{e}}(\boldsymbol{X}, \boldsymbol{F}) \quad \xrightarrow{\text{frame invariance}} \quad \widetilde{\Psi}^{\mathrm{e}}(\boldsymbol{X}, \boldsymbol{C}) \quad \xrightarrow{\text{isotropy}} \quad \psi^{\mathrm{e}}(\boldsymbol{X}, \boldsymbol{b}) \,. \tag{6.81}$$

The relations shown above state that frame invariance and isotropy allow the replacement of $\Psi^{\mathrm{e}}(\boldsymbol{X}, \boldsymbol{F})$ with $\psi^{\mathrm{e}}(\boldsymbol{X}, \boldsymbol{b})$, where the latter energy function can be represented in terms of the eigenvalues of $\boldsymbol{b}$. From multiplicative plasticity, we can replace $\boldsymbol{b}$ with $\boldsymbol{b}^{\mathrm{e}}$ and use either the principal logarithmic stretches or the invariants,

$$\psi^{\mathrm{e}}(\boldsymbol{X}, \boldsymbol{b}^{\mathrm{e}}) \quad \longrightarrow \quad \widehat{\psi}^{\mathrm{e}}(\boldsymbol{X}, \epsilon_1^{\mathrm{e}}, \epsilon_2^{\mathrm{e}}, \epsilon_3^{\mathrm{e}}) \quad \longrightarrow \quad \widetilde{\psi}^{\mathrm{e}}(\boldsymbol{X}, \epsilon_v^{\mathrm{e}}, \epsilon_s^{\mathrm{e}}) \,. \tag{6.82}$$

In the above representation, we assume that $\widetilde{\psi}^{\mathrm{e}}$ depends only on the first two strain invariants, as before.

In the finite deformation range, the geometric meanings of the strain invariants are as follows. The first invariant is the sum of the elastic logarithmic principal stretches,

$$\epsilon_v^{\mathrm{e}} = \sum_{A=1}^{3} \epsilon_A^{\mathrm{e}} = \sum_{A=1}^{3} \log(\lambda_A^{\mathrm{e}}) = \log(\lambda_1^{\mathrm{e}}\lambda_2^{\mathrm{e}}\lambda_3^{\mathrm{e}}) = \log(J^{\mathrm{e}})\,, \tag{6.83}$$

where $J^{\mathrm{e}} = \det(\boldsymbol{F}^{\mathrm{e}})$. From this, we derive the deviatoric principal logarithmic stretches

$$e_A^{\mathrm{e}} = \epsilon_A^{\mathrm{e}} - \frac{1}{3}\epsilon_v^{\mathrm{e}}\,, \qquad A = 1,2,3\,, \tag{6.84}$$

which represent the principal volume-preserving logarithmic stretches (Simo 1992) in the sense

$$e_A^{\mathrm{e}} = \log(\ell_A^{\mathrm{e}}) \quad \Longrightarrow \quad \ell_A^{\mathrm{e}} = (J^{\mathrm{e}})^{-1/3}\lambda_A^{\mathrm{e}}\,, \qquad A = 1,2,3\,. \tag{6.85}$$

Indeed, we see that $\ell_1^{\mathrm{e}}\ell_2^{\mathrm{e}}\ell_3^{\mathrm{e}} \equiv 1$. By analogy with equation (6.9), we can define the equivalent deviatoric strain invariant

$$\epsilon_s^{\mathrm{e}} = \sqrt{\frac{2}{3}}\Big(\sum_{A=1}^{3} e_A^{\mathrm{e}} e_A^{\mathrm{e}}\Big)^{1/2}\,. \tag{6.86}$$

Focusing on a stored energy function that reproduces a pressure-dependent elastic bulk modulus, we rewrite equation (6.31) as

$$\widetilde{\psi}^{\mathrm{e}}(\epsilon_v^{\mathrm{e}}, \epsilon_s^{\mathrm{e}}) = -P_0\widehat{c}_r \exp\left(\frac{\epsilon_{v0} - \epsilon_v^{\mathrm{e}}}{\widehat{c}_r}\right) + \frac{3}{2}\mu(\epsilon_s^{\mathrm{e}})^2 + \widetilde{\psi}_0\,, \tag{6.87}$$

where $P_0 < 0$ is the reference Kirchhoff pressure and $\widehat{c}_r$ is the elastic compressibility index in Kirchhoff stress space. We note at the outset that the above equation is not an acceptable model of elasticity for extreme strains. This can be seen easily if we substitute $\log(J^{\mathrm{e}})$ for ϵ_v^{e}. In order for $\widetilde{\psi}^{\mathrm{e}}$ to replicate the behavior at extreme strains, $\widetilde{\psi}^{\mathrm{e}} \to \infty$ as $J^{\mathrm{e}} \to 0$ (satisfied), and $\widetilde{\psi}^{\mathrm{e}} \to \infty$ as $J^{\mathrm{e}} \to \infty$ (not satisfied). The above form of $\widetilde{\psi}^{\mathrm{e}}$ is valid only within a certain range of elastic deformations, and where the stored energy is a convex function of the elastic logarithmic principal stretches.

Next, we turn to the yield function for the MCC model. The first two Kirchhoff stress invariants can be evaluated from the hyperelastic constitutive equations

$$P = \frac{\partial\widetilde{\psi}^{\mathrm{e}}}{\partial\epsilon_v^{\mathrm{e}}}\,, \qquad Q = \frac{\partial\widetilde{\psi}^{\mathrm{e}}}{\partial\epsilon_s^{\mathrm{e}}}\,. \tag{6.88}$$

The third invariant is Lode's angle θ (see Chapter 4), which is used along with the ellipticity ρ to evaluate the scaling function $\zeta = \zeta(\rho, \theta)$. Thus, the yield function in Kirchhoff stress space for the MCC model takes the form

$$\widetilde{f} = \zeta^2 \frac{Q^2}{M^2} + P(P - P_c) \leq 0\,, \tag{6.89}$$

where $P_c < 0$ is the Kirchhoff preconsolidation stress.

The principal Kirchhoff stresses can be determined from the chain rule

$$\tau_A = \frac{\partial \widehat{\psi}^{\mathrm{e}}}{\partial \epsilon^{\mathrm{e}}_A} = P\frac{\partial \epsilon^{\mathrm{e}}_v}{\partial \epsilon^{\mathrm{e}}_A} + Q\frac{\partial \epsilon^{\mathrm{e}}_s}{\partial \epsilon^{\mathrm{e}}_A}\,, \tag{6.90}$$

where

$$\frac{\partial \epsilon^{\mathrm{e}}_v}{\partial \epsilon^{\mathrm{e}}_A} = 1\,, \qquad \frac{\partial \epsilon^{\mathrm{e}}_s}{\partial \epsilon^{\mathrm{e}}_A} = \frac{2}{3} e^{\mathrm{e}}_A / \epsilon^{\mathrm{e}}_s\,. \tag{6.91}$$

From these, we can construct the full Kirchhoff stress tensor using the spectral representation

$$\boldsymbol{\tau} = \sum_{A=1}^{3} \tau_A \boldsymbol{n}^{(A)} \otimes \boldsymbol{n}^{(A)}\,, \tag{6.92}$$

where $\boldsymbol{n}^{(A)} \otimes \boldsymbol{n}^{(A)}$ are the spectral directions of $\boldsymbol{b}^{\mathrm{e}}$.

Finally, we consider the hardening law for the MCC model. The product decomposition of the deformation gradient produces the identity

$$J = \det(\boldsymbol{F}) = \det(\boldsymbol{F}^{\mathrm{e}})\det(\boldsymbol{F}^{\mathrm{p}}) = J^{\mathrm{e}} J^{\mathrm{p}}\,, \tag{6.93}$$

where $J^{\mathrm{e}} = \det(\boldsymbol{F}^{\mathrm{e}})$ and $J^{\mathrm{p}} = \det(\boldsymbol{F}^{\mathrm{p}})$. Taking the natural logarithms yields

$$\log(J) = \log(J^{\mathrm{e}}) + \log(J^{\mathrm{p}}) \qquad \Longleftrightarrow \qquad \epsilon_v = \epsilon^{\mathrm{e}}_v + \epsilon^{\mathrm{p}}_v\,. \tag{6.94}$$

The rate form of the above equation is

$$\frac{\dot{J}}{J} = \frac{\dot{J}^{\mathrm{e}}}{J^{\mathrm{e}}} + \frac{\dot{J}^{\mathrm{p}}}{J^{\mathrm{p}}} \qquad \Longleftrightarrow \qquad \dot{\epsilon}_v = \dot{\epsilon}^{\mathrm{e}}_v + \dot{\epsilon}^{\mathrm{p}}_v\,. \tag{6.95}$$

Since $J = \overline{v}/\overline{v}_{\mathrm{ref}}$ and $\dot{J} = \dot{\overline{v}}/\overline{v}_{\mathrm{ref}}$, where $\overline{v}_{\mathrm{ref}}$ is the fixed reference specific volume, the hardening law (6.24) extends to the finite deformation range in the form

$$\frac{\dot{J}}{J} = -\widehat{c}_c \frac{\dot{P}_c}{P_c}\,, \tag{6.96}$$

where $\widehat{c}_c > 0$ is the compressibility index in Kirchhoff space. To obtain the plastic part, we subtract the elastic rebound. According to equation (6.26), the elastic rebound is of the form

$$\frac{\dot{J}^{\mathrm{e}}}{J^{\mathrm{e}}} = -\widehat{c}_r \frac{\dot{P}_c}{P_c}\,. \tag{6.97}$$

Subtracting (6.97) from (6.96) yields the hardening law in the finite deformation range,

$$\dot{P}_c = -\frac{\dot{J}^{\mathrm{p}}/J^{\mathrm{p}}}{\widehat{c}_c - \widehat{c}_r} P_c = -\frac{\dot{\epsilon}^{\mathrm{p}}_v}{\widehat{c}_c - \widehat{c}_r} P_c\,. \tag{6.98}$$

The integral is a simple exponential function of the form

$$P_c = P_{c,n} \exp\left(-\frac{\epsilon_v^{\mathrm{p}} - \epsilon_{v,n}^{\mathrm{p}}}{\widehat{C}_c - \widehat{C}_r} \right) = P_{c,n} \exp\left(-\frac{\epsilon_v^{\mathrm{e\,tr}} - \epsilon_v^{\mathrm{e}}}{\widehat{C}_c - \widehat{C}_r} \right). \tag{6.99}$$

It may be worthwhile to verify the last part of the above equation, see Exercise 6.7.

Exercise 6.6. Consider a stored energy function of the form (cf. Exercise 5.5)

$$\widehat{\psi}^{\mathrm{e}} = \frac{1}{2}\lambda(\epsilon_1^{\mathrm{e}} + \epsilon_2^{\mathrm{e}} + \epsilon_3^{\mathrm{e}})^2 + \mu[(\epsilon_1^{\mathrm{e}})^2 + (\epsilon_2^{\mathrm{e}})^2 + (\epsilon_3^{\mathrm{e}})^2] ,$$

where $\epsilon_A^{\mathrm{e}} = \log(\lambda_A^{\mathrm{e}})$ for $A = 1, 2, 3$; and λ and μ are the Lamé parameters. Express this function in terms of $\log(J^{\mathrm{e}})$ and the principal volume-preserving logarithmic stretches ℓ_A^{e}, for $A = 1, 2, 3$. Then, show that this function predicts the correct behavior for extreme strains. The above expression for $\widehat{\psi}^{\mathrm{e}}$ is very intuitive to use, but, unfortunately, it is not a convex function of J^{e} (Simo 1992).

Exercise 6.7. In equation (6.99), show that $\epsilon_v^{\mathrm{p}} - \epsilon_{v,n}^{\mathrm{p}} = \epsilon_v^{\mathrm{e\,tr}} - \epsilon_v^{\mathrm{e}}$.

6.11 Plane Strain Compression of a Sand

In this section, we present a rare simulation of the mechanical behavior of a specimen of loose silica-concrete sand with variable density. Simulations involving variable density are uncommon, first, because density as a state variable is rarely integrated into the constitutive formulation, and, second, because one has to measure and quantify the spatial density variation within the domain of analysis to make the simulation meaningful. Current advances in digital imaging have made the latter process feasible, since we now have the technology to quantify the spatial variation of density noninvasively. In the example that follows, we use X-Ray Computed Tomography (CT) and digital image processing to quantify the spatial density variation in a rectangular specimen of sand. The simulation involves axially compressing this heterogeneous specimen under plane strain loading condition.

Figure 6.10 portrays the laboratory measured spatial density variation in the silica-concrete sand expressed in terms of specific volume. The higher the specific volume, the looser the sand. The black and white image is a negative image, i.e., darker regions delineate looser zones. Density is superimposed on two finite element meshes consisting of four-node quadrilateral elements integrated with a standard 2 × 2 rule. For reference, similar finite element simulations were conducted by Borja et al. (2013) using constant strain triangular elements. The refined mesh has an equally refined description of heterogeneity, making the mesh sensitivity study more challenging. The specimen is first subjected to an all-around confinement of 30 kPa, and then compressed vertically in plane strain under displacement control. The bottom side of the finite element mesh is supported on vertical rollers to simulate a sled used during actual testing. The top side is also supported on rollers except the upper left-hand corner is pinned to the support. Stress histories are all stored at the Gauss integration points. For the most part, stresses in the looser regions lie on the compression cap of the yield surface.

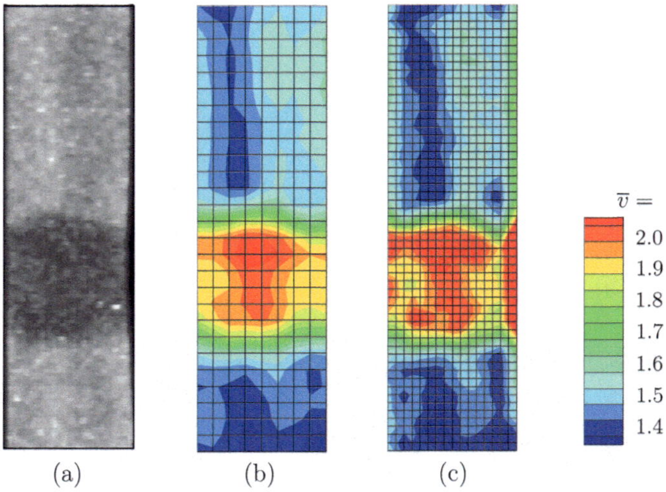

Fig. 6.10 Spatial variation of specific volume $\overline{v}$ in a rectangular specimen of silica-concrete sand: (a) X-Ray CT image; (b) coarser mesh with 252 nodes; (c) finer mesh with 952 nodes. The specimen is 137 mm tall by 40 mm wide by 80 mm deep (out-of-plane).

We employ the sand constitutive model described in Section 6.7 for the simulations. The complete set of material parameters can be found in Borja et al. (2013). We use a finite deformation formulation based on multiplicative plasticity similar to the development presented in the previous section for the MCC model. Movement of the compression cap is driven by the deviatoric component of the plastic strain. A distinctive feature of the model is a state parameter that allows the specific volume, as a state variable, to be 'detached' from the critical state line (Borja and Andrade 2006). This state parameter determines whether the stress point lies below or above the critical state line. Depending on the sign of this state parameter, the compression cap can either expand or shrink.

Figure 6.11 shows the deformed meshes and contours of deviatoric strains for the two finite element meshes after a vertical compression of 6 mm. Both meshes suggest the development of a shear band, a localized deformation style discussed further in Chapters 7 and 9. The resulting force-compression curves are portrayed in Fig. 6.12, indicating nearly identical responses predicted by the two meshes prior to the peak load. This suggests a lack of mesh sensitivity during the early part of loading up until the peak load. Beyond the peak load, the two curves diverge from each other, suggesting some mesh sensitivity issues afflicting the solution. Similar simulations presented by

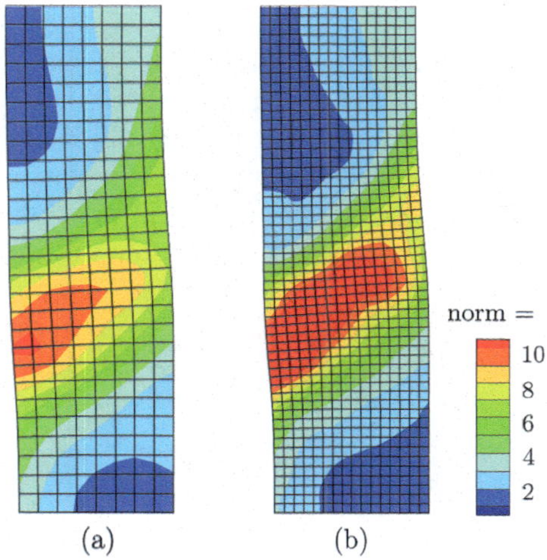

Fig. 6.11 Deformed meshes with contours of deviatoric strain at a total vertical compression of 6 mm: (a) coarser mesh; (b) finer mesh. Color bar is norm of deviatoric principal logarithmic stretches in percent.

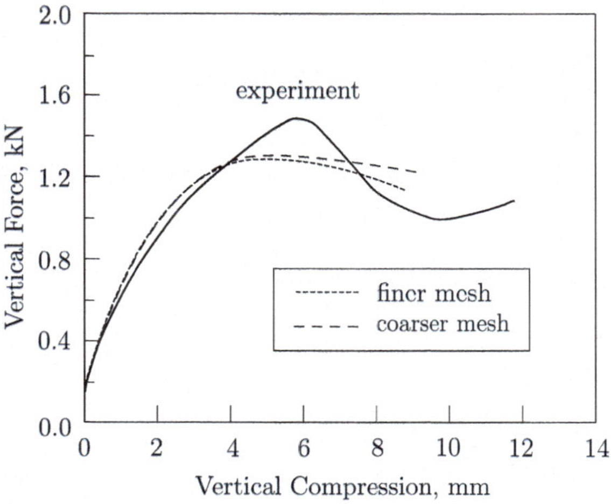

Fig. 6.12 Resultant force versus compression curves for plane strain compression example. Numerical solutions exhibit mesh sensitivity beyond a vertical compression of around 6 mm.

Borja et al. (2013) indicate a more dramatic manifestation of mesh sensitivity of the solution at post-peak when constant strain triangular elements are used. Chapter 9 addresses this aspect from the point of view of bifurcation of the solution typically associated with softening responses.

6.12 Summary

Hyperelasticity is a sound framework for accommodating nonlinear elasticity, since it ensures that no energy is generated or lost from a closed elastic loading cycle. Plasticity models can be embedded easily within the hyperelastic framework. For the class of hyperelastic-plastic models considered in this chapter, a return-mapping algorithm in the elastic strain space provides a robust framework for stress-point integration. The technique is compatible with the algorithm employed for stress-point integration in the finite deformation regime, where the return mapping is performed in the deformation space rather than in the stress space; more specifically, in the space defined by the elastic logarithmic principal stretches.

7 Discontinuities

7.1 Introduction

Material failure and damage typically involve some form of discontinuous deformation over a narrow zone. The mechanisms of deformation within this zone can be very complex, and they occur at multiple scales. Consider a geologic fault, as an example, which evolves from shearing across a series of échelon joints forming pockets of highly damaged rock. As shearing progresses, a fault core, or cataclasite zone, develops between surrounding less damaged zones made up of joints and sheared joints. As one moves farther away from the fault core, the rock becomes less and less damaged, until one finds the competent host rock that marks the end of the fault zone (Aydin et al. 2006). Most of the offset or slip between the surrounding rock occurs at the cataclasite zone where rocks could be reduced to a very fine-grained clay-like material due to all the slipping and grinding taking place (Chester et al. 1993). Because the fine-grained materials in the cataclasite zone may be considered frictional in nature, fault zones are also considered frictional. However, due to the multi-scale nature of deformation, the effective friction coefficient on a fault must be derived from upscaling the frictional properties of the fine-grained materials comprising the fault core. In general, the coefficient of friction of a fault is expected to vary with degree of damage on the host rock.

Figure 7.1 shows different levels of representation of a geologic fault, from a highly simplified representation consisting of one single crack, to a much more complex representation involving a gouge zone or fault core (black) containing granulized particles and surrounded by fractures and cracks. On a smaller scale, fractures and cracks are much narrower zones of discontinuity where adjacent surfaces either separate or slide past each other. But as the degree of damage increases, they could become part of a thicker fault core composed of discrete elements that do not have any semblance to the original cracks. Because of nearly overlapping definitions and variations in scale, qualitative descriptions of failure modes are quite artificial. However, we distinguish between two mechanisms of deformation: a continuum localized mode where the two sides of damage zone are in direct physical contact; and a separation mode where traction-free surfaces are created. Examples of continuum localized deformation include faults and deformation bands (Borja and Aydin 2004, Borja 2004, Scholz 1990); fractures and cracks are examples

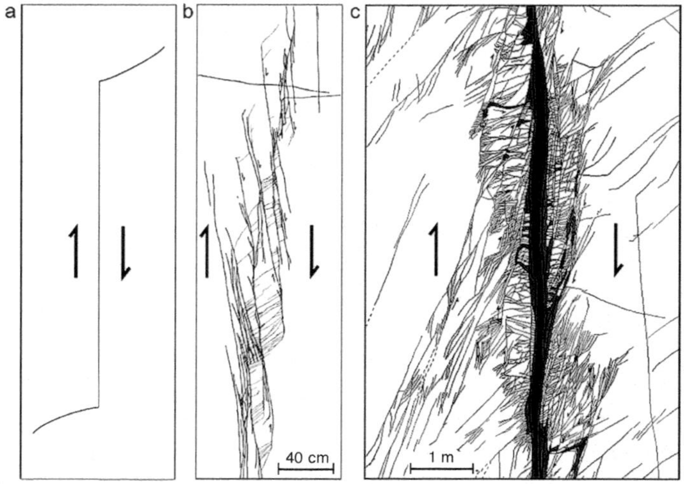

Fig. 7.1 Sketches depicting fault development by sliding along preexisting shear planes: (a) simplified representation showing a main fault and wing cracks; (b) shearing across a series of échelon joints; (c) detailed representation of the fault core (black) and surrounding damage zones. Image reproduced from Aydin et al. (2006).

of opening mode (Broek 1986). In this chapter, we use the finite element method as a tool to capture these two localized deformation modes.

Standard finite element techniques employ a conforming interpolation of displacement and must be 'enhanced' to accommodate discontinuous deformation. If the discontinuity is known right from the beginning, we can align the sides of the finite elements to conform with the discontinuity and employ classic nonlinear contact mechanics techniques (Laursen 2003, Wriggers 2006, Sanz et al. 2007). However, cracks can nucleate and propagate in random directions, in which case, it would not be possible to specify the sliding surfaces a priori. Adaptive remeshing has been used in the past to allow the element sides to adaptively define the geometry of a randomly propagating discontinuity. However, this approach is cumbersome to use, and problems arise when transferring information from the old mesh to the new mesh particularly in the presence of bulk plasticity.

This chapter highlights two finite element techniques where a jump in the displacement field is embedded into a fixed finite element mesh, called the background grid, to accommodate randomly propagating discontinuities. The first is a global enrichment technique based on the extended finite element method, where additional degrees of freedom are introduced into the existing nodes to interpolate the slip continuously. Higher resolution descriptions typically are achieved, particularly when combined with a crack-tip enrichment,

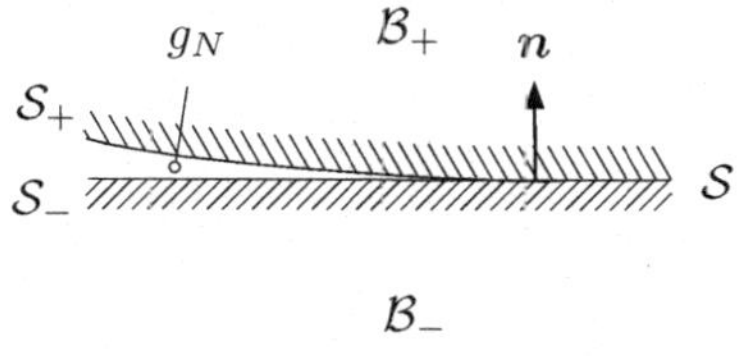

Fig. 7.2 Discontinuity $\mathcal{S}$ and gap function g_N

but the technique results in a larger system of equations to solve. The second is a local enrichment called the assumed enhanced strain method, where local degrees of freedom are added to the elements to enhance the displacement interpolation. In this technique, the additional degrees of freedom are eliminated on the element level and the global system of equations is unchanged. A robust contact algorithm is essential for either method, and below we outline important developments in the area of contact mechanics that are relevant to the two enhancement techniques.

7.2 Contact with Cohesion and Friction

Consider a discontinuity $\mathcal{S}$ defined by faces $\mathcal{S}_-$ and $\mathcal{S}_+$, which we call the master and slave faces, respectively. The corresponding sides are denoted by $\mathcal{B}_-$ and $\mathcal{B}_+$, see Fig. 7.2. Assuming that $\mathcal{S}$ is initially closed, the gap function at any point $\boldsymbol{x}$ on $\mathcal{S}$ may be obtained from the displacements of the contacting faces as

$$g_N = (\boldsymbol{u}_+ - \boldsymbol{u}_-) \cdot \boldsymbol{n} = \widetilde{\boldsymbol{u}} \cdot \boldsymbol{n}\,, \tag{7.1}$$

where $\boldsymbol{u}_-$ and $\boldsymbol{u}_+$ are the displacements of the master and slave faces, respectively, and $\boldsymbol{n}$ is the unit normal vector to $\mathcal{S}_-$ and pointing in the direction $\mathcal{S}_+$. The contact pressure between these two faces is given by

$$t_N = -\boldsymbol{\sigma} : (\boldsymbol{n} \otimes \boldsymbol{n})\,, \tag{7.2}$$

where $\boldsymbol{\sigma}$ is the Cauchy stress tensor at $\boldsymbol{x}$. The contact condition on $\mathcal{S}$ is described by the Kuhn-Tucker relations

$$t_N \geq 0\,, \qquad -g_N \leq 0\,, \qquad t_N g_N = 0\,. \tag{7.3}$$

We satisfy the no-contact condition when $-g_N < 0$ and $t_N = 0$, and the contact condition when $g_N = 0$ and $t_N > 0$.

Let $(\boldsymbol{m}_1, \boldsymbol{m}_2, \boldsymbol{n})$ form three mutually orthogonal unit vectors on $\mathcal{S}$ so that $\boldsymbol{m}_1$ and $\boldsymbol{m}_2$ are both tangent to $\mathcal{S}_-$. Under a contact condition the relative tangential displacement vector on $\mathcal{S}$ is given by

$$\boldsymbol{g}_T = \sum_{\alpha=1}^{2} \widetilde{\boldsymbol{u}} \cdot (\boldsymbol{m}_\alpha \otimes \boldsymbol{m}_\alpha) = \sum_{\alpha=1}^{2} (\widetilde{\boldsymbol{u}} \cdot \boldsymbol{m}_\alpha)\boldsymbol{m}_\alpha\,. \tag{7.4}$$

The tangential traction vector is

$$\boldsymbol{t}_T = \sum_{\alpha=1}^{2} (\boldsymbol{n} \cdot \boldsymbol{\sigma} \cdot \boldsymbol{m}_\alpha)\boldsymbol{m}_\alpha \,. \tag{7.5}$$

The norm of this vector, $\|\boldsymbol{t}_T\|$, is the resolved shear stress on $\mathcal{S}$.

We distinguish between stick and slip conditions. The criteria for stick and slip may be cast in the framework of classical plasticity theory using the standard cohesion-friction law

$$f = \|\boldsymbol{t}_T\| - \mu t_N - c \leq 0 \,, \tag{7.6}$$

where μ is the coefficient of friction and c is the cohesion. Employing a flow rule on this yield function defines the tangential relative velocity of sliding,

$$\dot{\boldsymbol{g}}_T = \dot{\zeta} \frac{\partial f}{\partial \boldsymbol{t}_T} = \dot{\zeta} \frac{\boldsymbol{t}_T}{\|\boldsymbol{t}_T\|} \,, \tag{7.7}$$

where $\dot{\zeta} = \|\dot{\boldsymbol{g}}_T\|$ is the slip speed. The flow rule then states that the slip velocity vector has the same direction as the tangential traction vector. The criteria for tangential stick/slip follows the standard Kuhn-Tucker conditions

$$\dot{\zeta} \geq 0 \,, \qquad f \leq 0 \,, \qquad \dot{\zeta} f = 0 \,. \tag{7.8}$$

We have a stick condition when $\dot{\zeta} = 0$ and $f < 0$, and a slip condition when $\dot{\zeta} > 0$ and $f = 0$.

The contact problem revolves around one of the most important material parameters in the scientific and engineering literature: the coefficient of friction μ. Wear changes the value of μ (Dieterich 1978), as do seismic rates (Di Toro et al. 2004, Hanks 1977). Laboratory tests on rough contacting surfaces sliding past each other have shown that the value of μ depends on the rate of sliding as well as on a state variable reflecting the maturity of contact. For slow rate of sliding, on the order of microns per second, a constitutive law in best agreement with experimental data is the Dieterich-Ruina friction law (Dieterich 1979, Ruina 1983), expressed as

$$\mu = \mu^* + A \log\left(\frac{V}{V^*}\right) + B \log\left(\frac{\theta}{\theta^*}\right) , \tag{7.9}$$

where V is slip speed, θ is a state variable, log is natural logarithm (log = ln) and A, B, μ^*, V^*, and θ^* are material parameters. Two of these material parameters (typically V^* and θ^*) can be assigned arbitrary values since they only serve as normalizing constants. The state variable θ is linked to the changing set of frictional contacts and wear on the material, and evolves according to the equation

$$\dot{\theta} = 1 - \frac{\theta V}{D_c} \,, \tag{7.10}$$

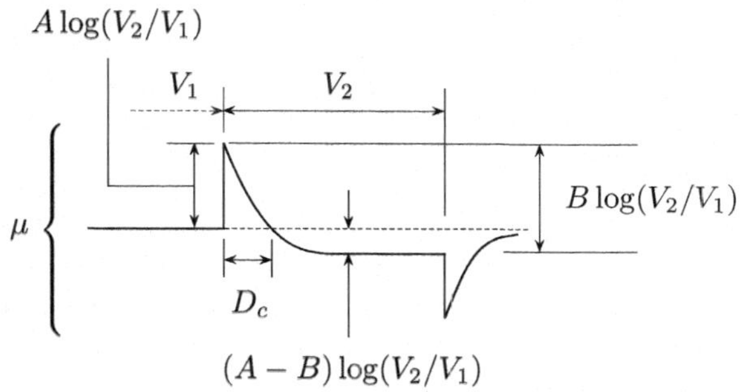

Fig. 7.3 Variation of coefficient of friction μ in a velocity stepping test ($V_2 > V_1$)

where D_c is a material parameter often referred to as the 'characteristic sliding distance' (Dieterich 1979). For seismic faulting, D_c is the critical slip distance over which strength breaks down during earthquake nucleation (Marone and Kilgore 1993).

Figure 7.3 depicts the physical significance of the constitutive law (7.9) and the characteristic slip D_c. At a constant sliding velocity V_1 the coefficient of friction is μ_1. After a velocity time stepping to a higher velocity V_2 the coefficient of friction instantaneously increases by an amount $A\log(V_2/V_1)$, and then decreases by an amount $B\log(V_2/V_1)$ for a net decrease of $(A-B)\log(V_2/V_1)$. Typically $B > A$, so a stepping to a faster speed results in net decrease in the coefficient of friction. The reverse is true for a stepping to a slower slip speed, i.e., the coefficient of friction instantaneously decreases first but stabilizes to a higher value than the initial one prior to velocity stepping, for a net increase in the coefficient of friction.

Equation (7.9) is singular when the slip speed is zero. By interpreting frictional resistance as a rate process, this equation can be regularized to accommodate zero slip speed as follows

$$\mu = A\sinh^{-1}\left[\frac{V}{2V^*}\exp\left(\frac{\mu^* + B\log(\theta/\theta^*)}{A}\right)\right]. \tag{7.11}$$

The above equation predicts $\mu = 0$ when $V = 0$, implying that, statistically, forward and backward jumps in a rate process cancel when there is no directional energy. Therefore, constitutive law (7.11) can be viewed as the complete Arrhenius equation with forward and backward jumps in a rate process, whereas equation (7.9) can be considered as a truncated version of the Arrhenius equation considering forward jumps only (Chester and Higgs 1992, Lapusta et al. 2000, Rice and Cocco 2007).

7.3 Lagrange Multipliers Method

There are three commonly used methods for imposing the contact and stick/slip conditions: (a) the Lagrange multipliers method, (b) the penalty method, and (c) the augmented Lagrangian method. The Lagrange multipliers method imposes the contact condition exactly; however, it is unwieldy to implement in the presence of cohesive-frictional contact. The penalty method accommodates cohesive-frictional contact, but does not impose the contact condition exactly. The augmented Lagrangian method uses the penalty method to mimic the Lagrange multipliers method, but requires an additional layer of iterations. We present the mathematical formulations for each of the three methods below.

Consider a quasistatic boundary-value problem over a domain $\mathcal{B}$ delimited by boundary $\partial\mathcal{B}$, such that $\partial\mathcal{B} = \overline{\partial\mathcal{B}_{\boldsymbol{t}} \cup \partial\mathcal{B}_{\boldsymbol{u}}}$ and $\varnothing = \partial\mathcal{B}_{\boldsymbol{t}} \cap \partial\mathcal{B}_{\boldsymbol{u}}$, where $\partial\mathcal{B}_{\boldsymbol{t}}$ and $\partial\mathcal{B}_{\boldsymbol{u}}$ are portions of the boundary where the tractions and displacements, respectively, are prescribed. We assume a well-defined discontinuity $\mathcal{S} \in \mathcal{B}$ containing a segment $\widetilde{\mathcal{S}} \in \mathcal{S}$ in contact mode, and a segment $\mathcal{S}\backslash\widetilde{\mathcal{S}}$ in opening mode. The strong form of the boundary-value problem is as follows: For all $\boldsymbol{x} \in \mathcal{B}$, find the displacement field $\boldsymbol{u}$ such that

$$\left.\begin{aligned} \boldsymbol{\nabla}\cdot\boldsymbol{\sigma} + \boldsymbol{f} &= \boldsymbol{0} \quad \text{in } \mathcal{B}\backslash\mathcal{S} \\ \boldsymbol{u} &= \widehat{\boldsymbol{u}} \quad \text{on } \partial\mathcal{B}_{\boldsymbol{u}} \\ \boldsymbol{\nu}\cdot\boldsymbol{\sigma} &= \widehat{\boldsymbol{t}} \quad \text{on } \partial\mathcal{B}_{\boldsymbol{t}} \\ \boldsymbol{n}\cdot\boldsymbol{\sigma} &= \boldsymbol{t}_{-} \quad \text{on } \widetilde{\mathcal{S}}_{-} \\ -\boldsymbol{n}\cdot\boldsymbol{\sigma} &= \boldsymbol{t}_{+} \quad \text{on } \widetilde{\mathcal{S}}_{+} \end{aligned}\right\} \tag{7.12}$$

subject to the constraints

$$\ell(\boldsymbol{x}) \equiv t_N(\boldsymbol{x}) > 0\,, \qquad -g_N(\boldsymbol{x}) \leq 0\,, \qquad \ell(\boldsymbol{x}) g_N(\boldsymbol{x}) = 0 \tag{7.13}$$

for all $\boldsymbol{x} \in \widetilde{\mathcal{S}}$, where $\boldsymbol{\nu}$ is the outward unit normal vector to $\partial\mathcal{B}_{\boldsymbol{t}}$. Equilibrium on the contact surface demands that

$$\boldsymbol{t}(\boldsymbol{x}) \equiv \boldsymbol{t}_{-}(\boldsymbol{x}) = -\boldsymbol{t}_{+}(\boldsymbol{x}) \qquad \forall\boldsymbol{x} \in \widetilde{\mathcal{S}}\,. \tag{7.14}$$

A new symbol $\ell = \ell(\boldsymbol{x})$ has been introduced to highlight that the contact pressure is now taken as the Lagrange multiplier.

To develop the weak form, we consider a set of trial functions

$$\boldsymbol{\mathcal{U}} = \{\boldsymbol{u}\,|\,\boldsymbol{u} \in H^1(\mathcal{B}),\ \boldsymbol{u} = \widehat{\boldsymbol{u}} \text{ on } \partial\mathcal{B}_{\boldsymbol{u}}\} \tag{7.15}$$

and a set of variations

$$\boldsymbol{\mathcal{V}} = \{\boldsymbol{\eta}\,|\,\boldsymbol{\eta} \in H^1(\mathcal{B}),\ \boldsymbol{\eta} = \boldsymbol{0} \text{ on } \partial\mathcal{B}_{\boldsymbol{u}}\}\,. \tag{7.16}$$

The 'classic' variational equation takes the form

$$\int_{\mathcal{B}\backslash\mathcal{S}} \boldsymbol{\nabla}^s\boldsymbol{\eta} : \boldsymbol{\sigma}\,\mathrm{d}V + \mathcal{G}_c = \int_{\mathcal{B}\backslash\mathcal{S}} \boldsymbol{\eta}\cdot\boldsymbol{f}\,\mathrm{d}V + \int_{\partial\mathcal{B}_t} \boldsymbol{\eta}\cdot\widehat{\boldsymbol{t}}\,\mathrm{d}A\,, \tag{7.17}$$

where

$$\mathcal{G}_c = \int_{\widetilde{\mathcal{S}}} \widetilde{\boldsymbol{\eta}} \cdot \boldsymbol{t} \,\mathrm{d}A \tag{7.18}$$

is a contact integral representing the virtual work done by the traction vector on $\widetilde{\mathcal{S}}$, and

$$\widetilde{\boldsymbol{\eta}} = \boldsymbol{\eta}_+ - \boldsymbol{\eta}_- \tag{7.19}$$

is the discontinuity in the variation on $\widetilde{\mathcal{S}}$.

For smooth contact the traction vector on the contact surface reduces to the form $\boldsymbol{t} = -\ell \boldsymbol{n}$, and the contact integral simplifies to

$$\mathcal{G}_c = -\int_{\widetilde{\mathcal{S}}} \widetilde{\boldsymbol{\eta}} \cdot \boldsymbol{n} \,\ell \,\mathrm{d}A\,. \tag{7.20}$$

We note that in this formulation, $\widetilde{\boldsymbol{\eta}}$ is not an independent variation. In the context of the standard finite element formulation, this requires that the element sides be aligned to the contacting faces. In Section 7.6, we shall present an extended finite element formulation in which $\widetilde{\boldsymbol{\eta}}$ is treated as an independent variation.

The integral $\mathcal{G}_c$ contains an unknown Lagrange multiplier ℓ. To complete the formulation of the problem, we must impose the constraint of no interpenetration on $\widetilde{\mathcal{S}}$. Let $\psi(\boldsymbol{x}) \geq 0$ denote the variation associated with the Lagrange multiplier $\ell(\boldsymbol{x})$. There are no boundary conditions for the functions ψ and ℓ, so we simply require that $(\psi, \ell) \in \mathcal{P}$, where $\mathcal{P}$ is a collection of nonnegative L_2 functions. Point-wise satisfaction of the no-interpenetration condition requires that $\widetilde{\boldsymbol{u}}(\boldsymbol{x}) \cdot \boldsymbol{n}(\boldsymbol{x}) = 0$ for all $\boldsymbol{x} \in \widetilde{\mathcal{S}}$. Converted into a weighted integral, we write

$$\int_{\widetilde{\mathcal{S}}} \psi(\boldsymbol{x}) \widetilde{\boldsymbol{u}}(\boldsymbol{x}) \cdot \boldsymbol{n}(\boldsymbol{x}) \,\mathrm{d}A = 0 \qquad \forall \psi \in \mathcal{P}\,. \tag{7.21}$$

Equations (7.17) and (7.21) constitute a two-field $(\boldsymbol{u}, \ell)$ mixed variational equations that must be solved simultaneously. The presence of the Lagrange multiplier ℓ results in an indefinite structure of the associated matrix problem (see Section 7.6).

7.4 Penalty and Augmented Lagrangian Methods

The unwieldy form of the variational equations resulting from the Lagrange multipliers formulation motivates a penalty regularization. In the penalty formulation, the Lagrange multiplier $\ell(\boldsymbol{x})$ is expressed in terms of the gap function g_N via a constitutive equation of the form

$$\ell(\boldsymbol{x}) = -k_N g_N(\boldsymbol{x}) \geq 0 \qquad \forall \boldsymbol{x} \in \widetilde{\mathcal{S}}\,, \tag{7.22}$$

where $k_N \gg 1$ is a prescribed penalty parameter. To appreciate the idea behind the penalty regularization, one may view the discontinuity $\mathcal{S}$ as having springs of finite stiffness k_N inserted between the contacting faces, which can only take compression. Since k_N is finite, some interpenetration may occur between the surfaces, so that $g_N < 0$. By prescribing a very large stiffness, however, the interpenetration can be made smaller and smaller, thus satisfying the constraint $g_N \approx 0$ in an approximate sense. With the above penalty regularization, the traction vector on the discontinuity takes the form

$$\boldsymbol{t} = \boldsymbol{t}_T + k_N(\widetilde{\boldsymbol{u}} \cdot \boldsymbol{n})\boldsymbol{n}\,. \tag{7.23}$$

We have thus eliminated the Lagrange multiplier ℓ from (7.17), and reduced the variational equation to the standard displacement form.

The tangential component of traction remains to be quantified. In developing an expression for the evolution of $\boldsymbol{t}_T$, we must consider both stick and slip conditions as well as the incremental nature of loading. A rate constitutive equation of the following form provides a starting point

$$\dot{\boldsymbol{t}}_T = k_T\Big(\dot{\boldsymbol{g}}_T - \dot{\zeta}\frac{\boldsymbol{t}_T}{\|\boldsymbol{t}_T\|}\Big)\,, \tag{7.24}$$

where k_T is a prescribed tangential penalty parameter. As $k_T \to \infty$, we recover either the stick condition, $\dot{\boldsymbol{g}}_T \to \boldsymbol{0}$, or the flow rule for the slip condition, $\dot{\boldsymbol{g}}_T \to \dot{\zeta}\boldsymbol{t}_T/\|\boldsymbol{t}_T\|$, since $\dot{\boldsymbol{t}}_T$ is bounded. Equation (7.24) is a classic elastic-plastic rate constitutive equation for which we can use the conventional return mapping scheme to develop a time-integrated version.

As usual, we assume that $(\boldsymbol{t}_T)_n$, $\Delta\boldsymbol{g}_T$, and $g_N < 0$ are given at the beginning of the load step. Here, a negative value for the gap function g_N implies that there is some interpenetration between the contacting surfaces due to the finite value of k_N. To determine $(\boldsymbol{t}_T)_{n+1} \equiv \boldsymbol{t}_T$, we first calculate an elastic predictor

$$\boldsymbol{t}_T^{\mathrm{tr}} = (\boldsymbol{t}_T)_n + k_T\Delta\boldsymbol{g}_T \tag{7.25}$$

and check if $\|\boldsymbol{t}_T^{\mathrm{tr}}\| + \mu k_N g_N - c \le 0$. If the inequality is true, we accept the predictor as the final value (stick condition). Otherwise, we use the backward implicit algorithm to correct for plastic sliding (slip condition),

$$\boldsymbol{t}_T = \boldsymbol{t}_T^{\mathrm{tr}} - k_T\Delta\zeta\frac{\boldsymbol{t}_T^{\mathrm{tr}}}{\|\boldsymbol{t}_T^{\mathrm{tr}}\|} = \Big(1 - k_T\frac{\Delta\zeta}{\|\boldsymbol{t}_T^{\mathrm{tr}}\|}\Big)\boldsymbol{t}_T^{\mathrm{tr}}\,. \tag{7.26}$$

The above equation indicates that the direction of the tangential traction vector is the same as the direction of the elastic predictor.

We can obtain the incremental slip $\Delta\zeta$ from the consistency condition, $f = 0$, which gives

$$\Delta\zeta = \frac{\|\boldsymbol{t}_T^{\mathrm{tr}}\| + \mu k_N g_N - c}{k_T}\,. \tag{7.27}$$

Inserting back into (7.26) and simplifying gives

$$\boldsymbol{t}_T = (c - \mu k_N g_N)\frac{\boldsymbol{t}_T^{\mathrm{tr}}}{\|\boldsymbol{t}_T^{\mathrm{tr}}\|}\,, \qquad g_N < 0\,. \tag{7.28}$$

The penalty method is simpler to implement than the Lagrange multipliers method, and it does not change the number of global degrees of freedom for the matrix problem. Furthermore, the method can handle the stick-slip condition quite easily. However, there are some shortcomings of the method as well. Apart from the fact that the penalty method only provides an approximate solution to the contact problem, the large values of the penalty parameters required to approximately impose the no-overlap condition results in an ill-conditioned coefficient matrix. This makes linear equation solving difficult particularly with an iterative solver such as the Newton-Krylov solver (White and Borja 2011).

The augmented Lagrangian method strives to strike a balance between the more accurate Lagrange multipliers method and the simpler penalty method (Simo and Laursen 1992). In the augmented Lagrangian method the Lagrange multiplier ℓ is written as

$$\ell(\boldsymbol{x}) = \overline{\ell}(\boldsymbol{x}) + \overline{k}_N\langle -g_N(\boldsymbol{x})\rangle\,, \qquad \forall \boldsymbol{x} \in \widetilde{\mathcal{S}}\,, \tag{7.29}$$

where $\overline{\ell}$ is a fixed approximation to the Lagrange multiplier, $\overline{k}_N$ is a penalty parameter, and the Macauley brackets $\langle\cdot\rangle$ describe a ramp function. We can make the value of $\overline{k}_N$ as large as the regular penalty parameter k_N, or somewhat smaller. In the latter case, the overlap between the contacting surfaces is expected to be larger, but the algorithm does provide a way to iteratively reduce the amount of this overlap.

Inserting (7.29) into (7.18), the contact integral becomes

$$\mathcal{G}_c = \int_{\widetilde{\mathcal{S}}} \widetilde{\boldsymbol{\eta}} \cdot \boldsymbol{t}_T \, \mathrm{d}A - \int_{\widetilde{\mathcal{S}}} (\widetilde{\boldsymbol{\eta}} \cdot \boldsymbol{n})[\overline{\ell} - \overline{k}_N(\widetilde{\boldsymbol{u}} \cdot \boldsymbol{n})] \, \mathrm{d}A\,. \tag{7.30}$$

Now, if we set $\overline{\ell} = 0$ and $\overline{k}_N = k_N$, then we recover the full penalty method. On the other hand, if we set $g_N = 0$, then $\overline{\ell}$ must be the exact Lagrange multiplier. The idea then is to make $\overline{\ell}$ as close to the exact Lagrange multiplier as possible so that we satisfy the condition $g_N \approx 0$ even if we use a small value of $\overline{k}_N$. Note that in the augmented Lagrangian method, $\overline{\ell}$ is a prescribed number for which the algorithm aims to find a reasonable value to minimize the amount of interpenetration.

Box 7.1 shows a flow chart of the augmented Lagrangian algorithm for finding the Lagrange multiplier $\overline{\ell}$ that yields an acceptable amount of penetration. Step 1 initializes the estimate of the Lagrange multiplier, which can be any nonnegative number including zero. An alternative initial guess would be the converged value of $\overline{\ell}$ from the previous load step. Step 2 iterates to find the displacement field $\boldsymbol{u}$, with $\overline{\ell}$ held fixed. Step 3 checks to ensure that the interpenetration is acceptable. If the overlap is still too large, then Step 4 updates $\overline{\ell}$ and continues with the iteration. The downside of the augmented Lagrange method is the additional layer of iterations necessary to find an acceptable value of Lagrange multiplier. This layer of iterations nests over another layer to solve the global nonlinear problem in Step 2. In short, there is price to pay for the accuracy and simplicity of the augmented Lagrangian technique.

Step 1. Initialize $\overline{\ell}^k = 0$, with iteration counter $k = 0$
Step 2. Solve $\boldsymbol{u}$ with $\mathcal{G}_\mathrm{c} = \mathcal{G}_\mathrm{c}(\overline{\ell}^k)$
Step 3. Check: Is $-g_N(\boldsymbol{x}) = -\widetilde{\boldsymbol{u}} \cdot \boldsymbol{n} \leq \mathrm{TOL}$ for all $\boldsymbol{x} \in \widetilde{\mathcal{S}}$?
Yes, augmentation convergence achieved and exit.
Step 4. No, set $\overline{\ell}^{k+1} = \langle \overline{\ell}^k - \overline{k}_N(\widetilde{\boldsymbol{u}} \cdot \boldsymbol{n}) \rangle$, $k \leftarrow k+1$,
and go to Step 2.

Box 7.1 Flow chart of the augmented Lagrangian method

7.5 Enriched Finite Elements

For obvious reason, it would not be possible to determine the path of propagation of a crack that has not even formed. Therefore, unless a discontinuity is predefined, we cannot structure the finite element mesh so that the sides of the elements are always aligned to the discontinuity. Conventional finite elements can only interpolate conforming displacement fields, i.e., smooth fields that are free of shocks such as a crack. If a discontinuity passes through the interior of these elements, the shape functions must be enhanced to accommodate this shock. This section deals with the enhancement of the finite element shape functions with a Heaviside function to accommodate the kinematics of a shock in the form of a displacement discontinuity.

Consider a mesh of finite elements in domain $\mathcal{B}$ containing a discontinuity $\mathcal{S}$ passing through the interior of finite elements within the subdomain $\mathcal{B}^h = \mathcal{B}^h_- \cup \mathcal{B}^h_+$ delineated by boundaries $\mathcal{S}^h_-$ and $\mathcal{S}^h_+$, see Fig. 7.4. The goal is to enrich the finite element shape functions within this subregion in order to resolve a discontinuous displacement field on $\mathcal{S}$. The desired enrichment has the form

$$\boldsymbol{u}(\boldsymbol{x}) = \overline{\boldsymbol{u}}(\boldsymbol{x}) + H_\mathcal{S}(\boldsymbol{x})\widetilde{\boldsymbol{u}}(\boldsymbol{x}) \,, \tag{7.31}$$

where $H_\mathcal{S}(\boldsymbol{x})$ is the Heaviside function defined as

$$H_\mathcal{S}(\boldsymbol{x}) = \begin{cases} 1 & \text{if } \boldsymbol{x} \in \mathcal{B}_+ \\ 0 & \text{if } \boldsymbol{x} \in \mathcal{B}_- \,. \end{cases} \tag{7.32}$$

In the above equation, $\boldsymbol{u}(\boldsymbol{x})$ is the displacement field, $\overline{\boldsymbol{u}}(\boldsymbol{x})$ is the continuous part of $\boldsymbol{u}$, $\widetilde{\boldsymbol{u}}(\boldsymbol{x})$ is the discontinuous part, or jump, $\mathcal{B}_-$ is the master side, and $\mathcal{B}_+$ is the slave side. Figure 7.5 shows the geometric meaning of the above displacement fields in one dimension. The open dots are nodal values of $\boldsymbol{u}$, and a displacement jump occurs on $\mathcal{S}$ between nodes A and B. The continuous displacement field $\overline{\boldsymbol{u}}$ is one without the jump, whereas $\widetilde{\boldsymbol{u}}$ defines the jump.

Alternatively, the displacement field can be re-parameterized as

$$\boldsymbol{u}(\boldsymbol{x}) = \boldsymbol{u}_c(\boldsymbol{x}) + M_\mathcal{S}(\boldsymbol{x})\widetilde{\boldsymbol{u}}(\boldsymbol{x}) \,. \tag{7.33}$$

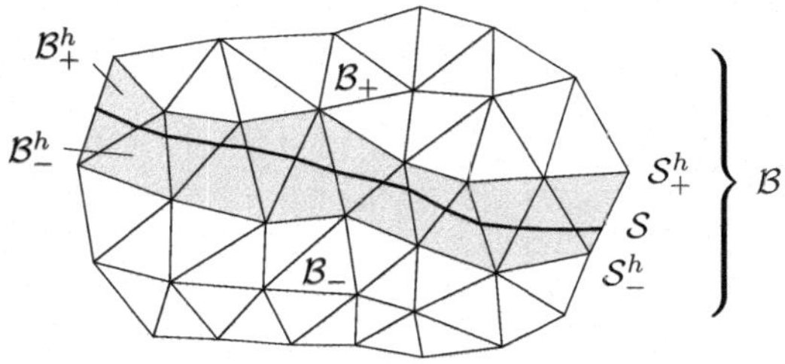

Fig. 7.4 Discontinuity $\mathcal{S}$ embedded in a fixed finite element mesh. Shaded region $\mathcal{B}^h = \mathcal{B}_-^h \cup \mathcal{B}_+^h$ is the support of $M_\mathcal{S}(\boldsymbol{x})$; $\mathcal{S}_-^h$ and $\mathcal{S}_+^h$ are the boundaries of this support.

In this re-parameterized version, the displacement field $\boldsymbol{u}_c(\boldsymbol{x})$ is the conforming displacement field. Outside the subregion $\mathcal{B}^h$ in Fig. 7.4, or outside the interval AB in Fig. 7.5, $\boldsymbol{u}_c(\boldsymbol{x}) = \boldsymbol{u}(\boldsymbol{x})$, but inside $\mathcal{B}^h$, or inside the interval AB, it ramps up in a conforming fashion. Note that $\boldsymbol{u}_c$ is not an approximation to the displacement jump on $\mathcal{S}$. To recover a concentrated displacement jump on $\mathcal{S}$, a blending term $M_\mathcal{S}(\boldsymbol{x})\widetilde{\boldsymbol{u}}(\boldsymbol{x})$ (shaded region in Fig. 7.5) must be added to the conforming displacement field. Blending is done only within the subregion $\mathcal{B}^h$, or within the interval AB, called the support of $M_\mathcal{S}(\boldsymbol{x})$ (Stakgold 1998).

The specific form for $M_\mathcal{S}(\boldsymbol{x})$ is

$$M_\mathcal{S}(\boldsymbol{x}) = H_\mathcal{S}(\boldsymbol{x}) - f^h(\boldsymbol{x}) , \tag{7.34}$$

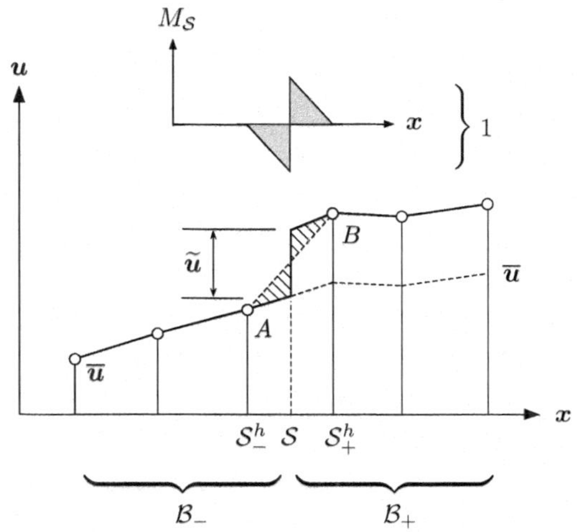

Fig. 7.5 Schematic representations of $\overline{\boldsymbol{u}}(\boldsymbol{x})$, $\widetilde{\boldsymbol{u}}(\boldsymbol{x})$, and $M_\mathcal{S}(\boldsymbol{x})$ for a solid with an embedded discontinuity. Line AB is the conforming displacement field $\boldsymbol{u}_c$.

where $f^h(\boldsymbol{x})$ is an arbitrary smooth blending function that satisfies the following requirements

$$f^h(\boldsymbol{x})\widetilde{\boldsymbol{u}}(\boldsymbol{x}) = \begin{cases} \boldsymbol{0} & \text{on } \mathcal{S}^h_- \cup (\mathcal{B}_- \backslash \mathcal{B}^h_-), \\ \widetilde{\boldsymbol{u}}(\boldsymbol{x}) & \text{on } \mathcal{S}^h_+ \cup (\mathcal{B}_+ \backslash \mathcal{B}^h_+). \end{cases} \tag{7.35}$$

Only the finite elements within the support of $M_{\mathcal{S}}(\boldsymbol{x})$ are enriched with the blending term. Outside the support of $M_{\mathcal{S}}(\boldsymbol{x})$, conventional finite element interpolation is used.

To further illustrate the geometric meaning of f^h, Figs. 7.6 and 7.7 show, respectively, the blending functions appropriate for constant-strain triangular (CST) and four-node quadrilateral finite elements. For a CST element cut by a discontinuity, the positive side of $\mathcal{S}$ could contain either one node or two nodes. For a quadrilateral element, the positive side of $\mathcal{S}$ could contain one, two, or three nodes (Borja and Regueiro 2001, Regueiro and Borja 2001). Irrespective of how the discontinuity is placed within the element, one can always associate $f^h(\boldsymbol{x})$ with the standard finite element basis shape functions. For example, for a CST element with one node A on the positive side of $\mathcal{S}$, the blending function is simply the finite element basis shape function N_A; for a CST element with two nodes A and B on the positive side, $f^h(\boldsymbol{x}) = N_A + N_B$, and so on.

Exercise 7.1. Develop expressions for the blending function f^h for an enriched quadrilateral element with one node, two nodes, and three nodes on the positive side of $\mathcal{S}$, see Fig. 7.7. Take $f^h = f^h(\xi, \eta)$, where $-1 \leq \xi, \eta \leq 1$ are the natural coordinates for the element.

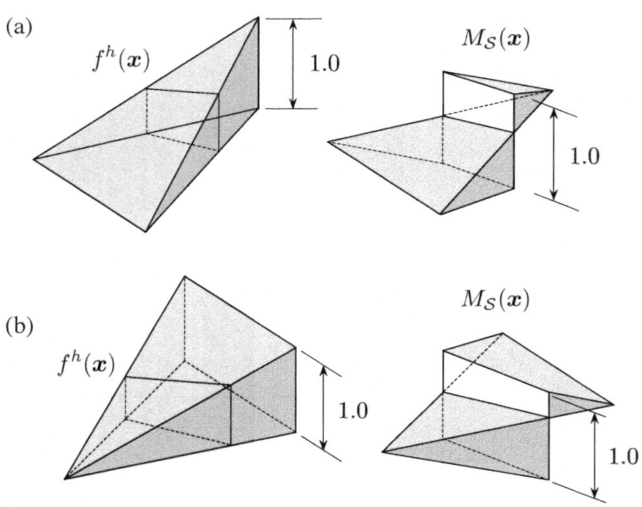

Fig. 7.6 Blending function $f^h(\boldsymbol{x})$ and jump function $M_{\mathcal{S}}(\boldsymbol{x})$ for enriched triangular finite elements: (a) one node in $\mathcal{B}_+$; (b) two nodes in $\mathcal{B}_+$

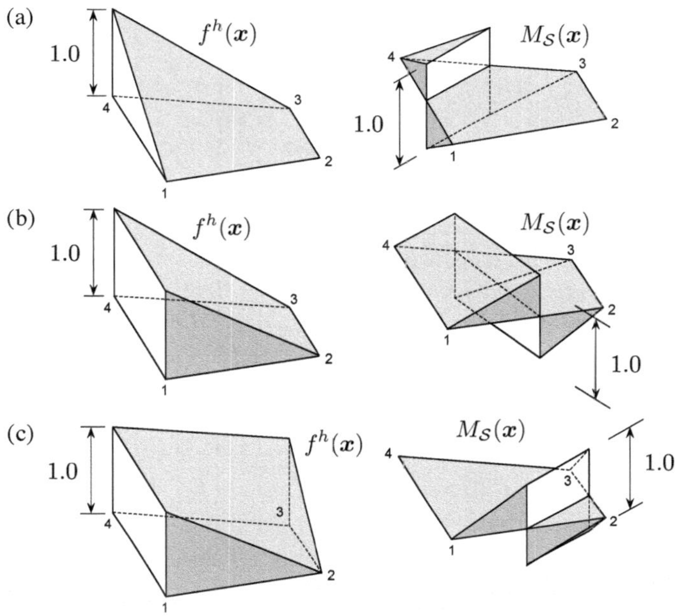

Fig. 7.7 Blending function $f^h(\boldsymbol{x})$ and jump function $M_{\mathcal{S}}(\boldsymbol{x})$ for enriched quadrilateral finite element: (a) one node in $\mathcal{B}_+$; (b) two nodes in $\mathcal{B}_+$; (c) three nodes in $\mathcal{B}_+$

Exercise 7.2. Consider a semi-infinite elastic plate defined in domain $-\infty < x < \infty$ and $0 \leq y \leq 1$, and assume a horizontal crack at $y = 1/2$. The bottom boundary of the plate is fixed at $y = 0$ while the top boundary moves horizontally according to the equation $u_x(y = 1) = v_x t$, where t is time and v_x is the constant horizontal velocity of the top boundary. The plate has a shear modulus of rigidity μ^e and the crack has a maximum shear resistance τ_Y. Develop expressions for the blending function $f^h(x, y)$, the discontinuous function $M_{\mathcal{S}}(x, y)$, and the velocity components $\overline{\boldsymbol{v}}(x, y)$ and $\widetilde{\boldsymbol{v}}(x, y)$ assuming the entire plate is the support of $M_{\mathcal{S}}$.

7.6 Extended Finite Element Method

A trial function for $\boldsymbol{u}(\boldsymbol{x})$ of the form shown in equation (7.33) suggests a similar form for the variation

$$\boldsymbol{\eta}(\boldsymbol{x}) = \boldsymbol{\eta}_c(\boldsymbol{x}) + M_{\mathcal{S}}(\boldsymbol{x})\widetilde{\boldsymbol{\eta}}(\boldsymbol{x}), \tag{7.36}$$

where $\boldsymbol{\eta}_c$ is the conforming part, and $\widetilde{\boldsymbol{\eta}}$ is the jump in the variation having the same physical meaning as the one defined in (7.19). This results in the variational equation

$$\int_{\mathcal{B}\setminus\mathcal{S}} [\nabla^s \boldsymbol{\eta}_c + H_{\mathcal{S}} \nabla^s \widetilde{\boldsymbol{\eta}} - \nabla^s (f^h \widetilde{\boldsymbol{\eta}})] : \boldsymbol{\sigma} \, \mathrm{d}V + \mathcal{G}_c$$
$$= \int_{\mathcal{B}\setminus\mathcal{S}} (\boldsymbol{\eta}_c + M_{\mathcal{S}} \widetilde{\boldsymbol{\eta}}) \cdot \boldsymbol{f} \, \mathrm{d}V + \int_{\partial \mathcal{B}_t} (\boldsymbol{\eta}_c + M_{\mathcal{S}} \widetilde{\boldsymbol{\eta}}) \cdot \widehat{\boldsymbol{t}} \, \mathrm{d}A \,, \quad (7.37)$$

where $\mathcal{G}_c$ is the same contact integral defined in (7.18).

We recall that in the classic form (7.18) of the contact integral, $\widetilde{\boldsymbol{\eta}}$ is not an independent variation. Furthermore, there is no blending term in the classic formulation because the discontinuities are located on the exterior boundaries, and so the trial and weighting functions define conforming fields all throughout the domain, i.e., $\boldsymbol{u} = \boldsymbol{u}_c$ and $\boldsymbol{\eta} = \boldsymbol{\eta}_c$. However, in the extended finite element formulation, $\boldsymbol{\eta}_c$ and $\widetilde{\boldsymbol{\eta}}$ are considered as two independent variations. Accordingly, we obtain two independent variational equations: The first is the regular part,

$$\int_{\mathcal{B}\setminus\mathcal{S}} \nabla^s \boldsymbol{\eta}_c : \boldsymbol{\sigma} \, \mathrm{d}V = \int_{\mathcal{B}\setminus\mathcal{S}} \boldsymbol{\eta}_c \cdot \boldsymbol{f} \, \mathrm{d}V + \int_{\partial \mathcal{B}_t} \boldsymbol{\eta}_c \cdot \widehat{\boldsymbol{t}} \, \mathrm{d}A \,. \quad (7.38)$$

The second is the enrichment part containing all the blending terms

$$\int_{\mathcal{B}^h\setminus\mathcal{S}} [H_{\mathcal{S}} \nabla^s \widetilde{\boldsymbol{\eta}} - \nabla^s (f^h \widetilde{\boldsymbol{\eta}})] : \boldsymbol{\sigma} \, \mathrm{d}V + \mathcal{G}_c$$
$$= \int_{\mathcal{B}^h\setminus\mathcal{S}} M_{\mathcal{S}} \widetilde{\boldsymbol{\eta}} \cdot \boldsymbol{f} \, \mathrm{d}V + \int_{\partial \mathcal{B}_t} M_{\mathcal{S}} \widetilde{\boldsymbol{\eta}} \cdot \widehat{\boldsymbol{t}} \, \mathrm{d}A \,. \quad (7.39)$$

Note that the region of integration for the enrichment part is limited to the support of $M_{\mathcal{S}}$, i.e., $\mathcal{B}^h \subset \mathcal{B}$ (Liu and Borja 2008).

The contact integral $\mathcal{G}_c$ contains either the Lagrange multiplier ℓ or simply the discontinuous displacement $\widetilde{\boldsymbol{u}}$ depending on whether one uses the Lagrange multipliers or penalty formulation. In either case, separate finite element interpolations are required for the conforming displacement field $\boldsymbol{u}_c$ and for the blending field $M_{\mathcal{S}}\widetilde{\boldsymbol{u}}$. Figure 7.8 portrays the enriched degrees

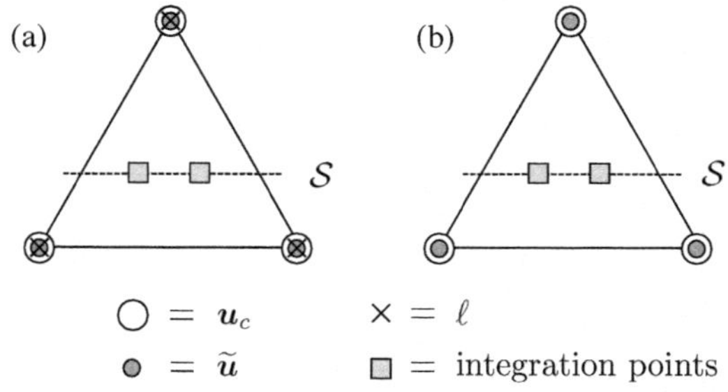

Fig. 7.8 Enriched CST elements: (a) regular nodes enriched with slip degrees of freedom $\widetilde{\boldsymbol{u}}$ and Lagrange multiplier ℓ; and (b) regular nodes enriched with slip degrees of freedom $\widetilde{\boldsymbol{u}}$

of freedom for a CST element. In the Lagrange multipliers formulation, the nodes are enriched with $(\widetilde{\boldsymbol{u}}, \ell)$ degrees of freedom, in addition to the $\boldsymbol{u}_c$ degrees of freedom; in a penalty regularization, the nodes are enriched only with $\widetilde{\boldsymbol{u}}$. The nodal values of $\widetilde{\boldsymbol{u}}$ may not be zero, but the expression $M_{\mathcal{S}}\widetilde{\boldsymbol{u}}$ is always zero when evaluated at the nodes. For a CST element, the slip and contact pressure are integrated on $\mathcal{S}$ using a two-point quadrature rule.

The conforming part of displacement field is approximated by standard C^0 shape functions $N_I(\boldsymbol{x})$,

$$\boldsymbol{u}_c(\boldsymbol{x}) = \sum_{I \in N_{\text{nodes}}} N_I(\boldsymbol{x})\boldsymbol{d}_I = \boldsymbol{N}\boldsymbol{d}, \tag{7.40}$$

where N_{nodes} is the set of total nodes and $\boldsymbol{d}_I$ is the regular displacement vector at node I. The discontinuous part is limited to the compact support of $M_{\mathcal{S}}(\boldsymbol{x})$, which is approximated by the same standard shape functions $N_I(\boldsymbol{x})$,

$$\widetilde{\boldsymbol{u}}(\boldsymbol{x}) = \sum_{I \in N_{\text{enr}}} N_I(\boldsymbol{x})\boldsymbol{a}_I = \hat{\boldsymbol{N}}\boldsymbol{a}, \tag{7.41}$$

where $\boldsymbol{a}_I$ is the vector of nodal enhancements and N_{enr} is the number of enriched nodes. In the Lagrange multipliers formulation, the Lagrange multiplier is interpolated using the same standard C^0 shape functions as

$$-\ell(\boldsymbol{x}) = \sum_{I \in N_{\text{enr}}} N_I(\boldsymbol{x})p_I = \bar{\boldsymbol{N}}\boldsymbol{p}, \tag{7.42}$$

where $\boldsymbol{p}$ is a column vector containing the nodal (negative) contact pressures at the enriched nodes. The negative sign follows from continuum mechanics convention that negative stresses are compressive, so the Lagrange multipliers are all negative numbers.

The displacement field $f^h(\boldsymbol{x})\widetilde{\boldsymbol{u}}(\boldsymbol{x})$ is given by the expression

$$f^h(\boldsymbol{x})\widetilde{\boldsymbol{u}}(\boldsymbol{x}) = \sum_{I \in N_{\text{enr}}} N_I(\boldsymbol{x})H_{\mathcal{S}}(\boldsymbol{x}_I)\boldsymbol{a}_I \,. \tag{7.43}$$

The discontinuous part of displacement field, $M_{\mathcal{S}}(\boldsymbol{x})\widetilde{\boldsymbol{u}}(\boldsymbol{x})$, is given by

$$M_{\mathcal{S}}(\boldsymbol{x})\widetilde{\boldsymbol{u}}(\boldsymbol{x}) = \sum_{I \in N_{\text{enr}}} \widetilde{N}_I(\boldsymbol{x})\boldsymbol{a}_I = \widetilde{\boldsymbol{N}}\boldsymbol{a}, \tag{7.44}$$

where

$$\widetilde{N}_I(\boldsymbol{x}) = N_I(\boldsymbol{x})[H_{\mathcal{S}}(\boldsymbol{x}) - H_{\mathcal{S}}(\boldsymbol{x}_I)] \,. \tag{7.45}$$

Note that the dimensions of $\hat{\boldsymbol{N}}$ and $\widetilde{\boldsymbol{N}}$ are the same. The discontinuous displacement on $\mathcal{S}$ is

$$\widetilde{\boldsymbol{u}}(\boldsymbol{x})|_{\mathcal{S}} = \sum_{I \in N_{\text{enr}}} N_I(\boldsymbol{x})|_{\mathcal{S}}\boldsymbol{a}_I \qquad \forall \boldsymbol{x} \in \mathcal{S} \,. \tag{7.46}$$

Since the term $M_{\mathcal{S}}\widetilde{\boldsymbol{u}}$ vanishes at the enriched nodes, the calculated conforming displacements at the enriched nodes are the actual displacements.

We now develop the matrix equation for the Lagrange multipliers formulation. Denoting the strain-displacement transformation array by the usual symbols $\boldsymbol{B} = \boldsymbol{\nabla}^s \boldsymbol{N}$ and $\widetilde{\boldsymbol{B}} = \boldsymbol{\nabla}^s \widetilde{\boldsymbol{N}}$, variational equation (7.38) results in the matrix equation

$$\boldsymbol{F}_{\text{INT}}(\boldsymbol{d}, \boldsymbol{a}) = \boldsymbol{F}_{\text{EXT}}\,, \tag{7.47}$$

where

$$\begin{aligned} \boldsymbol{F}_{\text{INT}} &= \int_{\mathcal{B}\setminus\mathcal{S}} \boldsymbol{B}^{\mathsf{T}} \boldsymbol{\sigma}\, \mathrm{d}V\,, \\ \boldsymbol{F}_{\text{EXT}} &= \int_{\mathcal{B}\setminus\mathcal{S}} \boldsymbol{N}^{\mathsf{T}} \boldsymbol{f}\, \mathrm{d}V + \int_{\partial\mathcal{B}_t} \boldsymbol{N}^{\mathsf{T}} \widehat{\boldsymbol{t}}\, \mathrm{d}A\,. \end{aligned} \tag{7.48}$$

Variational equation (7.39) yields the matrix equation

$$\boldsymbol{\mathcal{F}}_{\text{INT}}(\boldsymbol{d}, \boldsymbol{a}) + \boldsymbol{\mathcal{G}}_{\text{INT}}(\boldsymbol{a}, \boldsymbol{p}) + \boldsymbol{K}_{23}\boldsymbol{p} = \boldsymbol{\mathcal{F}}_{\text{EXT}}\,, \tag{7.49}$$

where

$$\begin{aligned} \boldsymbol{\mathcal{F}}_{\text{INT}} &= \int_{\mathcal{B}^h\setminus\mathcal{S}} \widetilde{\boldsymbol{B}}^{\mathsf{T}} \boldsymbol{\sigma}\, \mathrm{d}V\,, \\ \boldsymbol{\mathcal{G}}_{\text{INT}} &= \int_{\mathcal{S}} \hat{\boldsymbol{N}}^{\mathsf{T}} \boldsymbol{t}_T\, \mathrm{d}A\,, \qquad \hat{\boldsymbol{K}}_{23} = \int_{\widetilde{\mathcal{S}}} \hat{\boldsymbol{N}}^{\mathsf{T}} \boldsymbol{n} \bar{\boldsymbol{N}}\, \mathrm{d}A\,, \\ \boldsymbol{\mathcal{F}}_{\text{EXT}} &= \int_{\mathcal{B}^h\setminus\mathcal{S}} \widetilde{\boldsymbol{N}}^{\mathsf{T}} \boldsymbol{f}\, \mathrm{d}V + \int_{\partial\mathcal{B}_t} \widetilde{\boldsymbol{N}}^{\mathsf{T}} \widehat{\boldsymbol{t}}\, \mathrm{d}A\,. \end{aligned} \tag{7.50}$$

The no-interpenetration constraint (7.21) yields the matrix equation

$$\boldsymbol{K}_{32}\boldsymbol{a} = \boldsymbol{0}\,, \qquad \boldsymbol{K}_{32} = \hat{\boldsymbol{K}}_{23}^{\mathsf{T}}\,. \tag{7.51}$$

The system of equations to solve is nonlinear and must be linearized for iteration with Newton's method. The linearized version yields a coefficient matrix with a block structure of the form

$$\begin{bmatrix} \boldsymbol{K}_{11} & \boldsymbol{K}_{12} & \boldsymbol{0} \\ \boldsymbol{K}_{21} & \boldsymbol{K}_{22} & \boldsymbol{K}_{23} \\ \boldsymbol{0} & \boldsymbol{K}_{32} & \boldsymbol{0} \end{bmatrix} \begin{Bmatrix} \delta\boldsymbol{d} \\ \delta\boldsymbol{a} \\ \delta\boldsymbol{p} \end{Bmatrix} = \begin{Bmatrix} \delta\boldsymbol{F}_{\text{EXT}} \\ \delta\boldsymbol{\mathcal{F}}_{\text{EXT}} \\ \boldsymbol{0} \end{Bmatrix}, \tag{7.52}$$

where $\boldsymbol{K}_{11} = \partial\boldsymbol{F}_{\text{INT}}/\partial\boldsymbol{d}$, $\boldsymbol{K}_{12} = \partial\boldsymbol{F}_{\text{INT}}/\partial\boldsymbol{a}$, $\boldsymbol{K}_{21} = \partial\boldsymbol{\mathcal{F}}_{\text{INT}}/\partial\boldsymbol{d}$, $\boldsymbol{K}_{22} = \partial\boldsymbol{\mathcal{F}}_{\text{INT}}/\partial\boldsymbol{a} + \partial\boldsymbol{\mathcal{G}}_{\text{INT}}/\partial\boldsymbol{a}$, and $\boldsymbol{K}_{23} = \partial\boldsymbol{\mathcal{G}}_{\text{INT}}/\partial\boldsymbol{p} + \hat{\boldsymbol{K}}_{23}$. The zero-block appearing in the (3,3) position results in an indefinite coefficient matrix, similar to what is encountered in the mixed formulation for Stokes problem. Of interest is the unavoidable equal-order interpolation of slip $\widetilde{\boldsymbol{u}}$ and Lagrange multiplier ℓ, since we have the same parent element to enrich with both degrees of freedom. Equal-order interpolation of displacement and pressure is known to produce spurious oscillation in the pressure term. The next section deals with a stabilization scheme alleviating the contact pressure oscillation in the context of the Lagrange multipliers formulation.

Exercise 7.3. Consider a CST element with nodes denoted as A, B, and C. Assume that a discontinuity passes through the interior of this element so that node A is on the slave side, and nodes B and C are on the master side. Evaluate the values of the shape function $\widetilde{N}_I(\boldsymbol{x})$ at the nodes. Repeat the problem if nodes A and B are on the slave side, and node C is on the master side.

Exercise 7.4. A CST element with an embedded crack is enriched to accommodate a jump in displacement field. The coordinates of nodes defining the element are $a(0,0)$, $b(1,0)$, and $c(0,1)$. The crack is defined by the equation $y = x$. The nodal enrichments are $\boldsymbol{a}_a = \{\widetilde{u}_{xa}, \widetilde{u}_{ya}\}$, $\boldsymbol{a}_b = \{\widetilde{u}_{xb}, \widetilde{u}_{yb}\}$, and $\boldsymbol{a}_c = \{\widetilde{u}_{xc}, \widetilde{u}_{yc}\}$, so that $\widetilde{\boldsymbol{u}}(\boldsymbol{x}) = \sum_{A=1}^{3} N_A \boldsymbol{a}_A$. Determine the variation of displacement jump on the crack inside this element. At what point on the crack is the norm of the displacement jump maximum?

7.7 Stabilized Formulation

In the solution of Stokes equation it is known that the discrete velocity and pressure spaces must be chosen to satisfy the discrete LBB condition (Brezzi 1990)

$$\sup_{\boldsymbol{v}^h \in \boldsymbol{\mathcal{U}}^h} \frac{\int_\Omega \psi^h \nabla \cdot \boldsymbol{v}^h \, \partial\Omega}{\|\boldsymbol{v}^h\|_1} \geq C \|\psi^h\|_0 \qquad \forall \psi^h \in \mathcal{P}^h, \tag{7.53}$$

where $\boldsymbol{\mathcal{U}}^h$ and $\mathcal{P}^h$ are the discrete spaces of trial functions for velocity and pressure satisfying the essential boundary conditions, and $C > 0$ independent of the element length scale h. Many linear-pressure/linear-velocity interpolations do not satisfy this condition and lead to unstable approximations.

For the linear pair, Bochev and co-workers (Bochev et al. 2006, Bochev and Dohrmann 2006) demonstrated that a weaker condition for stability may be phrased as

$$\sup_{\boldsymbol{v}^h \in \boldsymbol{\mathcal{U}}^h} \frac{\int_\Omega \psi^h \nabla \cdot \boldsymbol{v}^h \, \partial\Omega}{\|\boldsymbol{v}^h\|_1} \geq C_1 \|\psi^h\|_0 - C_2 \|\psi^h - \Pi \psi^h\|_0 \quad \forall \, \psi^h \in \mathcal{P}^h, \tag{7.54}$$

where $\Pi : L_2(\Omega) \to R_0$ is a suitable projection operator operating on the pressure term, and with $C_1 > 0$ and $C_2 > 0$ independent of the element length scale h. The term $C_2 \|q^h - \Pi q^h\|_0$ quantifies the inherent deficiency in the linear pair and motivates a similar stabilization approach for the contact problem.

Figure 7.9 illustrates the instability generated by the equal-order pair interpolation for slip and contact pressure in the context of the extended finite

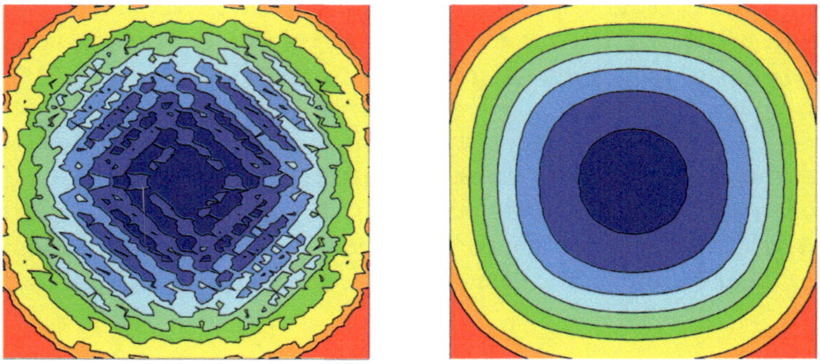

Fig. 7.9 Unstabilized (left) and stabilized (right) contact pressures on a pair of smooth cubes pressed against each other. After Liu and Borja (2010).

element method. In this example, an elastic cube is clamped at both its top and bottom ends and pressed in the vertical direction. The cube is discretized with three-dimensional tetrahedral elements, and a smooth horizontal plane is embedded midway between the top and bottom surfaces of the cube, cutting the interior of the tetrahedral elements. The Lagrange multipliers method is used to impose the contact condition. The left figure shows a contour of Lagrange multipliers exhibiting unphysical oscillation of the contact pressures, in contrast to the right figure which shows the correct, smooth solution for the contact pressure.

A stabilization methodology has been advocated by Liu and Borja (2010) to correct the deficiency inherent in the linear displacement-linear contact pressure interpolation with the extended finite element method. The methodology follows the same polynomial pressure projection stabilization advocated by Bochev and co-workers described above. The idea is to add stabilizing terms to the no-interpenetration condition (7.21). We expound on this subject below in the context of the Lagrange multipliers formulation. The readers may wish to consult Liu and Borja (2010) for similar discussions pertaining to the penalty and augmented Lagrangian methods.

We first assume a slip mode so that by stabilizing the contact pressure we also stabilize the shear component of traction. Note that the resolved shear stress is simply the contact pressure multiplied by the coefficient of friction (plus cohesion), so if the contact pressure is stabilized, the resolved shear stress also is stabilized. For the Lagrange multipliers method the stabilized form of (7.21) is

$$\int_{\widetilde{\mathcal{S}}} \psi \widetilde{\boldsymbol{u}} \cdot \boldsymbol{n} \, \mathrm{d}A - \int_{\widetilde{\mathcal{S}}} \frac{\kappa}{2M} (\psi - \Pi \psi)(\ell - \Pi \ell) \, \mathrm{d}A = 0 \,, \tag{7.55}$$

where M is a normalizing constant, $\kappa > 0$ is a specified stabilization parameter, and Π is the same projection operator introduced in (7.54). We can view the parameter M as a modulus or spring stiffness representing the pressure necessary to produce a unit indentation on the bulk material. The dimensionless parameter κ serves to tune the stabilization and has been normalized in the above equation so that $\kappa \approx 1.0$. We remark that there is a wide range of acceptable values of κ that work, and choosing a specific value for the problem at hand should not be a concern.

The projection operator Π must have a suitable range, and for CST elements it can be evaluated from the area mean of the contact pressure,

$$\Pi \ell_{A^e} = \frac{1}{A^e} \int_{A^e} \ell \, \mathrm{d}A \,. \tag{7.56}$$

Specifically, for CST elements we have

$$\Pi N_A = \frac{1}{3}\,, \qquad A = 1, 2, 3\,, \tag{7.57}$$

where N_A is the basis shape function for node A. The concept can be extended to 3D tetrahedral elements, i.e.,

$$\Pi N_A = \frac{1}{4}\,, \qquad A = 1, 2, 3, 4\,. \tag{7.58}$$

With this stabilization, (7.51) now becomes

$$\boldsymbol{K}_{32}\boldsymbol{a} + \boldsymbol{K}_{33}\boldsymbol{p} = \boldsymbol{0}\,, \tag{7.59}$$

where

$$\boldsymbol{K}_{33} = \int_{\tilde{\mathcal{S}}} \frac{\kappa}{2M} (\bar{\boldsymbol{N}} - \Pi \bar{\boldsymbol{N}})^{\mathsf{T}} (\bar{\boldsymbol{N}} - \Pi \bar{\boldsymbol{N}}) \, \mathrm{d}A \,. \tag{7.60}$$

Equation (7.52) becomes

$$\begin{bmatrix} \boldsymbol{K}_{11} & \boldsymbol{K}_{12} & \boldsymbol{0} \\ \boldsymbol{K}_{21} & \boldsymbol{K}_{22} & \boldsymbol{K}_{23} \\ \boldsymbol{0} & \boldsymbol{K}_{32} & \boldsymbol{K}_{33} \end{bmatrix} \begin{Bmatrix} \delta \boldsymbol{d} \\ \delta \boldsymbol{a} \\ \delta \boldsymbol{p} \end{Bmatrix} = \begin{Bmatrix} \delta \boldsymbol{F}_{\mathrm{EXT}} \\ \delta \boldsymbol{\mathcal{F}}_{\mathrm{EXT}} \\ \boldsymbol{0} \end{Bmatrix} . \tag{7.61}$$

Note that the (3,3) block is no longer a null submatrix, reflecting the effect of the stabilization. With the above technique, the solution recovers the smooth distribution of contact pressure as shown on the right contour of Fig. 7.9.

For a stick mode the tangential component of traction is an independent variable that does not depend on the normal contact pressure ℓ. Hence, stabilizing the contact pressure does not necessarily stabilize the tangential traction. In other words, both the contact pressure and tangential traction must now be stabilized. Let $\ell_T = \|\boldsymbol{t}_T\|$ denote the resolved shear stress, and $\boldsymbol{m} = \boldsymbol{t}_T/\ell_T$ denote the normalized direction. We treat ℓ_T as a second Lagrange multiplier so that the contact surface integral appearing in (7.18) becomes

$$\mathcal{G}_c = \int_{\tilde{\mathcal{S}}} \ell_T (\widetilde{\boldsymbol{\eta}} \cdot \boldsymbol{m}) \, \mathrm{d}A - \int_{\tilde{\mathcal{S}}} \ell (\widetilde{\boldsymbol{\eta}} \cdot \boldsymbol{n}) \, \mathrm{d}A \,. \tag{7.62}$$

The relevant constraints for a stick mode are

$$\ell(\boldsymbol{x}) > 0\,, \quad g_N(\boldsymbol{x}) = 0\,, \quad \ell_T(\boldsymbol{x}) > 0\,, \quad \widetilde{\boldsymbol{u}}(\boldsymbol{x}) \cdot \boldsymbol{m}(\boldsymbol{x}) = 0\,, \quad \forall \boldsymbol{x} \in \widetilde{\mathcal{S}}\,. \tag{7.63}$$

Denoting by ψ_T the weighting function for ℓ_T, the stabilized weighted integrals are

$$\int_{\widetilde{\mathcal{S}}} \psi \widetilde{\boldsymbol{u}} \cdot \boldsymbol{n} \, \mathrm{d}A - \int_{\widetilde{\mathcal{S}}} \frac{\kappa}{2M} (\psi - \Pi\psi)(\ell - \Pi\ell) \, \mathrm{d}A = 0\,, \tag{7.64}$$

$$\int_{\widetilde{\mathcal{S}}} \psi_T \widetilde{\boldsymbol{u}} \cdot \boldsymbol{m} \, \mathrm{d}A - \int_{\widetilde{\mathcal{S}}} \frac{\kappa}{2M} (\psi_T - \Pi\psi_T)(\ell_T - \Pi\ell_T) \, \mathrm{d}A = 0\,. \tag{7.65}$$

The corresponding stabilized finite element matrix equations are

$$\begin{bmatrix} \boldsymbol{K}_{11} & \boldsymbol{K}_{12} & \boldsymbol{0} & \boldsymbol{0} \\ \boldsymbol{K}_{21} & \boldsymbol{K}_{22} & \boldsymbol{K}_{23} & \boldsymbol{K}_{24} \\ \boldsymbol{0} & \boldsymbol{K}_{32} & \boldsymbol{K}_{33} & \boldsymbol{0} \\ \boldsymbol{0} & \boldsymbol{K}_{42} & \boldsymbol{0} & \boldsymbol{K}_{44} \end{bmatrix} \begin{Bmatrix} \delta \boldsymbol{d} \\ \delta \boldsymbol{a} \\ \delta \boldsymbol{p} \\ \delta \boldsymbol{q} \end{Bmatrix} = \begin{Bmatrix} \delta \boldsymbol{F}_{\mathrm{EXT}} \\ \delta \boldsymbol{\mathcal{F}}_{\mathrm{EXT}} \\ \boldsymbol{0} \\ \boldsymbol{0} \end{Bmatrix}, \tag{7.66}$$

where $\boldsymbol{q}$ is the nodal vector of tangential traction ℓ_T. The submatrices are the same as before, i.e., $\boldsymbol{K}_{11} = \partial \boldsymbol{F}_{\mathrm{INT}}/\partial \boldsymbol{d}$, $\boldsymbol{K}_{12} = \partial \boldsymbol{F}_{\mathrm{INT}}/\partial \boldsymbol{a}$, and $\boldsymbol{K}_{21} = \partial \boldsymbol{\mathcal{F}}_{\mathrm{INT}}/\partial \boldsymbol{d}$; except that, now, $\boldsymbol{K}_{22} = \partial \boldsymbol{\mathcal{F}}_{\mathrm{INT}}/\partial \boldsymbol{a}$, $\boldsymbol{K}_{44} = -\boldsymbol{K}_{33}$ (see (7.60)),

$$\boldsymbol{K}_{23} = \int_{\widetilde{\mathcal{S}}} \hat{\boldsymbol{N}}^{\mathsf{T}} \boldsymbol{n} \bar{\boldsymbol{N}} \, \mathrm{d}A\,, \qquad \boldsymbol{K}_{24} = \int_{\widetilde{\mathcal{S}}} \hat{\boldsymbol{N}}^{\mathsf{T}} \boldsymbol{m} \bar{\boldsymbol{N}} \, \mathrm{d}A\,, \tag{7.67}$$

$\boldsymbol{K}_{32} = \boldsymbol{K}_{23}$ and $\boldsymbol{K}_{42} = \boldsymbol{K}_{24}$. For a stick mode, this stabilization technique has been shown to circumvent pressure and shear stress oscillations (Liu and Borja 2010).

7.8 Strong Discontinuity

So far we have considered a discontinuous displacement field $\widetilde{\boldsymbol{u}}(\boldsymbol{x})$ that is continuously interpolated across element boundaries. Figure 7.10a shows a schematic representation of this field. The vector $\boldsymbol{a}$ contains the nodal values of $\widetilde{\boldsymbol{u}}$, which are used to interpolate the discontinuous displacement on $\mathcal{S}$. Of interest is a discontinuous displacement on $\mathcal{S}$ that is interpolated in a piecewise constant fashion. As shown in Fig. 7.10b, the upshot of a piecewise constant interpolation is that the displacement field $\widetilde{\boldsymbol{u}}$ is now discontinuous across element boundaries. However, the enrichment is purely local to the element, so it does not engender additional global degrees of freedom to solve.

The assumed enhanced strain technique revolves around the idea of a strong discontinuity. We remark that the displacement discontinuities described in the previous sections are all strong discontinuities, but they are treated in the context of an interface problem. In the assumed enhanced strain formulation, the displacement discontinuities are treated in the context of a

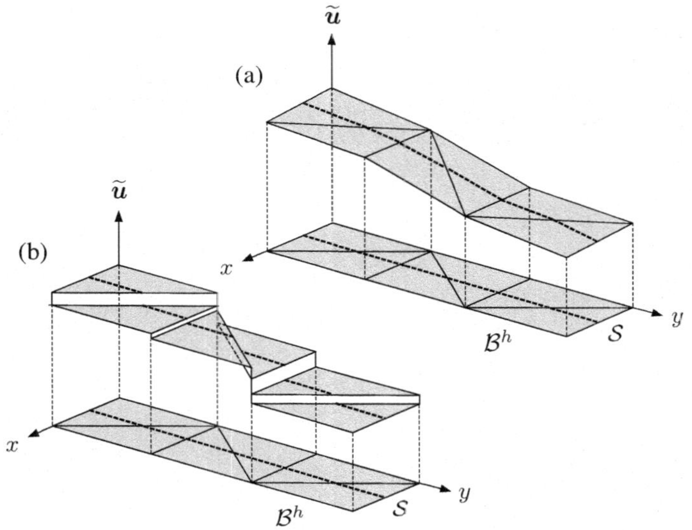

Fig. 7.10 Schematic representation of discontinuous displacement field $\widetilde{\boldsymbol{u}}$ on CST elements: (a) continuous interpolation; (b) piecewise constant interpolation

continuum problem. To this end, the Dirac delta function from theory of distributions (Stakgold 1998) plays a key role in the formulation. The Dirac delta function is a generalized function that is zero everywhere but has an infinitely sharp peak bounding a unit area. The delta function operates on another function $\phi(x)$ in the following way:

$$\int_{-\infty}^{\infty} \delta(x-\xi)\phi(x)\,\mathrm{d}x = \phi(\xi)\,, \tag{7.68}$$

i.e., it sifts the value of a regular function $\phi(x)$ at the location of the sharp peak.

Figure 7.11 shows a block with a so-called deformation band of thickness h. The displacement field in the neighborhood of the band varies according to the ramp function

$$\boldsymbol{u} = \begin{cases} \overline{\boldsymbol{u}}\,, & \text{if } \kappa \leq 0\,; \\ \overline{\boldsymbol{u}} + (\kappa/h)\widetilde{\boldsymbol{u}}|_{\mathcal{S}}\,, & \text{if } 0 \leq \kappa \leq h\,; \\ \overline{\boldsymbol{u}} + \widetilde{\boldsymbol{u}}\,, & \text{if } \kappa \geq h\,, \end{cases} \tag{7.69}$$

where $\widetilde{\boldsymbol{u}}|_{\mathcal{S}}$ is the displacement discontinuity on $\mathcal{S}$, and κ is a normal parameter perpendicular to the band. The above displacement field produces the infinitesimal strain field

$$\boldsymbol{\epsilon} = \begin{cases} \overline{\boldsymbol{\epsilon}}\,, & \text{outside the band}\,; \\ \overline{\boldsymbol{\epsilon}} + (1/h)(\widetilde{\boldsymbol{u}}|_{\mathcal{S}} \otimes \boldsymbol{n})\,, & \text{inside the band}\,, \end{cases} \tag{7.70}$$

where $\overline{\epsilon}(\boldsymbol{x}) = \nabla^s \overline{\boldsymbol{u}}(\boldsymbol{x})$ is a regular strain and the $O(1/h)$ term is the localized strain inside the band. Note that $\nabla \widetilde{\boldsymbol{u}}|_{\mathcal{S}} = \mathbf{0}$ since $\widetilde{\boldsymbol{u}}$ is assumed constant within each enhanced finite element. Therefore, we can drop the symbol $(\cdot)|_{\mathcal{S}}$ and simply use the abbreviated notation $\widetilde{\boldsymbol{u}} \equiv \widetilde{\boldsymbol{u}}|_{\mathcal{S}}$ since it is constant throughout the element. The condition $h > 0$ defines a weak discontinuity, whereas the limiting condition $h \to 0$ defines a strong discontinuity.

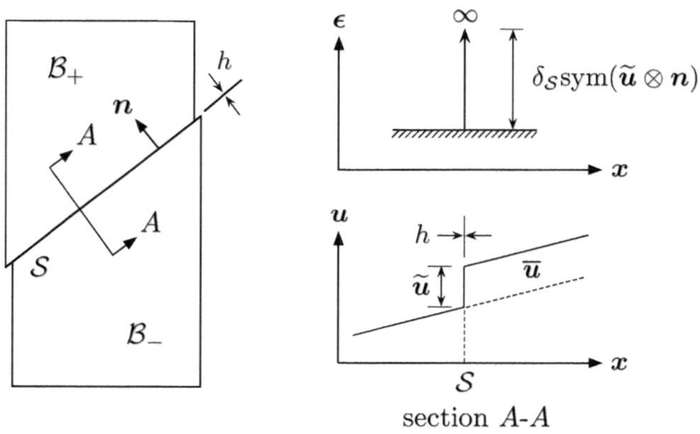

Fig. 7.11 Solid body with a deformation band of thickness h: The limiting condition $h \to 0$ defines a strong discontinuity

In the strong discontinuity limit, the displacement field in the neighborhood of the band can be represented with the aid of the Heaviside function as

$$\boldsymbol{u} = \overline{\boldsymbol{u}} + H_{\mathcal{S}}(\boldsymbol{x})\widetilde{\boldsymbol{u}}\,, \tag{7.71}$$

where

$$H_{\mathcal{S}}(\boldsymbol{x}) = \begin{cases} 1 & \text{if } \boldsymbol{x} \in \mathcal{B}_+ \\ 0 & \text{if } \boldsymbol{x} \in \mathcal{B}_-\,. \end{cases} \tag{7.72}$$

Noting that $\nabla H_{\mathcal{S}} = \delta_{\mathcal{S}}\boldsymbol{n}$, the resulting infinitesimal strain tensor inside the band is

$$\epsilon(\boldsymbol{x}) = \overline{\epsilon}(\boldsymbol{x}) + \delta_{\mathcal{S}}\mathrm{sym}(\widetilde{\boldsymbol{u}} \otimes \boldsymbol{n})\,, \tag{7.73}$$

where $\delta_{\mathcal{S}}$ is the delta function on $\mathcal{S}$.

Consider a strong discontinuity embedded in an elastic continuum. We use continuum theory of plasticity 'inside' the discontinuity and write the Cauchy stress rate tensor as

$$\dot{\boldsymbol{\sigma}} = \boldsymbol{c}^{\mathrm{e}} : (\dot{\epsilon} - \dot{\epsilon}^{\mathrm{p}}) = \boldsymbol{c}^{\mathrm{e}} : \left(\dot{\overline{\epsilon}} + \delta_{\mathcal{S}}\mathrm{sym}(\widetilde{\boldsymbol{v}} \otimes \boldsymbol{n}) - \dot{\lambda}\frac{\partial g}{\partial \boldsymbol{\sigma}}\right), \tag{7.74}$$

where g is the plastic potential function, $\widetilde{\boldsymbol{v}} = \dot{\widetilde{\boldsymbol{u}}}$, and $\dot{\lambda} > 0$ is the usual nonnegative consistency parameter. We readily note that the total strain rate $\dot{\epsilon}$ inside the discontinuity is very large, on the order δ, but $\dot{\boldsymbol{\sigma}}$ is a bounded function. Therefore, the consistency parameter $\dot{\lambda}$ representing the magnitude of plastic strain rate must also be very large, on the order δ,

$$\dot{\lambda} = \delta_S \dot{\lambda}_\delta \,, \tag{7.75}$$

where $\dot{\lambda}_\delta > 0$ is a regular bounded function.

Let f denote the yield function inside the strong discontinuity, and $f = 0$ denote the yield criterion. An example of an acceptable yield function is the simple cohesion-friction law given in (7.6). We emphasize at the outset that $f = 0$ may have very little similarity with the yield criterion for the bulk material. As noted in Section 7.1, the material inside the discontinuity typically is so damaged that it bears little semblance to the intact material from which it formed. Writing the consistency condition in terms of the yield criterion gives

$$\frac{\partial f}{\partial \boldsymbol{\sigma}} : \dot{\boldsymbol{\sigma}} - \dot{\lambda} H = 0 \,, \tag{7.76}$$

where H is the generalized plastic modulus inside the discontinuity. The first term in the above expression is a bounded function, whereas $\dot{\lambda}$ has a very large value on the order δ. For the consistency condition to be satisfied, the plastic modulus H must have a very small value, on the order $1/\delta$. This implies that the continuum response of the material inside the discontinuity must approach perfect plasticity. More specifically, we write

$$H = H_\delta / \delta_S \,, \tag{7.77}$$

where H_δ is a regular function. This yields the reduced consistency condition

$$\frac{\partial f}{\partial \boldsymbol{\sigma}} : \dot{\boldsymbol{\sigma}} - \dot{\lambda}_\delta H_\delta = 0 \,, \tag{7.78}$$

where both $\dot{\lambda}_\delta$ and H_δ are now regular functions. Note that whereas $\dot{\lambda}$ and H are continuum variables, $\dot{\lambda}_\delta$ and H_δ are interface variables. In fact, $\dot{\lambda}_\delta$ may be taken as the finite slip rate at the interface. In contrast, $\dot{\lambda}$ is the intense strain rate at the interface.

Substituting (7.74) into (7.78) produces two separate consistency conditions. The first condition pertains to the regular part,

$$\frac{\partial f}{\partial \boldsymbol{\sigma}} : \boldsymbol{c}^{\mathrm{e}} : \dot{\bar{\boldsymbol{\epsilon}}} - \dot{\lambda}_\delta H_\delta = 0 \,. \tag{7.79}$$

The second condition pertains to the singular part,

$$\frac{\partial f}{\partial \boldsymbol{\sigma}} : \boldsymbol{c}^{\mathrm{e}} : \left[\mathrm{sym}(\widetilde{\boldsymbol{v}} \otimes \boldsymbol{n}) - \dot{\lambda}_\delta \frac{\partial g}{\partial \boldsymbol{\sigma}} \right] = 0 \,, \tag{7.80}$$

from which we get

$$\mathrm{sym}(\widetilde{\boldsymbol{v}} \otimes \boldsymbol{n}) = \dot{\lambda}_\delta \frac{\partial g}{\partial \boldsymbol{\sigma}} \,. \tag{7.81}$$

We recall that g is the plastic potential function inside the discontinuity. Therefore, the singular part of the consistency condition requires that this function be of the form

$$g = \mathrm{sym}(\boldsymbol{m} \otimes \boldsymbol{n}) : \boldsymbol{\sigma} + g_0 \,, \tag{7.82}$$

where $\boldsymbol{m} = \widetilde{\boldsymbol{v}}/\|\widetilde{\boldsymbol{v}}\|$. Furthermore, we see that $\dot{\lambda}_\delta = \|\widetilde{\boldsymbol{v}}\|$, which means that if the material inside the discontinuity is undergoing isochoric deformation the plastic multiplier $\dot{\lambda}_\delta$ must be the slip rate. From equation (7.74), the Cauchy stress rate is now a bounded function and is given by

$$\dot{\boldsymbol{\sigma}} = \boldsymbol{c}^{\mathrm{e}} : \dot{\bar{\boldsymbol{\epsilon}}} \,. \tag{7.83}$$

This is the same Cauchy stress rate prevailing just outside the discontinuity where the host material is deforming elastically.

7.9 Assumed Enhanced Strain Method

There have been many different variations of the assumed enhanced strain formulation, but the main idea is the same: the element hosting a discontinuity is "softened" to capture the kinematics of a piecewise constant strong discontinuity. The magnitude of the discontinuity is statically condensed on the element level prior to global assembly, and the overall system of equations is unchanged. For constant strain triangular or constant strain tetrahedral elements, there is only one Gauss point required for numerical integration, so a static condensation on the element level is equivalent to a static condensation at the Gauss point level (Borja 2000). In what follows, we adopt the development used for the classic and extended finite element formulations to derive the relevant expressions leading to the assumed enhanced strain method.

Consider the classic variational form (7.17) along with expression (7.18) for the contact integral $\mathcal{G}_c$, and assume that $\boldsymbol{\eta}$ and $\widetilde{\boldsymbol{\eta}}$ are two independent variations. Substituting the Cauchy stress relation $\boldsymbol{t} = \boldsymbol{\sigma} \cdot \boldsymbol{n}$ into the contact integral, we get

$$\int_{\mathcal{B}\setminus\mathcal{S}} \boldsymbol{\nabla}^s \boldsymbol{\eta}_c : \boldsymbol{\sigma} \, \mathrm{d}V = \int_{\mathcal{B}\setminus\mathcal{S}} \boldsymbol{\eta}_c \cdot \boldsymbol{f} \, \mathrm{d}V + \int_{\partial\mathcal{B}_t} \boldsymbol{\eta}_c \cdot \widehat{\boldsymbol{t}} \, \mathrm{d}A \,, \tag{7.84}$$

$$\mathcal{G}_c = \int_{\mathcal{S}} \mathrm{sym}(\widetilde{\boldsymbol{\eta}} \otimes \boldsymbol{n}) : \boldsymbol{\sigma} \, \mathrm{d}A = 0 \,. \tag{7.85}$$

In the classic formulation $\boldsymbol{\eta}$ is a conforming variation, so in the variational equation above we have replaced $\boldsymbol{\eta}$ with the conforming variation $\boldsymbol{\eta}_c$.

The displacement field $\boldsymbol{u}(\boldsymbol{x})$ for any enriched element can be obtained from the re-parameterized form (7.33), which we rewrite below as (Borja and Regueiro 2001, Borja 2000)

$$\boldsymbol{u}(\boldsymbol{x}) = \boldsymbol{u}_c(\boldsymbol{x}) + [H_{\mathcal{S}}(\boldsymbol{x}) - f^h(\boldsymbol{x})]\widetilde{\boldsymbol{u}}(\boldsymbol{x}) \,. \tag{7.86}$$

We note that we have not used any blending term $f^h(\boldsymbol{x})$ for the variation $\widetilde{\boldsymbol{\eta}}$, which means that the trial and weighting functions have different forms. Therefore, we have a Petrov-Galerkin formulation. Accordingly, the infinitesimal strain field in the enriched element becomes

$$\boldsymbol{\epsilon}(\boldsymbol{x}) = \boldsymbol{\epsilon}_c(\boldsymbol{x}) - \text{sym}(\widetilde{\boldsymbol{u}} \otimes \boldsymbol{\nabla} f^h(\boldsymbol{x})) + \delta_{\mathcal{S}} \text{sym}(\widetilde{\boldsymbol{u}} \otimes \boldsymbol{n}), \tag{7.87}$$

where $\boldsymbol{\epsilon}(\boldsymbol{x}) = \boldsymbol{\nabla}^s \boldsymbol{u}(\boldsymbol{x})$ and $\boldsymbol{\epsilon}_c(\boldsymbol{x}) = \boldsymbol{\nabla}^s \boldsymbol{u}_c(\boldsymbol{x})$.

In the previous section we have shown that the singular component of the strain tensor is purely plastic, so the elastic component of strain must be the total strain minus the singular term, or

$$\boldsymbol{\epsilon}^{\text{e}} = \boldsymbol{\epsilon}_c - \text{sym}(\widetilde{\boldsymbol{u}} \otimes \boldsymbol{\nabla} f^h). \tag{7.88}$$

Note that the conforming strain $\boldsymbol{\epsilon}_c$ must be corrected by the blending term to recover the continuous elastic strain in the neighborhood of the discontinuity. Alternatively, we can write the displacement field $\boldsymbol{u}(\boldsymbol{x})$ for any enriched element in the original form (7.71), which uses the Heaviside function only, giving

$$\boldsymbol{\epsilon}(\boldsymbol{x}) = \overline{\boldsymbol{\epsilon}}(\boldsymbol{x}) + \delta_{\mathcal{S}} \text{sym}(\widetilde{\boldsymbol{u}} \otimes \boldsymbol{n}), \tag{7.89}$$

where $\overline{\boldsymbol{\epsilon}}(\boldsymbol{x}) = \boldsymbol{\nabla}^s \overline{\boldsymbol{u}}$. Comparing (7.87) with (7.89), we conclude that

$$\overline{\boldsymbol{\epsilon}} = \boldsymbol{\epsilon}_c - \text{sym}(\widetilde{\boldsymbol{u}} \otimes \boldsymbol{\nabla} f^h). \tag{7.90}$$

In other words, the elastic strain is nothing else but the continuous strain. This is in accord with the initial hypothesis that the host medium is elastic.

Figure 7.12 illustrates how the gradient $\boldsymbol{\nabla} f^h$ is calculated for an enriched CST element. An enriched CST element always contains a one-node side, which could lie either on the slave side (Fig. 7.12a) or the master side (Fig. 7.12b) of the discontinuity. Adopting the basic element shape function for f^h, the gradient $\boldsymbol{\nabla} f^h$ is always perpendicular to the two-node side. Taking $\boldsymbol{k}^e = (\boldsymbol{x}^* - \boldsymbol{x}_A)/h^e$, where h^e is the height of the triangle in the direction perpendicular to the two-node side, we readily see that $\boldsymbol{\nabla} f^h = \pm \boldsymbol{k}^e/h^e$, where the sign is selected so that $\boldsymbol{n} \cdot \boldsymbol{\nabla} f^h > 0$. If the two-node side is parallel to the discontinuity, then $\boldsymbol{\nabla} f^h = \boldsymbol{n}/h^e$. In general, the latter condition results in an optimal resolution of the kinematics of the discontinuity, but of course, it is not possible to predict the orientation of a propagating discontinuity. The least optimal condition is when $\boldsymbol{\nabla} f^h$ forms a sharp angle with $\boldsymbol{n}$. This problem can be alleviated by a minor adjustment of the nodal coordinates, see Borja (2000).

From the expression for the elastic component of the infinitesimal strain tensor, the Cauchy stress tensor can be calculated as

$$\boldsymbol{\sigma} = \boldsymbol{c}^{\text{e}} : \boldsymbol{\epsilon}_c - \boldsymbol{c}^{\text{e}} : \text{sym}(\widetilde{\boldsymbol{u}} \otimes \boldsymbol{\nabla} f^h). \tag{7.91}$$

It is useful to express the above equation following the usual format of return mapping. To this end, we impose the elastic constitutive equation at time t_n,

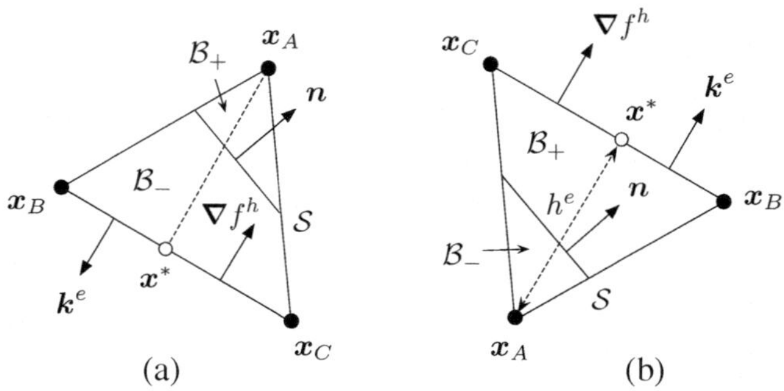

Fig. 7.12 CST elements with a discontinuity: (a) one node on the slave side of $\mathcal{S}$; (b) one node on the master side of $\mathcal{S}$

$$\boldsymbol{\sigma}_n = \boldsymbol{c}^{\mathrm{e}} : \boldsymbol{\epsilon}_{cn} - \boldsymbol{c}^{\mathrm{e}} : \mathrm{sym}(\widetilde{\boldsymbol{u}}_n \otimes \boldsymbol{\nabla} f^h) . \tag{7.92}$$

Subtracting one from the other yields

$$\boldsymbol{\sigma} - \boldsymbol{\sigma}_n = \boldsymbol{c}^{\mathrm{e}} : \Delta\boldsymbol{\epsilon}_c - \boldsymbol{c}^{\mathrm{e}} : \mathrm{sym}(\Delta\widetilde{\boldsymbol{u}} \otimes \boldsymbol{\nabla} f^h) , \tag{7.93}$$

where $\Delta\boldsymbol{\epsilon}_c = \boldsymbol{\epsilon}_c - \boldsymbol{\epsilon}_{cn}$ and $\Delta\widetilde{\boldsymbol{u}} = \widetilde{\boldsymbol{u}} - \widetilde{\boldsymbol{u}}_n$. Note that the jump direction does not change since the discontinuity is embedded in the solid (unless the solid is undergoing finite rotation, see Chapter 8).

We can rewrite (7.93) into a predictor-corrector format as follows

$$\boldsymbol{\sigma} = \boldsymbol{\sigma}^{\mathrm{tr}} - \Delta\lambda_\delta \boldsymbol{c}^{\mathrm{e}} : \mathrm{sym}(\boldsymbol{m} \otimes \boldsymbol{\nabla} f^h) , \tag{7.94}$$

where $\boldsymbol{\sigma}^{\mathrm{tr}} = \boldsymbol{\sigma}_n + \boldsymbol{c}^{\mathrm{e}} : \Delta\boldsymbol{\epsilon}_c$ is the Cauchy predictor stress calculated from the conforming part of the incremental strain, $\Delta\lambda_\delta = \|\Delta\widetilde{\boldsymbol{u}}\|$ is the slip increment, and $\boldsymbol{m} = \Delta\widetilde{\boldsymbol{u}}/\Delta\lambda_\delta$ is the unit slip direction. Since the discontinuity is embedded in the solid and the solid configuration does not change, following the hypothesis of infinitesimal deformation, the only unknown in the above constitutive equation is the slip increment $\Delta\lambda_\delta$. Note that the corrector part in the above predictor-corrector equation emanates from slip relaxation on the discontinuity and not a result of bulk plasticity, which does not enter in the present formulation.

To solve for the slip increment $\Delta\lambda_\delta$, we use the contact integral $\mathcal{G}_c$ given in equation (7.85) as an additional constraint. However, the expression for this contact integral must be modified for the case of piecewise constant slips due to the following reason. For piecewise constant slips, the contact integral may be broken up into individual element constraints due to the lack of connectivity of the slips, i.e.,

$$\int_{\mathcal{S}^e} \mathrm{sym}(\widetilde{\boldsymbol{\eta}}^e \otimes \boldsymbol{n}^e) : \boldsymbol{\sigma} \, \mathrm{d}A^e = 0 \qquad e = 1, 2, \ldots, n_{el} , \tag{7.95}$$

where $\mathcal{S}^e$ is the area of embedded surface within an enriched element volume $\mathcal{B}^e$ (in two dimensions, $\mathcal{S}^e$ is the length of embedded segment within an enriched element area $\mathcal{B}^e$), and n_{el} is the number of enriched elements. For a piecewise constant stress field, which is generated by constant strain triangular or constant strain tetrahedral elements, the integral takes the stronger form

$$\mathrm{sym}(\widetilde{\boldsymbol{\eta}}^e \otimes \boldsymbol{n}^e) : \boldsymbol{\sigma} = 0\,, \qquad e = 1, 2, \ldots, n_{el}\,. \tag{7.96}$$

The above equation is precisely the consistency condition for cohesionless contact ($c = 0$) when $\widetilde{\boldsymbol{\eta}}^e$ is constructed to be of the form

$$\widetilde{\boldsymbol{\eta}}^e = \boldsymbol{m}^e + \mu \boldsymbol{n}^e\,, \tag{7.97}$$

where μ is the coefficient of friction and $\boldsymbol{m}^e$ is a unit vector in the direction of the resolved shear stress. This can be verified by expanding

$$\mathrm{sym}(\boldsymbol{m}^e \otimes \boldsymbol{n}^e) : \boldsymbol{\sigma} + \mu\, \mathrm{sym}(\boldsymbol{n}^e \otimes \boldsymbol{n}^e) : \boldsymbol{\sigma} = \|\boldsymbol{t}_T\| - \mu t_N = 0\,, \tag{7.98}$$

and comparing the above expression with (7.6). However, if $c \neq 0$ the contact integral does not coincide with the consistency condition for constant strain elements. In a way, this is equivalent to the element not passing the patch test for constant strain fields (Hughes 1987).

The following modified expression for the contact integral passes the patch test for constant strain fields (Armero and Garikipati 1995, Simo and Oliver 1994)

$$\widetilde{\mathcal{G}}_c = \int_{\mathcal{B}^e} \widetilde{\gamma}^h : \boldsymbol{\sigma}\, \mathrm{d}V = 0\,, \qquad \widetilde{\gamma}^h = (C - \delta_{\mathcal{S}}) \widetilde{\boldsymbol{\eta}}^e \otimes \boldsymbol{n}^e\,. \tag{7.99}$$

This can be seen from the fact that the condition

$$\int_{\mathcal{B}^e} \widetilde{\gamma}^h\, \mathrm{d}V = \boldsymbol{0} \tag{7.100}$$

is identically satisfied by choosing $C = \mathcal{S}^e/\mathcal{B}^e$. Indeed, we can show that the above expression for $\widetilde{\mathcal{G}}_c$ reproduces a volume-integrated consistency condition.

The expression for $\widetilde{\mathcal{G}}_c$ when $C = \mathcal{S}^e/\mathcal{B}^e$ is

$$\widetilde{\mathcal{G}}_c = \frac{\mathcal{S}^e}{\mathcal{B}^e} \int_{\mathcal{B}^e} \mathrm{sym}(\widetilde{\boldsymbol{\eta}}^e \otimes \boldsymbol{n}^e) : \boldsymbol{\sigma}\, \mathrm{d}V - \int_{\mathcal{S}^e} \mathrm{sym}(\widetilde{\boldsymbol{\eta}}^e \otimes \boldsymbol{n}^e) : \boldsymbol{\sigma}\, \mathrm{d}A = 0\,. \tag{7.101}$$

However, the expression inside the second integral sign is nothing else but the cohesion c if we impose the yield criterion, cf. (7.6). Furthermore, c is either constant, or it may harden/soften with plastic slip, which is assumed constant within the element. Therefore, c must be constant throughout the element, and hence,

$$\frac{1}{\mathcal{B}^e} \mathrm{sym}(\widetilde{\boldsymbol{\eta}}^e \otimes \boldsymbol{n}^e) : \int_{\mathcal{B}^e} \boldsymbol{\sigma}\, \mathrm{d}V - c = 0\,. \tag{7.102}$$

For constant strain elements the above equation collapses to the consistency condition $f = 0$. For higher-order elements one needs to calculate the volume average (or area average, for 2D) of the stress term. Note from the above expression that $\mathcal{S}^e$ cancels out, so the above equation is true irrespective of the position of the discontinuity within an element.

Because the slips are piecewise constant, the contributions of the contact integral $\widetilde{\mathcal{G}}_c$ are purely local and may be statically condensed on the element level. This means that there are no additional degrees of freedom to solve other than the conforming nodal displacements. For a constant stress field, the required one-point Gauss integration provides an additional simplification. In this case, the consistency parameter $\Delta\lambda_\delta$ may be evaluated at the Gauss point level as

$$\Delta\lambda_\delta = \frac{\operatorname{sym}(\widetilde{\boldsymbol{\eta}}^e \otimes \boldsymbol{n}^e) : \boldsymbol{\sigma}^{\mathrm{tr}} - c}{\operatorname{sym}(\widetilde{\boldsymbol{\eta}}^e \otimes \boldsymbol{n}^e) : \boldsymbol{c}^e : \operatorname{sym}(\boldsymbol{m}^e \otimes \boldsymbol{\nabla} f^{h,e})} \,. \tag{7.103}$$

The final Cauchy stress tensor $\boldsymbol{\sigma}$ can then be evaluated directly from (7.94). As a final remark, linear jumps, such as the one proposed by Linder and Armero (2007), can also be accommodated by the strong discontinuity framework, although they are not covered in this book.

Exercise 7.5. Quasistatic and isothermal plastic deformations in single crystals arise from slips on specific crystallographic planes. A slip system β is defined by a unit normal vector $\boldsymbol{n}^{(\beta)}$ to the crystallographic plane and a unit slip direction $\boldsymbol{m}^{(\beta)}$. Forward and backward slips are treated separately. For a face-centered cubic crystal, there are 24 slip systems including forward and backward systems. Assume that a crystal is deforming in single slip (i.e., only one slip system is active) and that the yield function is given by

$$f = \operatorname{sym}(\boldsymbol{m} \otimes \boldsymbol{n}) : \boldsymbol{\sigma} - \tau_Y \le 0 \,,$$

where $\boldsymbol{\sigma}$ is the overall crystal stress, τ_Y is the resolved yield stress, and the slip tensor is $\boldsymbol{m} \otimes \boldsymbol{n} = \boldsymbol{m}^{(1)} \otimes \boldsymbol{n}^{(1)}$. Construct a contact integral $\widetilde{\mathcal{G}}_c$ satisfying the patch test and the consistency condition for this system.

7.10 Crack Tip Enrichment

Linear elastic fracture mechanics indicates that the strains at the crack tip are singular (Broek 1986). Thus, it is of interest to compare how the assumed enhanced strain, extended finite element method, and classic solutions resolve this singularity at the crack tip. To this end, we consider a 2×4 elastic rectangular block (consistent units throughout) with a 45° edge crack. The top and bottom boundaries are supported on rollers, and one corner node is pinned to prevent rigid-body horizontal translation. The block is compressed vertically, forcing the crack faces to rub against each other and slide. The elastic constants for the block are: Young's modulus $E = 10^5$ and Poisson's ratio $\nu = 0.30$. We assume a cohesion $c = 0$ and a coefficient of friction $\mu = 0.10$ on the crack; contact constraints are imposed by penalty regularization with $k_N = k_T = 10^7$.

The finite element mesh consists of 153 nodes and 256 CST elements arranged in a cross-diagonal pattern. For the two enrichment techniques (assumed enhanced strain and extended finite element methods) the crack passes through the element interior, but the elements are oriented so that their sides are nearly aligned to the crack. For the classic solution, the element sides define the crack itself, where each crack segment is represented by a pair of element faces. Figure 7.13 shows the position of the crack relative to the finite elements.

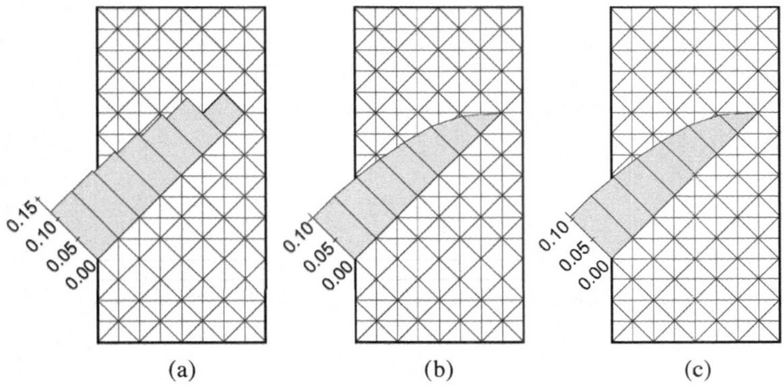

Fig. 7.13 Rectangular elastic block with a stationary edge crack: the block is compressed vertically in plane strain. Plots show the distribution of plastic slips along the crack: (a) assumed enhanced strain; (b) extended finite element; and (c) classic solutions. After Borja (2008).

Figures 7.13 and 7.14 portray the variations of slip calculated by the three finite element solutions. As expected, slip has a maximum value at the exposed face of the block and a minimum value at the crack tip. It is of interest to note that the classic and extended finite element solutions predict nearly the same slip distributions, although they are not one on top of the other in Fig. 7.14 because of the slightly different positions of the crack. However, both solutions predict smaller slips overall compared to the assumed enhanced strain solution. The latter solution exhibits the classical elliptical shape consistent with linear elastic fracture mechanics theory, whereas the other two solutions exhibit bullet-shaped slip distributions. Clearly, the continuous slip interpolation causes the deformation to lock in the vicinity of the crack tip.

Linear elastic fracture mechanics (Broek 1986, Belytschko and Black 1999) predicts the following near-tip displacement field for combined Modes I and II loading:

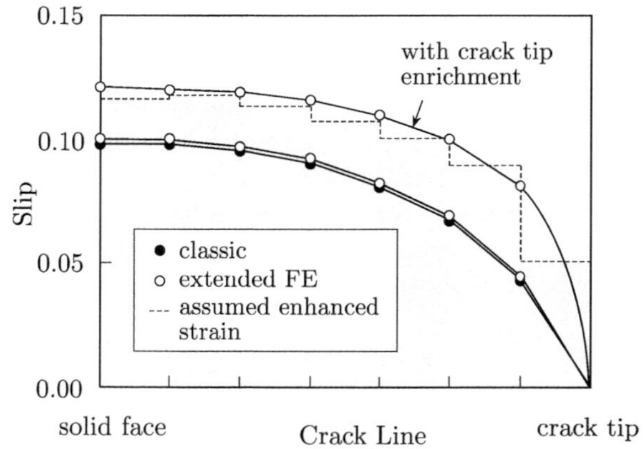

Fig. 7.14 Distribution of slip along a 45-degree edge crack with $c = 0$ and $\mu = 0.10$

$$\begin{aligned} u(r,\theta) &= \frac{K_{\rm I}}{2G}\sqrt{r/2G}\cos\left(\frac{\theta}{2}\right)\left[\kappa - 1 + 2\sin^2\left(\frac{\theta}{2}\right)\right] \\ &+ \frac{K_{\rm II}}{2G}\sqrt{r/2G}\sin\left(\frac{\theta}{2}\right)\left[\kappa + 1 + 2\cos^2\left(\frac{\theta}{2}\right)\right] \end{aligned} \tag{7.104}$$

for the x direction, and

$$\begin{aligned} v(r,\theta) &= \frac{K_{\rm I}}{2G}\sqrt{r/2G}\sin\left(\frac{\theta}{2}\right)\left[\kappa + 1 - 2\cos^2\left(\frac{\theta}{2}\right)\right] \\ &- \frac{K_{\rm II}}{2G}\sqrt{r/2G}\cos\left(\frac{\theta}{2}\right)\left[\kappa - 1 - 2\sin^2\left(\frac{\theta}{2}\right)\right] \end{aligned} \tag{7.105}$$

for the y direction, where

$$\kappa = \begin{cases} 3 - 4\nu, & \text{plane strain} \\ (3-\nu)/(1+\nu), & \text{plane stress} \end{cases} \tag{7.106}$$

is the Kolosov constant. In the above equations, $K_{\rm I}$ and $K_{\rm II}$ are stress intensity factors, r and θ are cylindrical coordinates originating from the crack tip, ν is Poisson's ratio, and G is the elastic shear modulus. The angle θ ranges from $-\pi$ to $+\pi$ ($\theta = 0$ is the crack line pointing away from the tip) so that $\sin(\theta/2)$ represents the discontinuous function on the crack. For pure Mode II loading the near-tip slip is given by

$$\zeta(r) = u(r,\pi) - u(r,-\pi) = \frac{K_{\rm II}}{G}(\kappa + 1)\sqrt{r/2G}\,. \tag{7.107}$$

The gradient of slip with respect to r is

$$\zeta'(r) = \frac{K_{\rm II}}{2G}\frac{\kappa + 1}{\sqrt{2Gr}}\,. \tag{7.108}$$

The above equation shows a $1/\sqrt{r}$ near-tip singularity of the gradient that cannot be captured by conventional finite element solutions. This explains why the near-tip deformations calculated by the classic and extended finite element solutions tend to lock. In contrast, slip predicted by the assumed enhanced strain solution shows an infinite gradient at the crack tip.

To handle the crack tip singularity, we enrich the description of the near-tip displacement field using a technique proposed by Belytschko and Black (1999). The idea is to introduce additional nodal degrees of freedom for the polygonal support of the crack tip to accommodate four shape functions

$$\begin{aligned}\gamma_i(r,\theta) &= \sqrt{r} f_i\,, \qquad f_1 = \cos\left(\frac{\theta}{2}\right), \qquad f_2 = \sin\left(\frac{\theta}{2}\right), \\ & \qquad f_3 = \sin\left(\frac{\theta}{2}\right)\sin\theta\,, \qquad f_4 = \cos\left(\frac{\theta}{2}\right)\sin\theta\,. \end{aligned} \tag{7.109}$$

The spans of the above shape functions are the same as the independent spans of equations (7.104) and (7.105), thus allowing the polygonal support to capture near-tip deformation. With the above crack tip enrichment in place, Fig. 18 shows an improved slip distribution calculated by the extended finite element solution utilizing both the Heaviside and crack-tip enrichments. Because of the enriched basis function near the crack tip, the problem of locking has been alleviated, and the solution now accommodates the near-tip singularity of the gradient.

7.11 Slip Weakening

Slip weakening is the degradation of strength of rocks during the initial stage of slip instability. The concept was motivated by the cohesive zone models for tensile fracture (Barenblatt 1962, Bilby et al. 1963, Dugdale 1960, Ida 1972, Palmer and Rice 1973). As the Earth's crust is deformed, the stress increases until it reaches a peak resistance. Rocks then fail either by developing a new fault zone or reactivating an old one. The shear strength decays to a lower level on those segments of the fault that slipped. Various slip weakening laws have been proposed and calibrated in the laboratory (Rummel et al. 1978, Okubo and Dietrich 1984, Wong 1982).

For modeling purposes, the intact rock and the fault that develops from it constitute two different materials and are represented by two different constitutive models. At the point of failure a discontinuity may be embedded into the failed element to simulate the nucleation of a fault. Then, the constitutive description for the element transitions accordingly. The constitutive model for the fault should provide a softening variable to facilitate a smooth transition from the pre-failure to the post-failure states. For cohesive-frictional materials, the cohesion of the fault is often chosen as this softening variable (Borja and Foster 2007).

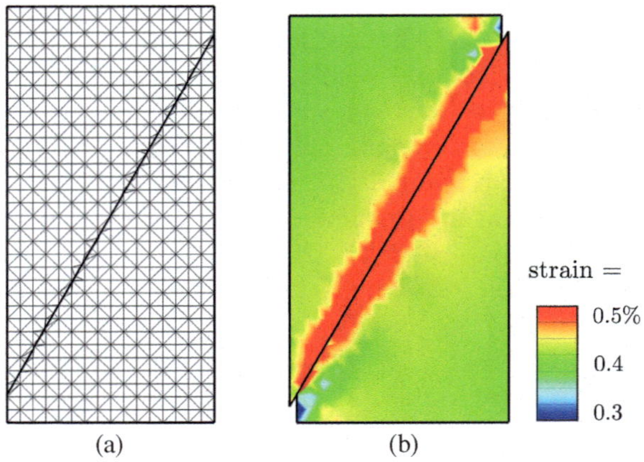

Fig. 7.15 Compression of a 0.04 m × 0.08 m block with a 60-degree frictional crack: (a) extended finite element mesh; and (b) cumulative bulk plastic strain. Bulk plasticity concentrates near the crack faces as the block slides along the frictional crack. Color bar is cumulative plastic strain in percent. After Liu and Borja (2009).

Figure 7.15a shows a rectangular block with a 60° frictional crack embedded in it. The crack is given an artificial cohesion that decays to zero due to slip weakening. The initial value of this cohesion is determined by equating the yield strength of the pre-failure model, taken as a non-associated Drucker-Prager model in this example, to the yield strength of the post-failure model at the failure stress. Over a characteristic distance D_c this cohesion drops to zero, allowing the coefficient of friction to take full control of the shear strength on the crack. The sides of the finite elements representing the block are not aligned to the crack, so it is not possible to use the classic formulation for this problem. Instead, we can employ either the assumed enhanced strain method (Foster et al. 2007) or the extended finite element technique (Liu and Borja 2009) and allow the crack to pass through the interior of the elements. Here, we use the latter technique.

To allow the frictional resistance on the crack to develop, a pressure of 250 MPa was applied on the two vertical faces of the block. Without this confining stress, the residual strength on the crack would be zero since there would be no contact pressure on it. Figure 7.16 shows the resulting load-compression plot. The failure stress is reached at a vertical displacement of around 0.8 mm, triggering a slip-weakening mechanism. To demonstrate the effect of variable friction, the simulation also includes a velocity stepping. The initial rate of vertical compression is 0.866 μm/s, which is equivalent to a slip speed of 1.0 μm/s at steady state. The rate of compression is then increased ten times, resulting in a slight drop in the value of the coefficient of friction ($\mu^* = 0.72$ in equation (7.9)).

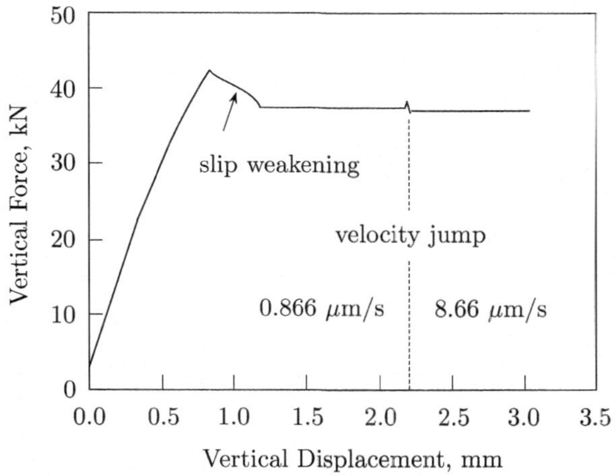

Fig. 7.16 Compression of a block with a 60-degree frictional crack: vertical force versus vertical displacement. After Liu and Borja (2009)

Figure 7.15b shows a contour of the cumulative bulk plastic strain at the conclusion of the simulation. Bulk plasticity concentrates in the region surrounding the crack, suggesting a potential for thickening of the failure zone. The displayed mechanism is in general agreement with the thickening of the fault core shown in the sketches of Fig. 7.1.

7.12 Summary

Cohesive-frictional sliding is an inelastic deformation concentrated within a narrow zone of negligible thickness. The classic formulation for an interface with negligible thickness may be employed for cohesive-frictional sliding if the geometry of the discontinuity is known a priori. For an evolving interface, such as a crack that nucleates and propagates as the solution progresses, enhanced finite element techniques including the extended finite element and assumed enhanced strain methods may be employed to allow the discontinuity to evolve on a fixed background grid. Bulk plasticity can also develop in the neighborhood of the discontinuity, in which case, the algorithm must be sufficiently robust to accommodate concurrent calculation of both distributed and localized plastic deformations within the domain of the problem.

8 Crystal Plasticity

8.1 Introduction

Metals and igneous rocks are most common materials with crystalline microstructures. Their elastoplastic properties are attributed to the existence of slip planes. The face-centered cubic (f.c.c.) structure has slip planes along (111) directions, while body-centered cubic (b.c.c.) crystals have slip planes in the (110) family. Because of their denser packing, materials with a f.c.c. structure, such as aluminum, copper, gold, and silver, tend to be more ductile than materials with a b.c.c. structure, such as iron, chromium, tungsten, and niobium. But packing alone does not determine the absolute ductility of a given material. Hexagonal close-packed (h.c.p.) and f.c.c. crystals both have coordination numbers of twelve, yet f.c.c. crystals have twelve slip planes while h.c.p. only has three, making slip in h.c.p. crystals more limited. Consequently, cadmium, zinc, titanium, beryllium, and other metals with a h.c.p. lattice structure tend to be more brittle than metals with a f.c.c. structure. Figures 8.1 and 8.2 compare the lattice arrangements of h.c.p. and f.c.c. crystals showing more limited slip mobility of the former compared to the latter.

Understanding the elasto-plastic behavior of single crystals is important for the prediction of the overall behavior of crystal aggregates. Quasistatic and isothermal plastic deformation of single crystals is due to slip on crystallographic planes, which occurs when the resolved shear stress on a critical slip system reaches a certain maximum value. The challenge lies in identifying the specific slip systems activated by a given increment of load, since the process typically involves selection from a pool of linearly dependent slip systems. In addition, 'hyperplanes' in stress space representing the yield surfaces for the slip systems intersect each other in a non-smooth fashion, producing edges and corners on which the gradient with respect to the stress tensor cannot be readily calculated.

In this chapter, we focus on multi-surface plasticity theory for single crystals and present theoretical and computational aspects of imposing linearly independent active constraints. We address the problem of identifying the independent active slip systems from a pool of linearly dependent ones using a robust 'ultimate' algorithm. The name has been coined for this technique because of its two desirable features: it is unconditionally convergent, and it is exact for an imposed crystal deformation varying as a ramp function.

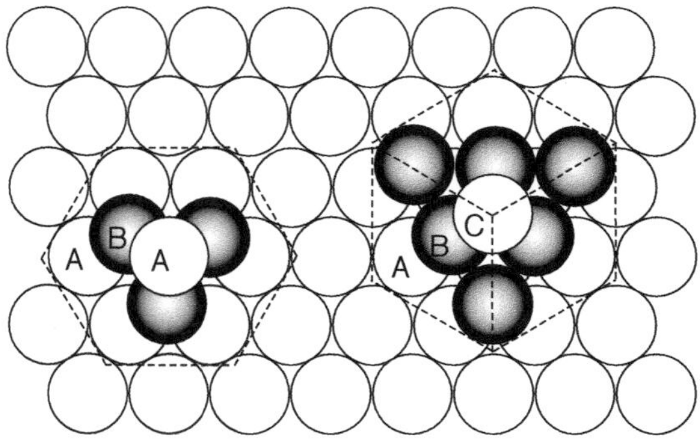

Fig. 8.1 H.c.p. (left) and f.c.c (right) arrangements of atoms. In a h.c.p. structure the positioning is ABABABA (every other layer is the same), while in a f.c.c. structure the positioning is ABCABCA (every third layer is the same). Dashed lines denote outline of each Bravais lattice.

Moreover, we address both the infinitesimal and finite deformation aspects of crystal plasticity. For the latter aspect, theory of distribution and strong discontinuity concepts are key ingredients of the mathematical formulation leading to the discrete micromechanics of elastoplastic crystals in the finite deformation range. In keeping with the overall format of the book, we only address the rate-independent aspects of crystal deformation, the reason being that rate-dependent regularization techniques abound in the literature, but a robust rate-independent formulation for elastoplastic crystals remains scarce.

8.2 Kinematics of Crystal Slip

We restrict the formulation to infinitesimal deformation for now and address the finite deformation aspects later in this chapter. Consider the usual additive decomposition of the infinitesimal strain rate tensor,

$$\dot{\boldsymbol{\epsilon}} = \dot{\boldsymbol{\epsilon}}^{\mathrm{e}} + \dot{\boldsymbol{\epsilon}}^{\mathrm{p}}\,, \tag{8.1}$$

where the elastic part $\dot{\boldsymbol{\epsilon}}^{\mathrm{e}}$ emanates from the crystal lattice deformation, and the plastic part $\dot{\boldsymbol{\epsilon}}^{\mathrm{p}}$ arises from slips on crystallographic planes. Denoting by $\boldsymbol{n}^{(\beta)}$ the unit normal to a crystallographic plane containing the β-slip system, $\boldsymbol{m}^{(\beta)}$ the direction of plastic slip, and $\dot{\gamma}^{(\beta)}$ the rate of crystallographic glide strain, the velocity in the direction of slip is written as

$$\boldsymbol{v}^{(\beta)} = \dot{\gamma}^{(\beta)}(\boldsymbol{x}\cdot\boldsymbol{n}^{(\beta)})\boldsymbol{m}^{(\beta)}\,. \tag{8.2}$$

Figure 8.3 shows the geometrical significance of the glide strain. In this figure $\dot{\gamma}^{(\beta)}$ is meant to be small, but it has been magnified to delineate the geometry

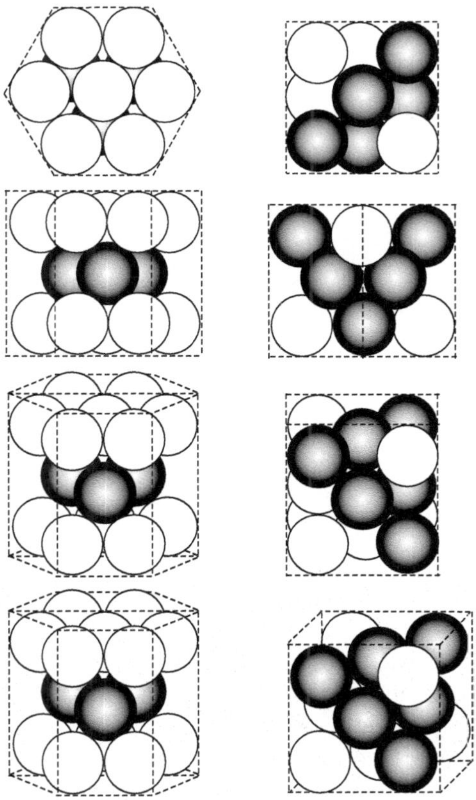

Fig. 8.2 Rotated views of h.c.p. (left) and f.c.c (right) arrangement of atoms

of deformation. The velocity gradient contributed by glide strain β can be evaluated from the expression

$$\nabla \boldsymbol{v}^{(\beta)} = \dot{\gamma}^{(\beta)} \boldsymbol{m}^{(\beta)} \otimes \boldsymbol{n}^{(\beta)} , \tag{8.3}$$

where $\boldsymbol{m}^{(\beta)} \otimes \boldsymbol{n}^{(\beta)}$ is the Schmid tensor (Schmid and Boas 1935). Summing over all active crystallographic slips results in the following expression for the plastic strain rate

$$\dot{\boldsymbol{\epsilon}}^{\mathrm{p}} = \sum_{\beta \text{ active}} \dot{\gamma}^{(\beta)} \boldsymbol{\alpha}^{(\beta)} , \tag{8.4}$$

where

$$\boldsymbol{\alpha}^{(\beta)} = \frac{1}{2} (\boldsymbol{m}^{(\beta)} \otimes \boldsymbol{n}^{(\beta)} + \boldsymbol{n}^{(\beta)} \otimes \boldsymbol{m}^{(\beta)}) . \tag{8.5}$$

Note that $\boldsymbol{m}^{(\beta)} \cdot \boldsymbol{n}^{(\beta)} = 0$, so the slips produce a volume-preserving deformation.

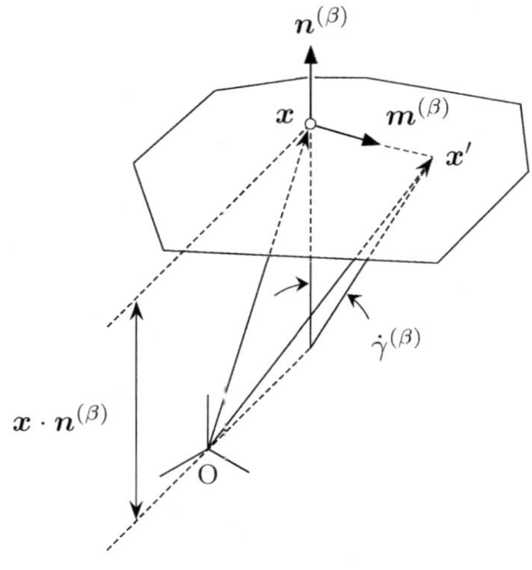

Fig. 8.3 Geometric meaning of plastic slip in a single crystal

8.3 Constitutive Framework

Consider a crystal with $2N$ potentially active slip systems, including forward and backward slip systems. There can be no more than N active slip systems, since if a forward slip direction is active then the backward direction must be latent. For example, a f.c.c. crystal possesses eight (111) planes in the unit cell and three $\langle 110 \rangle$ possible slip directions on each plane, so $N = 12$ for this crystal structure. Let $\tau_Y^{(\beta)}$ represent the yield stress for each slip system. By definition, the system is potentially active if at least one of the following conditions is satisfied:

$$f^{(\beta)} = \begin{cases} \boldsymbol{\sigma} : \boldsymbol{\alpha}^{(\beta)} - \tau_Y^{(\beta)} = 0\,, & \beta = 1, 2, \ldots, N \\ -\boldsymbol{\sigma} : \boldsymbol{\alpha}^{(\beta)} - \tau_Y^{(\beta)} = 0\,, & \beta = N+1, N+2, \ldots, 2N\,. \end{cases} \tag{8.6}$$

These yield conditions delineate $2N$ hyperplanes in the general stress space, and delimit the elastic region

$$\mathsf{E} = \{(\boldsymbol{\sigma}, \tau_Y^{(\beta)}) \in \mathbb{R}^6 \times \mathbb{R}^{2N} \,|\, f^{(\beta)} < 0\,,\ \beta = 1, 2, \ldots, 2N\}\,. \tag{8.7}$$

A more compact expression for the yield surfaces is

$$f^{(\beta)} = |\boldsymbol{\sigma} : \boldsymbol{\alpha}^{(\beta)}| - \tau_Y = 0\,, \qquad \beta = 1, 2, \ldots N\,, \tag{8.8}$$

where the absolute value sign accounts for both forward and backward slips.

Depending on the position of the stress point, two or more yield constraints could activate and trigger multiple components of plastic flow. Assuming an associative flow rule for the plastic flow, the plastic strain rate can be written as (Koiter 1960)

$$\dot{\boldsymbol{\epsilon}}^{\mathrm{p}} = \sum_{\beta=1}^{2N} \dot{\gamma}^{(\beta)} \frac{\partial f^{(\beta)}}{\partial \boldsymbol{\sigma}} = \sum_{\beta=1}^{2N} \dot{\gamma}^{(\beta)} \boldsymbol{\alpha}^{(\beta)}, \tag{8.9}$$

where $\boldsymbol{\alpha}^{(\beta)} = -\boldsymbol{\alpha}^{(\beta-N)}$ for $N < \beta \leq 2N$. The plastic multipliers $\dot{\gamma}^{(\beta)}$ must satisfy the classic Kuhn-Tucker conditions (Budiansky and Wu 1962)

$$\dot{\gamma}^{(\beta)} \geq 0, \qquad f^{(\beta)} \leq 0, \qquad \dot{\gamma}^{(\beta)} f^{(\beta)} = 0 \tag{8.10}$$

for all β. The system is active if $\dot{\gamma}^{(\beta)} > 0$ and $f^{(\beta)} = 0$, and inactive if $\dot{\gamma}^{(\beta)} = 0$ and $f^{(\beta)} < 0$. From this expression for the plastic strain rate tensor produced by the glide strains, we can write the elastic rate constitutive equation as

$$\dot{\boldsymbol{\sigma}} = \boldsymbol{c}^{\mathrm{e}} : \left(\dot{\boldsymbol{\epsilon}}^{\mathrm{e}} - \sum_{\beta} \dot{\gamma}^{(\beta)} \boldsymbol{\alpha}^{(\beta)} \right), \tag{8.11}$$

where $\boldsymbol{c}^{\mathrm{e}}$ is the crystal lattice elasticity tensor. Although the summation for the plastic strain rate tensor is carried out over all β's, only the active systems contribute to the sum.

A sufficiently general hardening law for elastoplastic crystals is given by the expression

$$\dot{\tau}_Y^{(\beta)} = \sum_{\xi=1}^{2N} h_{\beta\xi} \dot{\gamma}^{(\xi)}, \tag{8.12}$$

where $h_{\beta\xi}$ is a matrix of plastic hardening moduli whose elements generally depend on the history of deformation (Peirce et al. 1982). If the backward systems are all taken into consideration, a hardening law of this general form requires a total of $2N \times 2N$ plastic moduli, which is a very large number of parameters for most crystal structures. Assuming that the matrix $[h_{\beta\xi}]$ is diagonal, we can write

$$h_{\beta\xi} = h_{\beta} \delta_{\beta\xi} \qquad (\text{no sum}), \tag{8.13}$$

where $\delta_{\beta\xi}$ is the Kronecker delta, see Koiter (1953). Substituting in (8.12) results in the hardening law

$$\dot{\tau}_Y^{(\beta)} = h_{\beta} \dot{\gamma}^{(\beta)} \qquad (\text{no sum}), \tag{8.14}$$

where a total of $2N$ plastic moduli are still required. However, if we assume $h_{\beta\xi} = h$, then we obtain a simple hardening law due to Taylor (1938)

$$\dot{\tau}_Y^{(\beta)} = h \sum_{\xi=1}^{2N} \dot{\gamma}^{(\xi)}. \tag{8.15}$$

According to the above law, an equal increment of hardening is generated on all slip systems, including active and latent systems. With only one plastic modulus required, namely, h, the Taylor hardening has become the most

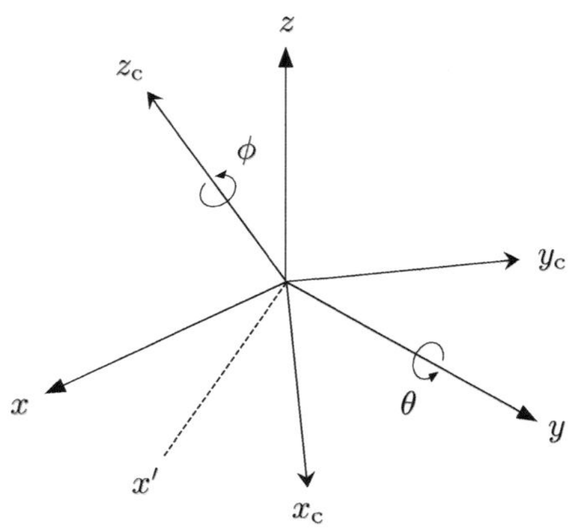

Fig. 8.4 Euler angles defining crystal axes (x_c, y_c, z_c) relative to the fixed system (x, y, z)

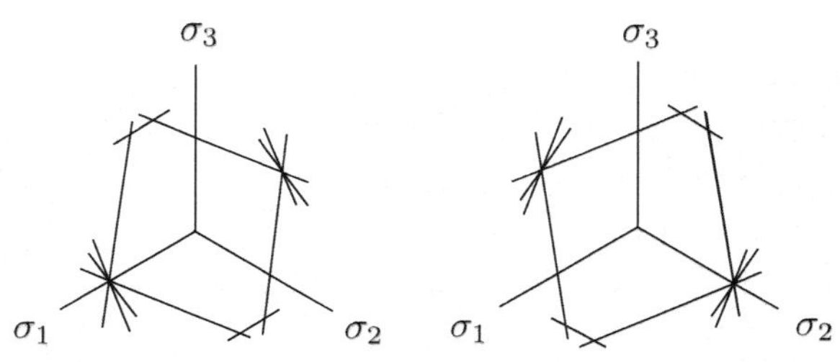

Fig. 8.5 Yield polyhedra for f.c.c. crystal with Euler angles $\theta = 0$ and $\phi = 18°$ (left) and $\theta = 72°$ and $\phi = 90°$ (right). After Borja and Wren (1993).

widely used law for crystals. With this law, forward and backward slips generate the same rate of hardening.

Crystal orientations can be described by the Euler angles between a fixed reference frame (x, y, z) and a crystal reference frame (x_c, y_c, z_c), as shown in Fig. 8.4. The Euler angles consist of a positive (right-hand rule) rotation of θ about the y-axis, followed by a positive rotation of ϕ about the z_c-axis. Figure 8.5 compares the yield polyhedra for a crystal with two different orientations relative to fixed principal stress axes. Polyhedra are the intersections of the yield hyperplanes $|\boldsymbol{\sigma} : \boldsymbol{\alpha}^{(\beta)}| = \tau_Y$, for $\beta = 1, \ldots, N$, with the π-plane. Some of the yield planes plot one on top of the other, while others cluster in a corner. In light of this, redundant constraints are to be expected when two or more slip systems activate.

8.4 Ultimate Algorithm

A numerical technique called 'ultimate algorithm' may be used to identify a group of linearly independent active systems from all potentially active constraints. The algorithm derives its name from the properties that: (a) it is unconditionally convergent, and (b) it is exact for imposed crystal deformation varying as a ramp function. In a nutshell, the ultimate algorithm divides the imposed crystal deformation in increments and tracks the sequence of slip system activation one by one. It is possible to identify different sets of linearly independent constraints, but the resulting overall crystal stress is demonstrably unique.

To illustrate the algorithm, we first define the set of active constraints as

$$\mathcal{J}_{\text{act}} = \{\beta \in \{1, 2, \ldots, N\} \mid f^{(\beta)} = 0 \text{ and } \dot{\gamma}^{(\beta)} \geq 0\} . \tag{8.16}$$

Linearly independent slip systems are then selected from the set $\mathcal{J}_{\text{act}}$, and collected in the subset $\overline{\mathcal{J}}_{\text{act}} \subset \mathcal{J}_{\text{act}}$. By definition, active slip systems are linearly independent if

$$\sum_{\beta \in \overline{\mathcal{J}}_{\text{act}}} \dot{\gamma}^{(\beta)} \boldsymbol{\alpha}^{(\beta)} = \mathbf{0} \quad \text{implies} \quad \dot{\gamma}^{(\beta)} = 0 \quad \forall \beta \in \overline{\mathcal{J}}_{\text{act}} , \tag{8.17}$$

see Nowinski (1981). The remaining slip systems in the set $\mathcal{J}_{\text{act}} \backslash \overline{\mathcal{J}}_{\text{act}}$ constitute all redundant constraints. The tensors $\boldsymbol{\alpha}^{(\beta)}$ for all $\beta \in \overline{\mathcal{J}}_{\text{act}}$ are said to form linearly independent bases for the plastic strain rate $\dot{\boldsymbol{\epsilon}}^{\text{p}}$. However, there can be only six independent bases for the symmetric tensor $\dot{\boldsymbol{\epsilon}}^{\text{p}}$. Furthermore, $\text{tr}(\dot{\boldsymbol{\epsilon}}^{\text{p}}) = 0$, which represents an additional constraint on the plastic strain rate. Therefore, the set $\overline{\mathcal{J}}_{\text{act}}$ can have no more than five elements.

To demonstrate how the ultimate algorithm works, consider first how a primary slip system activates. Writing (8.1) in discrete form gives

$$\Delta \boldsymbol{\epsilon} = \Delta \boldsymbol{\epsilon}^{\text{e}} + \Delta \boldsymbol{\epsilon}^{\text{p}} . \tag{8.18}$$

Assume that the strain increment is applied as a ramp function, i.e.,

$$\Delta \boldsymbol{\epsilon}(\tau) = \kappa \Delta \boldsymbol{\epsilon} , \qquad \kappa = \tau / \Delta t , \tag{8.19}$$

where $0 \leq \tau \leq \Delta t$. By 'sliding' τ from zero to Δt, or κ from zero to one, we can track the sequence of activation of slip systems one by one.

Now, assume that $|\boldsymbol{\alpha}^{(\beta)} : \boldsymbol{\sigma}_n| - \tau_{Y,n} < 0$ for all slip systems initially, which means that the initial stress point lies inside the elastic region. Applying a strain increment $\kappa \Delta \boldsymbol{\epsilon}$ given by (8.19) results in a stress evolution of the form

$$\boldsymbol{\sigma}(t) = \boldsymbol{\sigma}_n + \int_0^{\kappa \Delta t} \mathbf{c}^{\text{e}} : \dot{\boldsymbol{\epsilon}} \, \text{d}\tau = \boldsymbol{\sigma}_n + \kappa \mathbf{c}^{\text{e}} : \Delta \boldsymbol{\epsilon} , \tag{8.20}$$

provided that κ is small enough that no slip system activates. The goal is to search for the smallest value of κ that activates the most critical slip system. This entails evaluating κ for all slip systems and constructing the set

$$\Psi_1 = \{\kappa^{(\beta)} \in R_+ |\, \psi^{(\beta)}\boldsymbol{\alpha}^{(\beta)} : \boldsymbol{\sigma}(t) - \tau_{Y,n} = 0\}\,, \tag{8.21}$$

where $\psi^{(\beta)} = \text{sign}[\boldsymbol{\alpha}^{(\beta)} : \boldsymbol{\sigma}(t)]$. If $\kappa^{(\beta)} < 1$ for some β, then the critical slip system β_1 is the constraint that yields the smallest value of $\kappa^{(\beta)}$. This in turn determines the subincrement $\kappa\Delta\boldsymbol{\epsilon}$ that activates the initial operative (primary) system.

To identify the secondary slip system, we assume that there is an active primary slip system β_1 initially. Provided that the primary slip system does not deactivate and that no other slip system activates, the evolution of the crystal stress may be described by the equation

$$\begin{aligned} \boldsymbol{\sigma}(t) &= \boldsymbol{\sigma}_n + \int_0^{\kappa\Delta t} \boldsymbol{c}^{\mathrm{e}} : (\dot{\boldsymbol{\epsilon}} - \dot{\boldsymbol{\epsilon}}^{\mathrm{p}})\,\mathrm{d}t \\ &= \boldsymbol{\sigma}_n + \boldsymbol{c}^{\mathrm{e}} : (\kappa\Delta\boldsymbol{\epsilon} - \Delta\gamma^{(\beta_1)}\psi^{(\beta_1)}\boldsymbol{\alpha}^{(\beta_1)})\,. \end{aligned} \tag{8.22}$$

For a constant plastic modulus h the stresses must satisfy the consistency condition

$$\psi^{(\beta_1)}\boldsymbol{\alpha}^{(\beta_1)} : \boldsymbol{\sigma}(t) - (\tau_{Y,n} + h\Delta\gamma^{(\beta_1)}) = 0\,, \tag{8.23}$$

which gives the incremental slip on the primary system

$$\Delta\gamma^{(\beta_1)} = \frac{\psi^{(\beta_1)}\boldsymbol{\alpha}^{(\beta_1)} : \boldsymbol{c}^{\mathrm{e}}}{\mu_{\mathrm{c}} + h} : \kappa\Delta\boldsymbol{\epsilon}\,, \tag{8.24}$$

where μ_{c} is the elastic shear modulus of the crystal lattice. Once again, we can evaluate $\kappa^{(\beta)}$ for all slip systems and construct the set

$$\Psi_2 = \{\kappa^{(\beta)} \in R_+ |\psi^{(\beta)}\boldsymbol{\alpha}^{(\beta)} : \boldsymbol{\sigma}(t) - (\tau_{Y,n} + h\Delta\gamma^{(\beta_1)}) = 0\}\,. \tag{8.25}$$

If $\kappa^{(\beta)} < 1$ for some $\beta \in \{1, 2, \ldots, N\}\backslash\beta_1$, then the secondary slip system is the constraint β_2 that yields the smallest value of $\kappa^{(\beta)}$.

A similar idea may be employed to generalize the algorithm to the multislip case. Consider m linearly independent active systems, $\beta_1, \ldots, \beta_m$, and assume that the systems remain active as the incremental deformation $\kappa\Delta\boldsymbol{\epsilon}$ is applied. The evolution of the crystal stress during this period is given by

$$\boldsymbol{\sigma}(t) = \boldsymbol{\sigma}_n + \boldsymbol{c}^{\mathrm{e}} : \left(\kappa\Delta\boldsymbol{\epsilon} - \sum_{i=1}^{m} \Delta\gamma^{(\beta_i)}\psi^{(\beta_i)}\boldsymbol{\alpha}^{(\beta_i)}\right). \tag{8.26}$$

By imposing a total of m independent consistency conditions, we can determine the slips:

$$\psi^{(\beta_i)}\boldsymbol{\alpha}^{(\beta_i)} : \boldsymbol{\sigma}(t) - \left(\tau_{Y,n} + h\sum_{i=1}^{m} \Delta\gamma^{(\beta_i)}\right) = 0\,, \qquad i = 1, \ldots, m\,, \tag{8.27}$$

which gives

$$\Delta\gamma^{(\beta_i)} = \kappa \sum_{j=1}^{m} g_{ij}^{-1} \psi^{(\beta_j)} \boldsymbol{\alpha}^{(\beta_j)} : \boldsymbol{c}^{\mathrm{e}} : \Delta\boldsymbol{\epsilon}\,, \tag{8.28}$$

where

$$g_{ij} = \begin{cases} \mu_{\mathrm{c}} + h\,, & \text{if } i = j\,, \\ 2\mu_{\mathrm{c}} \psi^{(\beta_i)} \psi^{(\beta_j)} \boldsymbol{\alpha}^{(\beta_i)} : \boldsymbol{\alpha}^{(\beta_j)} + h\,, & \text{otherwise.} \end{cases} \tag{8.29}$$

Since the slips are linearly independent, the matrix $[g_{ij}]$ must be invertible. From the set

$$\begin{aligned} \Psi_{m+1} &= \Big\{ \kappa^{(\beta)} \in R_+ | \psi^{(\beta)} \boldsymbol{\alpha}^{(\beta)} : \boldsymbol{\sigma}(t) \\ &\quad - \Big(\tau_{Y,n} + \kappa^{(\beta)} h \sum_{i=1}^{m} \sum_{j=1}^{m} g_{ij}^{-1} \psi^{(\beta_j)} \boldsymbol{\alpha}^{(\beta_j)} : \boldsymbol{c}^{\mathrm{e}} : \Delta\boldsymbol{\epsilon} \Big) = 0 \Big\}\,, \end{aligned} \tag{8.30}$$

we can determine the next critical slip system.

Box 8.1 shows a flow chart of the ultimate algorithm for identifying linearly independent active constraints. To form the set $\overline{\mathcal{J}}_{\mathrm{act}}$, the g_{ij}'s are calculated from (8.29) and assembled into an array. A simple $\boldsymbol{LDU}$ factorization of this array identifies the redundant constraints from the zero elements of $\boldsymbol{D}$. Similar numerical techniques are presented by Anand and Kothari (1996) and Miehe and Schröder (2001). The new slips are calculated from the equation

$$\Delta\widetilde{\gamma}^{(\beta_i)} = \sum_{j=1}^{m} g_{ij}^{-1} \psi^{(\beta_j)} \boldsymbol{\alpha}^{(\beta_j)} : \boldsymbol{c}^{\mathrm{e}} : \Delta\boldsymbol{\epsilon}\,. \tag{8.31}$$

There is no guarantee that an active slip system will remain active when other systems activate, so Step 4 of Box 8.1 checks for the systems that deactivate from the sign of the calculated slips.

Step 1. If $\mathcal{J}_{\mathrm{act}} = \varnothing$, set $\overline{\mathcal{J}}_{\mathrm{act}} = \varnothing$, $\min \Psi_{n+1} \gg 1$, and go to Step 6
Step 2. Select $\overline{\mathcal{J}}_{\mathrm{act}} \subset \mathcal{J}_{\mathrm{act}}$
Step 3. Compute $\Delta\widetilde{\gamma}^{(\beta)}$ for all $\beta \in \overline{\mathcal{J}}_{\mathrm{act}}$
Step 4. If $\Delta\widetilde{\gamma}^{(\beta)} < 0$, drop $\Delta\widetilde{\gamma}^{(\beta)}_{\mathrm{min}}$ from $\overline{\mathcal{J}}_{\mathrm{act}}$ and go to Step 3
Step 5. Compute $\kappa^{(\beta)}$ for all $\beta \notin \overline{\mathcal{J}}_{\mathrm{act}}$ and assemble Ψ_{n+1}
Step 6. Set $\kappa^{(\beta_{m+1})} = \min(1, \min \Psi_{n+1})$ and update
$\boldsymbol{\sigma}_n \leftarrow \boldsymbol{\sigma}_n + \boldsymbol{c}^{\mathrm{e}} : (\kappa^{(\beta_{m+1})} \Delta\boldsymbol{\epsilon} - \sum_{\beta \in \overline{\mathcal{J}}_{\mathrm{act}}} \Delta\widetilde{\gamma}^{(\beta)} \psi^{(\beta)} \boldsymbol{\alpha}^{(\beta)})$
$\tau_{Y,n} \leftarrow \tau_{Y,n} + \kappa^{(\beta_{m+1})} h \sum_{\beta \in \overline{\mathcal{J}}_{\mathrm{act}}} \Delta\widetilde{\gamma}^{(\beta)}$
$\Delta\boldsymbol{\epsilon} \leftarrow \Delta\boldsymbol{\epsilon} - \kappa^{(\beta_{m+1})} \Delta\boldsymbol{\epsilon}$
$\overline{\mathcal{J}}_{\mathrm{act}} \leftarrow \overline{\mathcal{J}}_{\mathrm{act}} \cup \beta_{m+1}$
Step 8. If $\kappa^{(\beta_{m+1})} < 1$, go to Step 2.
Step 9. Otherwise, return to calling program.

Box 8.1 Ultimate algorithm for infinitesimal crystal plasticity

8.5 Uniqueness of Crystal Stress

With a large number of slip systems to select from, it is not possible to construct a unique combination of linearly independent systems. However, we can always show that the resulting overall crystal stress is unique and does not depend on the selected systems.

To demonstrate the uniqueness of the overall crystal stress, we collect all potentially active systems in the set $\mathcal{J}_{\rm act} = \{1, 2, \ldots, p, p+1, \ldots, p+q\}$, where $\psi^{(\beta)}\boldsymbol{\alpha}^{(\beta)} : \boldsymbol{\sigma}_n - \tau_{Y,n} = 0$ and $\psi^{(\beta)}\boldsymbol{\alpha}^{(\beta)} : \boldsymbol{c}^{\rm e} : \Delta\boldsymbol{\epsilon} \geq 0$ for all $\beta \in \mathcal{J}_{\rm act}$. The inequality means that the incremental elastic stress predictor $\boldsymbol{c}^{\rm e} : \Delta\boldsymbol{\epsilon}$ forms an acute angle with the gradient $\psi^{(\beta)}\boldsymbol{\alpha}^{(\beta)}$. We take $\overline{\mathcal{J}}_{\rm act} = \{1, 2, \ldots, p\} \subset \mathcal{J}_{\rm act}$ as the set of all active linearly independent slip systems. As noted earlier, this set results from an arbitrary selection of independent slip systems and is not unique. Thus, the set $\mathcal{J}_{\rm act} \backslash \overline{\mathcal{J}}_{\rm act} = \{p+1, \ldots, p+q\}$ form the redundant constraints. Since the slips are all positive, we have

$$\sum_{\eta=1}^{m} g_{\beta\eta}^{-1}\psi^{(\eta)}\boldsymbol{\alpha}^{(\eta)} : \boldsymbol{c}^{\rm e} : \Delta\boldsymbol{\epsilon} \geq 0 \qquad \forall \beta, \eta \in \overline{\mathcal{J}}_{\rm act}\,. \tag{8.32}$$

We now apply an incremental strain $\kappa\Delta\boldsymbol{\epsilon}$ and activate some slip systems in the crystal. The discrete consistency condition is

$$\kappa\psi^{(\beta)}\boldsymbol{\alpha}^{(\beta)} : \boldsymbol{c}^{\rm e} : \Delta\boldsymbol{\epsilon} = \sum_{\eta\in\mathcal{J}_{\rm act}} g_{\beta\eta}\Delta\gamma^{(\eta)} \qquad \forall \beta \in \overline{\mathcal{J}}_{\rm act}\,. \tag{8.33}$$

By freezing the slips on the redundant constraints, we form a non-singular matrix $g_{\beta\eta}$ that can be inverted to yield

$$\Delta\gamma^{(\beta)} = \kappa \sum_{\eta\in\overline{\mathcal{J}}_{\rm act}} g_{\beta\eta}^{-1}\psi^{(\eta)}\boldsymbol{\alpha}^{(\eta)} : \boldsymbol{c}^{\rm ep} : \Delta\boldsymbol{\epsilon} \geq 0\,. \tag{8.34}$$

Next, we show that the slips calculated from the linearly independent slip systems, together with the frozen slips for the redundant constraints, satisfy the consistency conditions for the redundant constraints. This is easily verified by writing

$$\begin{aligned}
\psi^{(\beta)}\boldsymbol{\alpha}^{(\beta)} : \Big(\boldsymbol{\sigma}_n \;&+\; \kappa\boldsymbol{c}^{\rm e} : \Delta\boldsymbol{\epsilon} - \boldsymbol{c}^{\rm e} : \sum_{\eta\in\overline{\mathcal{J}}_{\rm act}} \psi^{(\eta)}\Delta\gamma^{(\eta)}\boldsymbol{\alpha}^{(\eta)}\Big) \\
&= \tau_{Y,n} + \kappa\psi^{(\beta)}\boldsymbol{\alpha}^{(\beta)} : \boldsymbol{c}^{\rm e} : \Delta\boldsymbol{\epsilon} \\
&\quad - \sum_{\eta\in\overline{\mathcal{J}}_{\rm act}} 2\mu_{\rm c}\psi^{(\beta)}\psi^{(\eta)}\boldsymbol{\alpha}^{(\beta)} : \boldsymbol{\alpha}^{(\eta)}\Delta\gamma^{(\eta)} \\
&= \tau_{Y,n} + h \sum_{\eta\in\overline{\mathcal{J}}_{\rm act}} \Delta\gamma^{(\eta)}
\end{aligned} \tag{8.35}$$

for all $\beta \in \mathcal{J}_{\rm act}\backslash\overline{\mathcal{J}}_{\rm act}$, where $2\mu_{\rm c}\psi^{(\beta)}\psi^{(\eta)}\boldsymbol{\alpha}^{(\beta)} : \boldsymbol{\alpha}^{(\eta)} = g_{\beta\eta} - h$ for a Taylor-hardening crystal. By freezing the slips on all the redundant constraints, we see that the consistency conditions are identically satisfied on all active slip systems. The redundant systems do not contribute to the plastic strain; instead, they simply undergo neutral loading.

To show that the overall crystal stress is unique, we write the plastic strain increment in the general form

$$\Delta\boldsymbol{\epsilon} = \sum_{\beta\in\overline{\mathcal{J}}_{\text{act}}} \Delta\gamma^{(\beta)}\psi^{(\beta)}\boldsymbol{\alpha}^{(\beta)} + \sum_{\mathcal{J}_{\text{act}}\backslash\overline{\mathcal{J}}_{\text{act}}} \Delta\gamma^{(\beta)}\psi^{(\beta)}\boldsymbol{\alpha}^{(\beta)} . \tag{8.36}$$

Inverting (8.33) to obtain

$$\begin{aligned} \Delta\gamma^{(\beta)} &= \kappa \sum_{\eta\in\overline{\mathcal{J}}_{\text{act}}} g_{\beta\eta}^{-1}\psi^{(\eta)}\boldsymbol{\alpha}^{(\eta)} : \boldsymbol{c}^{\text{e}} : \Delta\boldsymbol{\epsilon} \\ &- \sum_{\eta\in\overline{\mathcal{J}}_{\text{act}}} \sum_{\xi\in\mathcal{J}\backslash\overline{\mathcal{J}}_{\text{act}}} g_{\beta\eta}^{-1} g_{\eta\xi}\Delta\gamma^{(\xi)} \qquad \forall\beta\in\overline{\mathcal{J}}_{\text{act}} , \end{aligned} \tag{8.37}$$

and substituting into (8.36), we obtain

$$\begin{aligned} \Delta\boldsymbol{\epsilon}^{\text{p}} &= \kappa \sum_{\beta\in\overline{\mathcal{J}}_{\text{act}}} \sum_{\eta\in\overline{\mathcal{J}}_{\text{act}}} (g_{\beta\eta}^{-1}\psi^{(\eta)}\boldsymbol{\alpha}^{(\eta)} : \boldsymbol{c}^{\text{e}} : \Delta\boldsymbol{\epsilon})\psi^{(\beta)}\boldsymbol{\alpha}^{(\beta)} \\ &- \sum_{\beta\in\overline{\mathcal{J}}_{\text{act}}} \sum_{\eta\in\overline{\mathcal{J}}_{\text{act}}} \sum_{\xi\in\mathcal{J}_{\text{act}}\backslash\overline{\mathcal{J}}_{\text{act}}} (g_{\beta\eta}^{-1} g_{\eta\xi}\Delta\gamma^{(\xi)})\psi^{(\beta)}\boldsymbol{\alpha}^{(\beta)} \\ &+ \sum_{\beta\in\mathcal{J}_{\text{act}}\backslash\overline{\mathcal{J}}_{\text{act}}} \Delta\gamma^{(\beta)}\psi^{(\beta)}\boldsymbol{\alpha}^{(\beta)} . \end{aligned} \tag{8.38}$$

Combining (8.33) and (8.34) yields the relation

$$\psi^{(\beta)}\boldsymbol{\alpha}^{(\beta)} = \sum_{\eta\in\overline{\mathcal{J}}_{\text{act}}} \sum_{\xi\in\overline{\mathcal{J}}_{\text{act}}} g_{\beta\eta} g_{\eta\xi}^{-1}\psi^{(\xi)}\boldsymbol{\alpha}^{(\xi)} \qquad \forall\beta\in\mathcal{J}_{\text{act}}\backslash\overline{\mathcal{J}}_{\text{act}} . \tag{8.39}$$

After substituting into (8.38), we note that the last two summation terms cancel, leaving an expression for the plastic strain increment that depends solely on the arbitrarily selected linearly independent systems, i.e.,

$$\Delta\boldsymbol{\epsilon}^{\text{p}} = \kappa \sum_{\beta\in\overline{\mathcal{J}}_{\text{act}}} \sum_{\eta\in\overline{\mathcal{J}}_{\text{act}}} (g_{\beta\eta}^{-1}\psi^{(\eta)}\boldsymbol{\alpha}^{(\eta)} : \boldsymbol{c}^{\text{e}} : \Delta\boldsymbol{\epsilon})\psi^{(\beta)}\boldsymbol{\alpha}^{(\beta)} . \tag{8.40}$$

This means that the elements of the set $\overline{\mathcal{J}}_{\text{act}}$ form a complete span and uniquely determine the crystal stress.

As an illustration, we consider a f.c.c. crystal with Euler angles $\theta = 0$ and $\phi = 18°$; Young's modulus $E = 1500$, Poisson's ratio $\nu = 0.33$, initial yield stress $\tau_{Y0} = 1.0$ (consistent units are used throughout). The crystal is subjected to two deformation histories: extension and simple shear. In the extension test the crystal is initially stress-free and stretched axially along the fixed z axis under prescribed strains $\epsilon_{zz} = \epsilon_{zz}(t)$, $\epsilon_{xx} = \epsilon_{yy} = -\epsilon_{zz}(t)/2$, and all other components of the strain tensor are zero. In the simple shear test, the crystal is also initially stress-free and sheared to prescribed strains $\epsilon_{yz}(t) = \epsilon_{zy}(t)$, while all other strain components are zero.

Figure 8.6 shows the stress-strain curves for $h = 0$ (perfect plasticity) and $h = 10$ (Taylor-hardening). The perfectly plastic crystal shows a horizontal plateau as expected, whereas the Taylor-hardening crystal shows a constant tangent modulus after activating all independent slip systems. In the extension test, a duplex system is encountered at initial yield. Further stretching activates two more linearly independent systems, and then a fifth one is detected at a higher strain. In the simple shear test, the five linearly independent systems are detected one by one. The resulting stress-strain responses are the same irrespective of the linearly independent slip systems selected.

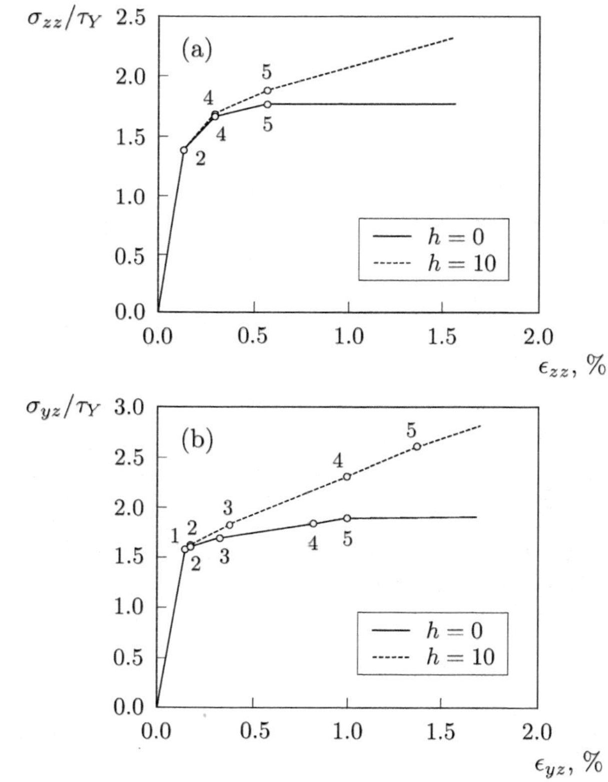

Fig. 8.6 Stress-point simulations on a f.c.c. crystal with Euler angles $\theta = 0$ and $\phi = 18°$: (a) isochoric extension test, and (b) isochoric simple shear test. Smaller-font numbers denote independent slip systems detected.

8.6 Large Deformation

Various formulations for crystalline materials experiencing large elastoplastic deformation have been proposed in the literature (Asaro and Rice 1977, Asaro 1983, Hill and Rice 1972, Rice 1971). In this chapter, we use the multiplicative plasticity framework developed in Chapter 5 to incorporate finite deformation effects into the crystal plasticity formulation.

In formulating a crystal plasticity theory within the framework of multiplicative plasticity, the following aspects must be considered. First, the lattice deformation is purely elastic, and plastic deformation is due solely to slips on crystallographic planes. Second, as the crystal deforms, the lattice stretches and rotates, so the embedded slip systems also stretch and rotate with the crystal lattice. Third, the lattice rotation is not the same as the crystal rotation, because the crystal rotation also includes a portion contributed by the plastic slips. Figure 8.7 differentiates between the lattice rotation (ABC') and crystal rotation (ABC) for a simplified crystal structure containing one slip system. The figure resembles a finite element with an embedded strong discontinuity, a subject discussed thoroughly in Chapter 7. The similarity between the two problems, that of a finite element with a strong discontinuity and a crystal lattice with an embedded slip plane, suggests that the strong discontinuity concept could very well be used for the crystal plasticity problem. We explore this idea in the following developments.

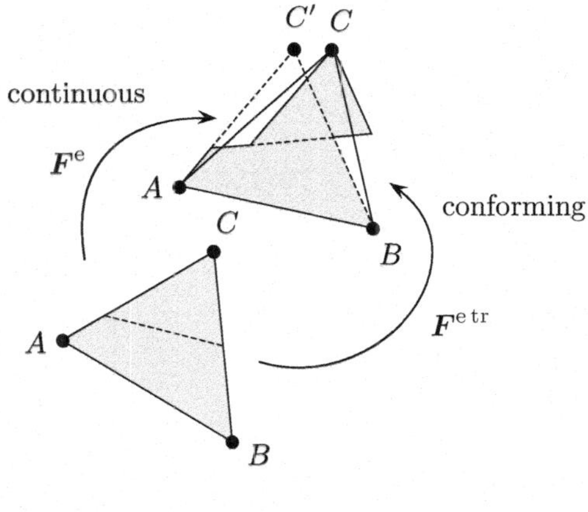

Fig. 8.7 Deformation of crystal lattice (continuous map, ABC') versus overall deformation of crystal (conforming map, ABC)

Consider a crystal with one active slip system. For convenience, we regularize the slip plane and represent it as a layer $\mathcal{D}_0$ with thickness h_0 in the reference configuration. The local deformation in the neighborhood of the slip plane is then given by the ramp function

$$\boldsymbol{\phi} = \begin{cases} \overline{\boldsymbol{\phi}}\,, & \text{if } c \leq 0\,, \\ \overline{\boldsymbol{\phi}} + c\widetilde{\boldsymbol{\phi}}/h_0\,, & \text{if } 0 \leq c \leq h_0\,, \\ \overline{\boldsymbol{\phi}} + \widetilde{\boldsymbol{\phi}}\,, & \text{if } c \geq h_0\,, \end{cases} \tag{8.41}$$

where $\overline{\boldsymbol{\phi}}$ is the deformation of the lattice in the neighborhood of the slip plane, and $\widetilde{\boldsymbol{\phi}}$ is the deformation jump over the thickness h_0. The variable c is a normal parameter given by the expression

$$c = (\boldsymbol{X} - \boldsymbol{Y}) \cdot \boldsymbol{N}\,, \tag{8.42}$$

where $\boldsymbol{Y}$ is any point on the master side of $\mathcal{D}_0$, and $\boldsymbol{N}$ is a unit normal vector in the direction of the slave side (note that the choice of master and slave sides is arbitrary). The corresponding deformation gradients are

$$\boldsymbol{F} = \frac{\partial \boldsymbol{\phi}}{\partial \boldsymbol{X}} = \begin{cases} \boldsymbol{F}^{\mathrm{e}} & \text{outside } \mathcal{D}_0\,, \\ \boldsymbol{F}^{\mathrm{e}} + ([\![\boldsymbol{\phi}]\!] \otimes \boldsymbol{N})/h_0 & \text{inside } \mathcal{D}_0\,, \end{cases} \tag{8.43}$$

where $\boldsymbol{F}^{\mathrm{e}} = \partial\overline{\boldsymbol{\phi}}/\partial\boldsymbol{X}$. The superscript "e" in $\boldsymbol{F}^{\mathrm{e}}$ stands for 'elastic' and suggests that the lattice deformation is purely elastic.

The velocity field takes similar forms,

$$\boldsymbol{V} = \begin{cases} \overline{\boldsymbol{V}}\,, & \text{if } c \leq 0\,, \\ \overline{\boldsymbol{V}} + c\widetilde{\boldsymbol{V}}/h_0\,, & \text{if } 0 \leq c \leq h_0\,, \\ \overline{\boldsymbol{V}} + \widetilde{\boldsymbol{V}}\,, & \text{if } c \geq h_0\,, \end{cases} \tag{8.44}$$

where $\overline{\boldsymbol{V}} = \dot{\overline{\boldsymbol{\phi}}}$ and $\widetilde{\boldsymbol{V}} = \dot{\widetilde{\boldsymbol{\phi}}}$. The time derivatives of $\boldsymbol{F}$ are

$$\dot{\boldsymbol{F}} = \begin{cases} \dot{\boldsymbol{F}}^{\mathrm{e}} & \text{outside } \mathcal{D}_0\,, \\ \dot{\boldsymbol{F}}^{\mathrm{e}} + (\widetilde{\boldsymbol{V}} \otimes \boldsymbol{N})/h_0 & \text{inside } \mathcal{D}_0\,, \end{cases} \tag{8.45}$$

where $\dot{\boldsymbol{F}} = \partial\boldsymbol{V}/\partial\boldsymbol{X}$ and $\dot{\boldsymbol{F}}^{\mathrm{e}} = \partial\overline{\boldsymbol{V}}/\partial\boldsymbol{X}$. The Sherman-Morrison formula provides the inverse of $\boldsymbol{F}$,

$$\boldsymbol{F}^{-1} = \begin{cases} \boldsymbol{F}^{\mathrm{e}-1} & \text{outside } \mathcal{D}_0\,, \\ \boldsymbol{F}^{\mathrm{e}-1} - (\boldsymbol{\Phi} \otimes \boldsymbol{N} \cdot \boldsymbol{F}^{\mathrm{e}-1})/(h_0 + \boldsymbol{N} \cdot \boldsymbol{\Phi}) & \text{inside } \mathcal{D}_0\,, \end{cases} \tag{8.46}$$

where $\boldsymbol{\Phi} = \boldsymbol{F}^{\mathrm{e}-1} \cdot \widetilde{\boldsymbol{\phi}}$ is the elastic pull-back of $\widetilde{\boldsymbol{\phi}}$. Applying Nanson's formula on any differential area on the slip plane yields the relation (Borja 2002b)

$$\frac{\boldsymbol{n}}{h} = \frac{\boldsymbol{N} \cdot \boldsymbol{F}^{-1}}{h_0} = \frac{\boldsymbol{N} \cdot \boldsymbol{F}^{\mathrm{e}-1}}{h_0 + \boldsymbol{N} \cdot \boldsymbol{\Phi}}\,, \tag{8.47}$$

where $\boldsymbol{n}$ is the unit normal vector to the slip plane in the current configuration, and h is the thickness of $\mathcal{D}_0$ in the current configuration. We readily see that the elastic component of deformation gradient is responsible for the rotation of the unit normal vector from $\boldsymbol{N}$ to $\boldsymbol{n}$. The velocity gradient is given by the relation $\boldsymbol{l} = \dot{\boldsymbol{F}} \cdot \boldsymbol{F}^{-1}$, so we obtain

$$\boldsymbol{l} = \begin{cases} \boldsymbol{l}^{\mathrm{e}} & \text{outside } \mathcal{D}_0\,, \\ \boldsymbol{l}^{\mathrm{e}} + (\overset{\triangledown}{\boldsymbol{\phi}} \otimes \boldsymbol{n})/h & \text{inside } \mathcal{D}_0\,, \end{cases} \tag{8.48}$$

where

$$\boldsymbol{l}^{\mathrm{e}} = \dot{\boldsymbol{F}}^{\mathrm{e}} \cdot \boldsymbol{F}^{\mathrm{e}-1}\,, \qquad \overset{\triangledown}{\boldsymbol{\phi}} = \dot{\widetilde{\boldsymbol{\phi}}} - \boldsymbol{l}^{\mathrm{e}} \cdot \widetilde{\boldsymbol{\phi}}\,. \tag{8.49}$$

Note that $\overset{\triangledown}{\boldsymbol{\phi}}$ contains a convected part, which makes it an objective rate co-rotational with $\widetilde{\boldsymbol{\phi}}$.

Let $\boldsymbol{M}$ and $\boldsymbol{N}$ define the primary slip system in the reference configuration, where $\boldsymbol{M}$ is a unit vector in the direction of slip. We assume at the outset that $\boldsymbol{M} \perp \boldsymbol{N}$. We have shown earlier that the elastic deformation gradient $\boldsymbol{F}^{\mathrm{e}}$ performs a covariant transformation on $\boldsymbol{N}$, so it must perform a contravariant transformation on $\boldsymbol{M}$, i.e.,

$$\widetilde{\boldsymbol{\phi}} := \boldsymbol{F}^{\mathrm{e}} \cdot \gamma \boldsymbol{M}\,, \tag{8.50}$$

where γ is the cumulative slip reckoned with respect to the reference configuration. The combined expression $\gamma\boldsymbol{M}$ then represents the slip vector measured with respect to the reference configuration. Since the reference configuration is fixed, then $\boldsymbol{M}$ is also fixed. Therefore, using the relation $\dot{\boldsymbol{F}}^{\mathrm{e}} = \boldsymbol{l}^{\mathrm{e}} \cdot \boldsymbol{F}^{\mathrm{e}}$, we get

$$\dot{\widetilde{\boldsymbol{\phi}}} = \boldsymbol{F}^{\mathrm{e}} \cdot \dot{\gamma}\boldsymbol{M} + \dot{\boldsymbol{F}}^{\mathrm{e}} \cdot \gamma\boldsymbol{M} = \dot{g}\boldsymbol{m} + \boldsymbol{l}^{\mathrm{e}} \cdot \widetilde{\boldsymbol{\phi}}\,, \tag{8.51}$$

where

$$\boldsymbol{m} = \boldsymbol{F}^{\mathrm{e}} \cdot \boldsymbol{M} / \|\boldsymbol{F}^{\mathrm{e}} \cdot \boldsymbol{M}\| \tag{8.52}$$

is the unit slip vector and $\dot{g} = \dot{\gamma}\|\boldsymbol{F}^{\mathrm{e}} \cdot \boldsymbol{M}\|$ is the associated slip rate in the current configuration. The co-rotational jump rate simplifies to the form

$$\overset{\triangledown}{\boldsymbol{\phi}} = \dot{g}\boldsymbol{m}\,. \tag{8.53}$$

In the strong discontinuity limit, both h_0 and h approach zero. We can then use theory of distributions once again and write the time derivative of the deformation gradient ‘inside’ the strong discontinuity as

$$\dot{\boldsymbol{F}} = \dot{\boldsymbol{F}}^{\mathrm{e}} + \delta_{\mathcal{D}}(\dot{\widetilde{\boldsymbol{\phi}}} \otimes \boldsymbol{N})\,, \tag{8.54}$$

where $\delta_{\mathcal{D}}$ is the usual Dirac delta function. Similarly, we can write the velocity gradient ‘inside’ the strong discontinuity as

$$\boldsymbol{l} = \boldsymbol{l}^{\mathrm{e}} + \delta_{\mathcal{D}}(\dot{g}\boldsymbol{m} \otimes \boldsymbol{n})\,. \tag{8.55}$$

This additive decomposition of the velocity gradient identifies the singular term as the plastic component, i.e.,

$$\boldsymbol{l}^{\mathrm{p}} = \delta_{\mathcal{D}}(\dot{g}\boldsymbol{m} \otimes \boldsymbol{n}) . \tag{8.56}$$

In addition, from the relation $\boldsymbol{l}^{\mathrm{e}} = \dot{\boldsymbol{F}}^{\mathrm{e}} \cdot \boldsymbol{F}^{\mathrm{e}-1}$ established in (8.49), and from the multiplicative decomposition of deformation gradient, we obtain

$$\boldsymbol{l} = \dot{\boldsymbol{F}}^{\mathrm{e}} \cdot \boldsymbol{F}^{\mathrm{e}-1} + \boldsymbol{F}^{\mathrm{e}} \cdot (\dot{\boldsymbol{F}}^{\mathrm{p}} \cdot \boldsymbol{F}^{\mathrm{p}-1}) \cdot \boldsymbol{F}^{\mathrm{e}-1} . \tag{8.57}$$

We identify the second term on the right-hand side as nothing but $\boldsymbol{l}^{\mathrm{p}}$, and so we get

$$\dot{\boldsymbol{F}}^{\mathrm{p}} \cdot \boldsymbol{F}^{\mathrm{p}-1} = \delta_{\mathcal{D}}(\boldsymbol{F}^{\mathrm{e}-1} \cdot \dot{g}\boldsymbol{m}) \otimes (\boldsymbol{n} \cdot \boldsymbol{F}^{\mathrm{e}}) . \tag{8.58}$$

This last equation has a form similar to the one presented by Rice (1971).

Equation (8.56) suggests that the consistency parameter is the delta function $\delta_{\mathcal{D}}$ with a regular plastic multiplier $\dot{g}$. If we take the yield function as

$$f = |(\boldsymbol{m} \otimes \boldsymbol{n}) : \boldsymbol{\tau}| - \tau_Y \equiv \psi(\boldsymbol{m} \otimes \boldsymbol{n}) : \boldsymbol{\tau} - \tau_Y \leq 0 , \tag{8.59}$$

where

$$\psi := \operatorname{sign}((\boldsymbol{m} \otimes \boldsymbol{n}) : \boldsymbol{\tau}) \tag{8.60}$$

is the sign function and τ_Y is the yield stress on the slip system, then the expression for $\boldsymbol{l}^{\mathrm{p}}$ follows the associative flow rule on the yield function in the sense that the plastic flow direction is the Schmid tensor $(\boldsymbol{m} \otimes \boldsymbol{n})$ itself. As usual, we can decompose the velocity gradient $\boldsymbol{l}^{\mathrm{p}}$ into symmetric and skew-symmetric parts as

$$\boldsymbol{d}^{\mathrm{p}} = \delta_{\mathcal{D}}\psi \operatorname{sym}(\dot{g}\boldsymbol{m} \otimes \boldsymbol{n}) , \qquad \boldsymbol{\omega}^{\mathrm{p}} = \delta_{\mathcal{D}}\psi \operatorname{skw}(\dot{g}\boldsymbol{m} \otimes \boldsymbol{n}) . \tag{8.61}$$

Since $\boldsymbol{m}$, $\boldsymbol{n}$, and $\boldsymbol{\tau}$ all depend on the elastic deformation gradient $\boldsymbol{F}^{\mathrm{e}}$ alone, then we can write the consistency condition on the yield function as

$$\dot{f} = \frac{\partial f}{\partial \boldsymbol{F}^{\mathrm{e}}} : \dot{\boldsymbol{F}}^{\mathrm{e}} - \delta_{\mathcal{D}}\dot{g}H = 0 , \tag{8.62}$$

where H is the generalized continuum plastic modulus. Following the strong discontinuity formulation of Chapter 7, we observe that the first term is a regular function whereas the second term is singular, unless H is of the form

$$H = h_{\mathcal{D}}/\delta_{\mathcal{D}} , \tag{8.63}$$

where $h_{\mathcal{D}}$ is a regular function representing the interface plastic modulus. This results in $H \to 0$, or perfect plasticity in the continuum sense, and $h_{\mathcal{D}}$ = regular function, or hardening/softening plasticity in the interface sense. The regular component $h_{\mathcal{D}}$ plays the role of the Taylor hardening modulus in subsequent discussions. In this case, the consistency condition simplifies to

$$\dot{f} = \frac{\partial f}{\partial \boldsymbol{F}^{\mathrm{e}}} : \dot{\boldsymbol{F}}^{\mathrm{e}} - \dot{g}h_{\mathcal{D}} = 0 . \tag{8.64}$$

Note that the variables are now all regular functions.

8.7 Multiscale Fields

The deformation fields described in the previous section are fine-scale fields—they delineate the elastic deformation of the crystal lattice and the finite slips on the crystallographic planes. The fine-scale fields must be upscaled to the crystal level to obtain the coarse-scale deformation field, which is a conforming field describing the overall deformation of the crystal. Interestingly, the theoretical developments related to the enriched finite elements presented in Chapter 7 apply equally well to the crystal plasticity problem. A glimpse of the similarity of the two problems is evident from Fig. 8.7, in which the triangles may be interpreted either as an enriched finite element or as a crystal volume. If one takes the latter point of view, then the domain delineated in Fig. 8.7 represents a representative elementary volume (REV) over which the crystal deformation is homogenized. Figure 8.8 shows a simplified one-dimensional representation of the crystal REV. The continuous deformation field delineates the fine-scale deformation of the crystal lattice; the slip is the fine-scale displacement jump. Over the REV range, the fine-scale fields are homogenized to obtain the conforming crystal deformation.

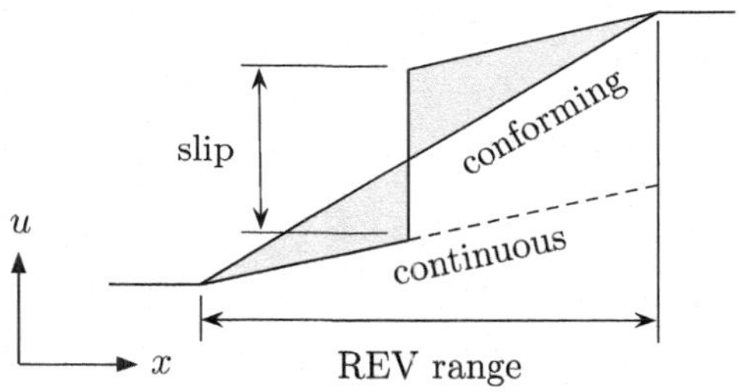

Fig. 8.8 One-dimensional representation of continuous and conforming displacement fields over a representative elementary volume (REV) for a crystal deforming in single slip

Consider the following fine-scale displacement field in the neighborhood of a slip plane:

$$\boldsymbol{u}(\boldsymbol{X}) = \overline{\boldsymbol{u}}(\boldsymbol{X}) + \widetilde{\boldsymbol{u}}(\boldsymbol{X}) H_{\mathcal{D}}(\boldsymbol{X}) , \qquad H_{\mathcal{D}}(\boldsymbol{X}) = \begin{cases} 1 & \text{if } \boldsymbol{X} \in \mathcal{D}_+ \\ 0 & \text{if } \boldsymbol{X} \in \mathcal{D}_- \end{cases} \tag{8.65}$$

where $\mathcal{D}_\pm$ represents regions in the crystal lattice on the opposite sides of the glide plane. As usual, we take the $(-)$ side as the master side and the $(+)$ side as the slave side. The displacement field $\overline{\boldsymbol{u}}$ is the continuous part of $\boldsymbol{u}$ and defines the elastic deformation of the crystal lattice, whereas $\widetilde{\boldsymbol{u}}$ is the slip given by

$$\widetilde{\boldsymbol{u}}(\boldsymbol{X}) = \zeta(\boldsymbol{X}) \boldsymbol{m}(\boldsymbol{X}) = \zeta \boldsymbol{F}^{\mathrm{e}} \cdot \boldsymbol{M} / \| \boldsymbol{F}^{\mathrm{e}} \cdot \boldsymbol{M} \| , \tag{8.66}$$

where

$$\zeta = \gamma \|\boldsymbol{F}^{\mathrm{e}} \cdot \boldsymbol{M}\| \tag{8.67}$$

is the cumulative slip co-rotational with the crystallographic plane.

To obtain a macroscopically smooth coarse-scale displacement field, we re-parameterize the fine-scale displacement field as follows,

$$\boldsymbol{u}(\boldsymbol{X}) = \boldsymbol{u}_c(\boldsymbol{X}) + \widetilde{\boldsymbol{u}}(\boldsymbol{X}) M_{\mathcal{D}}(\boldsymbol{X}), \tag{8.68}$$

where $\boldsymbol{u}_c$ is the coarse-scale displacement field, and

$$M_{\mathcal{D}} = H_{\mathcal{D}}(\boldsymbol{X}) - G(\boldsymbol{X}) \tag{8.69}$$

is the same blending function introduced in Chapter 7 in the context of the enriched finite element formulation. In the present case, $G(\boldsymbol{X})$ is an arbitrary smooth ramp function that varies from zero to one over a unit thickness of the crystal volume in the direction of $\boldsymbol{N}$. The actual form of the ramp function is immaterial to the formulation since we only need to extract a deformation gradient from it, i.e., $G(\boldsymbol{X})$ will disappear in the formulation. In keeping with the strong discontinuity discussion, we assume that the displacement jump $\widetilde{\boldsymbol{u}}$ is spatially constant on the slave side of the slip plane.

The continuous deformation is the total deformation minus the deformation jump, i.e.,

$$\overline{\boldsymbol{x}} = \boldsymbol{X} + \overline{\boldsymbol{u}}. \tag{8.70}$$

The conforming deformation delineates the homogenized deformation of the crystal,

$$\boldsymbol{x}_c = \boldsymbol{X} + \boldsymbol{u}_c. \tag{8.71}$$

The continuous and conforming deformations are related by the equation

$$\overline{\boldsymbol{x}} = \boldsymbol{x}_c - \widetilde{\boldsymbol{u}}(\boldsymbol{X}) G(\boldsymbol{X}). \tag{8.72}$$

Let $\boldsymbol{F}^{\mathrm{e}} = \partial \overline{\boldsymbol{x}} / \partial \boldsymbol{X}$ and $\boldsymbol{F}^{\mathrm{e\,tr}} = \partial \boldsymbol{x}_c / \partial \boldsymbol{X}$. Since $\widetilde{\boldsymbol{u}}$ is spatially constant on the slave side of the slip plane, we get

$$\boldsymbol{F}^{\mathrm{e}} = \boldsymbol{F}^{\mathrm{e\,tr}} - \zeta \boldsymbol{m} \otimes \frac{\partial G}{\partial \boldsymbol{X}} = \boldsymbol{F}^{\mathrm{e\,tr}} - \zeta \boldsymbol{m} \otimes \boldsymbol{N}. \tag{8.73}$$

Note that ζ is now taken as a measure of slip per unit distance, making it a dimensionless strain-like quantity, like the glide strain rate $\dot{\gamma}$ in Fig. 8.3. Imposing the last equation at time t_n gives

$$\boldsymbol{F}_n^{\mathrm{e}} = \boldsymbol{F}_n^{\mathrm{e\,tr}} - \zeta_n \boldsymbol{m}_n \otimes \frac{\partial G}{\partial \boldsymbol{X}} = \boldsymbol{F}_n^{\mathrm{e\,tr}} - \zeta_n \boldsymbol{m}_n \otimes \boldsymbol{N}, \tag{8.74}$$

where $\boldsymbol{F}_n^{\mathrm{e}} = \partial \overline{\boldsymbol{x}}_n / \partial \boldsymbol{X}$ and $\boldsymbol{F}_n^{\mathrm{e\,tr}} = \partial \boldsymbol{x}_{cn} / \partial \boldsymbol{X}$. The superscript "e" indicates an elastic deformation field for the lattice, whereas the superscript "e tr" suggests a trial predictor state in the absence of plastic slip. We shall refer to $\boldsymbol{F}^{\mathrm{e\,tr}}$ as the coarse-scale deformation gradient for the crystal.

8.8 Discrete Formulation

The discrete formulation time-integrates the stress and deformation variables over discrete time intervals. The assumption is that all variables at time t_n are known, and the goal is to calculate their values at time $t > t_n$. The relative deformation gradients at time $t > t_n$ can be written as

$$\boldsymbol{f}^{\mathrm{e}} = \frac{\partial \overline{\boldsymbol{x}}}{\partial \overline{\boldsymbol{x}}_n} = \boldsymbol{F}^{\mathrm{e}} \cdot \boldsymbol{F}_n^{\mathrm{e}-1}\,, \qquad \boldsymbol{f}^{\mathrm{e\,tr}} = \frac{\partial \boldsymbol{x}_c}{\partial \boldsymbol{x}_{cn}} = \boldsymbol{F}^{\mathrm{e\,tr}} \cdot \boldsymbol{F}_n^{\mathrm{e\,tr}-1}\,. \tag{8.75}$$

We can use the Sherman-Morrison formula to get the inverse

$$\boldsymbol{F}^{\mathrm{e}-1} = \boldsymbol{F}^{\mathrm{e\,tr}-1} + \frac{\zeta}{1-\zeta\beta}\boldsymbol{F}^{\mathrm{e\,tr}-1} \cdot \boldsymbol{m} \otimes \boldsymbol{N} \cdot \boldsymbol{F}^{\mathrm{e\,tr}-1}\,, \tag{8.76}$$

and

$$\boldsymbol{F}_n^{\mathrm{e}-1} = \boldsymbol{F}_n^{\mathrm{e\,tr}-1} + \frac{\zeta_n}{1-\zeta_n\beta_n}\boldsymbol{F}_n^{\mathrm{e\,tr}-1} \cdot \boldsymbol{m}_n \otimes \boldsymbol{N} \cdot \boldsymbol{F}_n^{\mathrm{e\,tr}-1}\,, \tag{8.77}$$

where $\beta = \boldsymbol{N} \cdot \boldsymbol{F}^{\mathrm{e\,tr}-1} \cdot \boldsymbol{m}$ and $\beta_n = \boldsymbol{N} \cdot \boldsymbol{F}_n^{\mathrm{e\,tr}-1} \cdot \boldsymbol{m}_n$. From equation (8.75), we find

$$\boldsymbol{f}^{\mathrm{e}} = \boldsymbol{f}^{\mathrm{e\,tr}} - \overset{\triangledown}{\boldsymbol{m}} \otimes \boldsymbol{N} \cdot \boldsymbol{F}_n^{\mathrm{e\,tr}-1}\,, \tag{8.78}$$

where

$$\overset{\triangledown}{\boldsymbol{m}} = \frac{\zeta}{1-\zeta_n\beta_n}\boldsymbol{m} - \frac{\zeta_n}{1-\zeta_n\beta_n}\boldsymbol{f}^{\mathrm{e\,tr}} \cdot \boldsymbol{m}_n \tag{8.79}$$

is the incremental slip over the relevant time interval.

The foregoing formulation shows that the elastic relative deformation gradient $\boldsymbol{f}^{\mathrm{e}}$ is a function of the current slip direction $\boldsymbol{m}$, which in turn is a function of $\boldsymbol{f}^{\mathrm{e}}$. This implicit relation is difficult solve analytically, unless we recognize the strain-driven format of the algorithm and the role played by the coarse-scale deformation gradient in the algorithmic formulation. In a strain-driven format, we are given the elastic left Cauchy-Green deformation tensor $\boldsymbol{b}_n^{\mathrm{e}}$ at time t_n, and the coarse-scale displacement increment $\Delta \boldsymbol{u}_c = \boldsymbol{u}_c - \boldsymbol{u}_{cn}$. The objective is to find the elastic left Cauchy-Green deformation tensor $\boldsymbol{b}^{\mathrm{e}}$ at any time $t > t_n$. The elastic predictors are defined by either of the following coarse-scale deformation gradients

$$\boldsymbol{F}^{\mathrm{e\,tr}} = \boldsymbol{F}_n^{\mathrm{e\,tr}} + \frac{\partial \Delta \boldsymbol{u}_c}{\partial \boldsymbol{X}}\,, \qquad \boldsymbol{f}^{\mathrm{e\,tr}} = \boldsymbol{1} + \frac{\partial \Delta \boldsymbol{u}_c}{\partial \boldsymbol{x}_{cn}}\,. \tag{8.80}$$

The slip direction $\boldsymbol{m}$ transforms according to (8.52), or, equivalently, according to the relation

$$\boldsymbol{m} = \boldsymbol{f}^{\mathrm{e}} \cdot \boldsymbol{m}_n / \|\boldsymbol{f}^{\mathrm{e}} \cdot \boldsymbol{m}_n\|\,. \tag{8.81}$$

Therefore, we obtain

$$\boldsymbol{f}^{\mathrm{e}} \cdot \boldsymbol{m}_n = \|\boldsymbol{f}^{\mathrm{e}} \cdot \boldsymbol{m}_n\| \boldsymbol{m} = \frac{1}{1-\zeta_n\beta_n}\boldsymbol{f}^{\mathrm{e\,tr}} \cdot \boldsymbol{m}_n - \frac{\zeta\beta_n}{1-\zeta_n\beta_n}\boldsymbol{m}\,. \tag{8.82}$$

We see that the vector formed by $\boldsymbol{f}^{\mathrm{e\,tr}} \cdot \boldsymbol{m}_n$ is parallel to the vector $\boldsymbol{m}$. Therefore, we can use the alternative expression

$$\boldsymbol{m} = \boldsymbol{f}^{\mathrm{e\,tr}} \cdot \boldsymbol{m}_n / \|\boldsymbol{f}^{\mathrm{e\,tr}} \cdot \boldsymbol{m}_n\| = \boldsymbol{F}^{\mathrm{e\,tr}} \cdot \boldsymbol{M} / \|\boldsymbol{F}^{\mathrm{e\,tr}} \cdot \boldsymbol{M}\| . \tag{8.83}$$

Similarly,

$$\boldsymbol{m}_n = \boldsymbol{F}_n^{\mathrm{e\,tr}} \cdot \boldsymbol{M} / \|\boldsymbol{F}_n^{\mathrm{e\,tr}} \cdot \boldsymbol{M}\| . \tag{8.84}$$

Because the coarse-scale deformation gradient $\boldsymbol{F}^{\mathrm{e\,tr}}$ determines the new orientation of $\boldsymbol{M}$, it follows that $\beta, \beta_n \propto \boldsymbol{M} \cdot \boldsymbol{N} \equiv 0$. Therefore, $\boldsymbol{f}^{\mathrm{e}}$ can be written in a more simplified form

$$\boldsymbol{f}^{\mathrm{e}} = \boldsymbol{f}^{\mathrm{e\,tr}} - \Delta\zeta \boldsymbol{m} \otimes \boldsymbol{N} \cdot \boldsymbol{F}_n^{\mathrm{e\,tr}-1} , \tag{8.85}$$

where

$$\Delta\zeta = \zeta - \zeta_n \|\boldsymbol{f}^{\mathrm{e\,tr}} \cdot \boldsymbol{m}_n\| \tag{8.86}$$

is the co-rotational slip increment. The foregoing development suggests that for single-slip system the push-forward of the continuous field on the slip system is the same as the push-forward of the conforming field on this same slip system. Figure 8.7 elucidates this conclusion pictorially: the lattice is deformed to configuration ABC', whereas the crystal is deformed to configuration ABC. However, the push-forward of either deformation gradient on the slip system is the same.

8.9 Stress-Point Integration

From the elastic relative deformation gradient $\boldsymbol{f}^{\mathrm{e}}$, we can time-integrate the elastic left Cauchy-Green deformation tensor $\boldsymbol{b}^{\mathrm{e}}$ as follows,

$$\boldsymbol{b}^{\mathrm{e}}(t > t_n) = \boldsymbol{f}^{\mathrm{e}} \cdot \boldsymbol{b}_n^{\mathrm{e}} \cdot \boldsymbol{f}^{\mathrm{e}\mathsf{T}} . \tag{8.87}$$

This expression for $\boldsymbol{b}^{\mathrm{e}}$ is exact, as can be verified by differentiating it with respect to time to obtain the original rate equation

$$\dot{\boldsymbol{b}}^{\mathrm{e}} = \boldsymbol{l}^{\mathrm{e}} \cdot \boldsymbol{b}^{\mathrm{e}} + \boldsymbol{b}^{\mathrm{e}} \cdot \boldsymbol{l}^{\mathrm{e}\mathsf{T}} , \qquad \boldsymbol{l}^{\mathrm{e}} = \dot{\boldsymbol{f}}^{\mathrm{e}} \cdot \boldsymbol{f}^{\mathrm{e}-1} . \tag{8.88}$$

Therefore, we can proceed with the sequence of calculations as follows: for a given $\Delta \boldsymbol{u}_c$, calculate the corresponding $\boldsymbol{b}^{\mathrm{e}}$.

Next, we write $\boldsymbol{b}^{\mathrm{e}}$ in spectral form as

$$\boldsymbol{b}^{\mathrm{e}} = \sum_{A=1}^{3} \lambda_A^2 \boldsymbol{n}^{(A)} \otimes \boldsymbol{n}^{(A)} , \tag{8.89}$$

where the λ_A^2's are the principal values of $\boldsymbol{b}^{\mathrm{e}}$ and the $\boldsymbol{n}^{(A)}$'s are the corresponding principal directions. We recall from Chapter 5 that the square roots of λ_A^2 are the elastic principal stretches, and $\epsilon_A = \log(\lambda_A)$ are the corresponding elastic logarithmic principal stretches. From the assumed isotropy in the elastic response, we can also write the symmetric Kirchhoff stress tensor in spectral form as

$$\boldsymbol{\tau} = \sum_{A=1}^{3} \tau_A \boldsymbol{n}^{(A)} \otimes \boldsymbol{n}^{(A)}, \qquad \tau_A = \frac{\partial \widehat{\psi}^{\mathrm{e}}}{\partial \epsilon_A}, \tag{8.90}$$

where $\widehat{\psi}^{\mathrm{e}} = \widehat{\psi}^{\mathrm{e}}(\epsilon_1, \epsilon_2, \epsilon_3)$ is the same stored energy function introduced in Chapter 5. A constraint on the Kirchhoff stress $\boldsymbol{\tau}$ is that it must satisfy the yield condition (8.59).

The following is a summary of the stress-point integration algorithm for a crystal deforming in single slip. Assume that the primary slip system is active at the beginning of the load increment so that

$$\psi_n(\boldsymbol{m}_n \otimes \boldsymbol{n}_n) : \boldsymbol{\tau}_n - \tau_{Yn} = 0, \tag{8.91}$$

where ψ_n is the sign function, $\tau_{Yn} = \tau_{Y0} + \zeta_n h_{\mathcal{D}}$, and τ_{Y0} is the initial yield stress. Introducing the sign function so that the vector $\psi \boldsymbol{m}$ always defines a forward slip direction, the expression for $\boldsymbol{f}^{\mathrm{e}}$ takes the form (cf. (8.85))

$$\boldsymbol{f}^{\mathrm{e}} = \boldsymbol{f}^{\mathrm{e\,tr}} - \Delta\zeta \psi \boldsymbol{m} \otimes \boldsymbol{N} \cdot \boldsymbol{F}_n^{\mathrm{e\,tr}-1}. \tag{8.92}$$

If $\Delta\zeta$ is known, we can sequentially calculate $\boldsymbol{f}^{\mathrm{e}} \to \boldsymbol{b}^{\mathrm{e}} \to \boldsymbol{\tau}$. However, $\Delta\zeta$ is unknown and must be calculated to satisfy the discrete consistency condition

$$r(\Delta\zeta) = \psi(\boldsymbol{m} \otimes \boldsymbol{n}) : \boldsymbol{\tau} - (\tau_{Yn} + \Delta\zeta h_{\mathcal{D}}) \to 0. \tag{8.93}$$

We have a nonlinear scalar equation in the slip increment $\Delta\zeta$, which we can solve by Newton's method. Noting that $\boldsymbol{m}_n$, $\boldsymbol{n}_n$, $\boldsymbol{\tau}_n$, and $\Delta\boldsymbol{u}_c$ are fixed during the local iteration, the tangent operator is calculated as

$$r'(\Delta\zeta) = \psi[(\boldsymbol{m} \otimes \boldsymbol{n}) : \boldsymbol{\tau}'(\Delta\zeta)] - h_{\mathcal{D}}. \tag{8.94}$$

The derivative of $\boldsymbol{\tau}$ is

$$\boldsymbol{\tau}'(\Delta\zeta) = \frac{1}{2}\boldsymbol{\varphi}^{\mathrm{e}} : \frac{\partial \boldsymbol{b}^{\mathrm{e}}}{\partial \Delta\zeta} = \boldsymbol{\alpha}^{\mathrm{e}} : \left(\frac{\partial \boldsymbol{f}^{\mathrm{e}}}{\partial \Delta\zeta} \cdot \boldsymbol{f}^{\mathrm{e}-1}\right), \tag{8.95}$$

where

$$\frac{\partial \boldsymbol{f}^{\mathrm{e}}}{\partial \Delta\zeta} = -\psi \boldsymbol{m} \otimes \boldsymbol{N} \cdot \boldsymbol{F}_n^{\mathrm{e\,tr}-1}, \tag{8.96}$$

and

$$\frac{\partial \boldsymbol{f}^{\mathrm{e}}}{\partial \Delta\zeta} \cdot \boldsymbol{f}^{\mathrm{e}-1} = -\psi \boldsymbol{m} \otimes \boldsymbol{N} \cdot \boldsymbol{F}^{\mathrm{e\,tr}-1}. \tag{8.97}$$

Therefore, we obtain

$$\boldsymbol{\tau}'(\Delta\zeta) = -\boldsymbol{\alpha}^{\mathrm{e}} : \left(\psi \boldsymbol{m} \otimes \boldsymbol{N} \cdot \boldsymbol{F}^{\mathrm{e\,tr}-1}\right). \tag{8.98}$$

In general, the iterative algorithm described above converges very fast, typically to machine precision in three iterations or less. The calculated incremental slip $\Delta\zeta$ is exact, as is the elastic relative deformation gradient $\boldsymbol{f}^{\mathrm{e}}$.

8.10 Multislip Systems

In a more general case, the coarse-scale displacement increment $\Delta \boldsymbol{u}_c$ may be large enough to trigger two or more slip systems. Some of these systems may be linearly dependent, in which case, a special filtering algorithm, such as the ultimate algorithm, may be employed to eliminate the redundant constraints. For the small strain case, we recall that tracking the individual slip systems is an explicit process because the elastoplastic tangential moduli are constant. However, for the finite deformation case, tracking the slip systems is an iterative process. In general, an iterative solution is employed to: (a) detect the next active slip system(s), and (b) track the evolution of the current slip systems. The discussions below are specific to a duplex system, but it is quite easy to see how the developments can be extended to more than two slip systems.

Consider a coarse-scale displacement increment imposed as a ramp function varying according to $\kappa \Delta \boldsymbol{u}_c$, where κ is a pseudo-time variable, similar in meaning to the same variable introduced in Section 8.4, which varies from $[0, 1]$. The corresponding coarse-scale deformation gradients are

$$\boldsymbol{F}^{\text{e tr}} = \boldsymbol{F}_n^{\text{e tr}} + \kappa \frac{\partial \Delta \boldsymbol{u}_c}{\partial \boldsymbol{X}}, \qquad \boldsymbol{f}^{\text{e tr}} = \boldsymbol{1} + \kappa \frac{\partial \Delta \boldsymbol{u}_c}{\partial \boldsymbol{x}_{cn}}. \tag{8.99}$$

Let α denote the primary slip system, which is assumed to be active right at the beginning of the load increment. We obtain the slip direction $\boldsymbol{m}^{(\alpha)}$ from the contravariant transformation

$$\boldsymbol{m}^{(\alpha)} = \boldsymbol{f}^{\text{e tr}} \cdot \boldsymbol{m}_n^{(\alpha)} / \| \boldsymbol{f}^{\text{e tr}} \cdot \boldsymbol{m}_n^{(\alpha)} \|, \tag{8.100}$$

and the unit normal $\boldsymbol{n}^{(\alpha)}$ from the covariant transformation

$$\boldsymbol{n}^{(\alpha)} = \boldsymbol{n}_n^{(\alpha)} \cdot \boldsymbol{f}^{\text{e tr}-1} / \| \boldsymbol{n}_n^{(\alpha)} \cdot \boldsymbol{f}^{\text{e tr}-1} \|. \tag{8.101}$$

For a crystal deforming in single slip, the elastic component of relative deformation gradient varies according to

$$\boldsymbol{f}^{\text{e}} = \boldsymbol{f}^{\text{e tr}} - \Delta \zeta^{(\alpha)} \boldsymbol{m}^{(\alpha)} \otimes \boldsymbol{N}^{(\alpha)} \cdot \boldsymbol{F}_n^{\text{e tr}-1}, \tag{8.102}$$

where

$$\Delta \zeta^{(\alpha)} = \zeta^{(\alpha)} - \zeta_n^{(\alpha)} \| \boldsymbol{f}^{\text{e tr}} \cdot \boldsymbol{m}_n^{(\alpha)} \| \tag{8.103}$$

is the slip increment. From this, we can update the elastic left Cauchy-Green deformation tensor as

$$\boldsymbol{b}^{\text{e}} = \boldsymbol{f}^{\text{e}} \cdot \boldsymbol{b}_n^{\text{e}} \cdot \boldsymbol{f}^{\text{e}\mathsf{T}}, \tag{8.104}$$

and then calculate the Kirchhoff stress tensor $\boldsymbol{\tau}$ from the hyperelastic constitutive equation. The discrete consistency condition determines the incremental slip $\Delta \zeta^{(\alpha)}$,

$$\psi^{(\alpha)} \boldsymbol{m}^{(\alpha)} \otimes \boldsymbol{n}^{(\alpha)} : \boldsymbol{\tau} - (\tau_{Yn} + \Delta \zeta^{(\alpha)} h_{\mathcal{D}}) = 0. \tag{8.105}$$

This sequence of calculations results in an exact solution for the incremental slip provided that the pseudo-time variable κ is not large enough to trigger a secondary slip.

Now, suppose we want to determine the value of $\kappa \leq 1$ that triggers a secondary slip. The search for the next critical slip system entails testing all inactive systems for yielding and identifying the specific system that requires the smallest value of κ to activate. The algorithm for calculating the value of κ for each inactive system begins with a push-forward on all slip vectors. The coarse-scale relative deformation gradient $\boldsymbol{f}^{\mathrm{e\,tr}}$ induces an exact push-forward on the primary slip system; therefore, it induces the same push-forward on all inactive systems, i.e.,

$$\boldsymbol{m}^{(\beta)} = \boldsymbol{f}^{\mathrm{e\,tr}} \cdot \boldsymbol{m}_n^{(\beta)} / \| \boldsymbol{f}^{\mathrm{e\,tr}} \cdot \boldsymbol{m}_n^{(\beta)} \| \tag{8.106}$$

and

$$\boldsymbol{n}^{(\beta)} = \boldsymbol{n}_n^{(\beta)} \cdot \boldsymbol{f}^{\mathrm{e\,tr}-1} / \| \boldsymbol{n}_n^{(\beta)} \cdot \boldsymbol{f}^{\mathrm{e\,tr}-1} \| \tag{8.107}$$

for all inactive β. The value of κ required for yielding of the next critical system β must satisfy the yield condition

$$\psi^{(\beta)} \boldsymbol{m}^{(\beta)} \otimes \boldsymbol{n}^{(\beta)} : \boldsymbol{\tau} - (\tau_{Yn} + \Delta\zeta^{(\alpha)} h_{\mathcal{D}}) = 0\,. \tag{8.108}$$

Equations (8.105) and (8.108) can be solved simultaneously for κ and $\Delta\zeta^{(\alpha)}$ using a simple Newton iteration.

Next, consider a duplex system (α, β) subjected to a conforming displacement field $\boldsymbol{u}_c$. The corresponding fine-scale displacement field is given by

$$\boldsymbol{u} = \overline{\boldsymbol{u}} + \sum_{\circ=(\alpha,\beta)} \widetilde{\boldsymbol{u}}^{\circ} H_{\mathcal{D}}^{\circ}\,, \qquad H_{\mathcal{D}}^{\circ}(\boldsymbol{X}) = \begin{cases} 1 & \text{if } \boldsymbol{X} \in \mathcal{D}_+^{\circ} \\ 0 & \text{if } \boldsymbol{X} \in \mathcal{D}_-^{\circ} \end{cases}\,, \tag{8.109}$$

where the symbol '$\circ$' pertains either to system α or β, and $\mathcal{D}_{\pm}^{\circ}$ are either the slave or master sides of the corresponding slip planes. The elastic components of deformation gradient are

$$\boldsymbol{F}^{\mathrm{e}} = \boldsymbol{F}^{\mathrm{e\,tr}} - \sum_{\circ=(\alpha,\beta)} \zeta^{\circ} \psi^{\circ} \boldsymbol{m}^{\circ} \otimes \boldsymbol{N}^{\circ} \tag{8.110}$$

for any time $t > t_n$, and

$$\boldsymbol{F}_n^{\mathrm{e}} = \boldsymbol{F}_n^{\mathrm{e\,tr}} - \sum_{\circ=(\alpha,\beta)} \zeta_n^{\circ} \psi_n^{\circ} \boldsymbol{m}_n^{\circ} \otimes \boldsymbol{N}^{\circ} \tag{8.111}$$

at time t_n. The following sign functions ensure that all slips are forward slips,

$$\psi^{\circ} = \mathrm{sign}[(\boldsymbol{m}^{\circ} \otimes \boldsymbol{n}^{\circ}) : \boldsymbol{\tau}]\,, \qquad \circ = (\alpha, \beta)\,. \tag{8.112}$$

The deformation gradient $\boldsymbol{F}_n^{\mathrm{e}}$ is known, so it can be inverted to yield

$$\boldsymbol{f}^{\mathrm{e}} = \boldsymbol{F}^{\mathrm{e\,tr}} \cdot \boldsymbol{F}_n^{\mathrm{e}-1} - \sum_{\circ=(\alpha,\beta)} \zeta^\circ \psi^\circ \boldsymbol{m}^\circ \otimes \boldsymbol{N}^\circ \cdot \boldsymbol{F}_n^{\mathrm{e}-1}, \tag{8.113}$$

where

$$\boldsymbol{m}^\circ = \boldsymbol{f}^{\mathrm{e}} \cdot \boldsymbol{m}_n^\circ / \|\boldsymbol{f}^{\mathrm{e}} \cdot \boldsymbol{m}_n^\circ\|, \qquad \circ = (\alpha, \beta). \tag{8.114}$$

We view equation (8.113) as a system of nine scalar equations in eleven scalar unknowns, namely, the nine elements of $\boldsymbol{f}^{\mathrm{e}}$ and the two slips $\zeta^{(\alpha)}$ and $\zeta^{(\beta)}$. The consistency conditions provide the two remaining equations, which we write in residual form as

$$R_\circ = (\psi^\circ \boldsymbol{m}^\circ \otimes \boldsymbol{n}^\circ) : \boldsymbol{\tau} - \Big(\tau_{Yn} + \sum_{\circ=a,b} \Delta\zeta^\circ h_{\mathcal{D}}\Big) \to 0, \tag{8.115}$$

where

$$\Delta\zeta^\circ = \zeta^\circ - \zeta_n^\circ \|\boldsymbol{f}^{\mathrm{e}} \cdot \boldsymbol{m}_n^\circ\|, \qquad \circ = (\alpha, \beta). \tag{8.116}$$

The algorithm described above is exact, but it entails a large number of unknowns to solve. This motivates the development of a simplified stress-point integration algorithm for multislip systems.

The first step in developing a simplified integration algorithm for multislip systems is to replace the continuous deformation gradient with the conforming deformation gradient to induce a push-forward on the vector $\boldsymbol{M}$:

$$\boldsymbol{m}^\circ = \boldsymbol{F}^{\mathrm{e\,tr}} \cdot \boldsymbol{M}^\circ / \|\boldsymbol{F}^{\mathrm{e\,tr}} \cdot \boldsymbol{M}^\circ\|, \qquad \circ = (\alpha, \beta). \tag{8.117}$$

The second step involves a recursive inversion of $\boldsymbol{F}_n^{\mathrm{e}}$ using the Sherman-Morrison formula and dropping the higher-order terms. Let

$$\boldsymbol{F}_{n\alpha}^{\mathrm{e}} = \boldsymbol{F}_n^{\mathrm{e\,tr}} - \zeta_n^{(\alpha)} \psi_n^{(\alpha)} \boldsymbol{m}_n^{(\alpha)} \otimes \boldsymbol{N}^{(\alpha)}; \tag{8.118}$$

then

$$\boldsymbol{F}_n^{\mathrm{e}} = \boldsymbol{F}_{n\alpha}^{\mathrm{e\,tr}} - \zeta_n^{(\beta)} \psi_n^{(\beta)} \boldsymbol{m}_n^{(\beta)} \otimes \boldsymbol{N}^{(\beta)}. \tag{8.119}$$

The respective inverses are

$$\boldsymbol{F}_{n\alpha}^{\mathrm{e}-1} = \boldsymbol{F}_n^{\mathrm{e\,tr}-1} + \zeta_n^{(\alpha)} \boldsymbol{F}_n^{\mathrm{e\,tr}-1} \cdot \psi_n^{(\alpha)} \boldsymbol{m}_n^{(\alpha)} \otimes \boldsymbol{N}^{(\alpha)} \cdot \boldsymbol{F}_n^{\mathrm{e\,tr}-1} \tag{8.120}$$

and

$$\boldsymbol{F}_n^{\mathrm{e}-1} = \boldsymbol{F}_{n\alpha}^{\mathrm{e\,tr}-1} + \zeta_n^{(\beta)} \boldsymbol{F}_{n\alpha}^{\mathrm{e\,tr}-1} \cdot \psi_n^{(\beta)} \boldsymbol{m}_n^{(\beta)} \otimes \boldsymbol{N}^{(\beta)} \cdot \boldsymbol{F}_{n\alpha}^{\mathrm{e\,tr}-1}. \tag{8.121}$$

Expanding the expressions above gives a polynomial equation for $\boldsymbol{f}^{\mathrm{e}}$ in $\zeta^{(\alpha)}$, $\zeta^{(\beta)}$, $\zeta_n^{(\alpha)}$, and $\zeta_n^{(\beta)}$. Ignoring the higher-order terms of the polynomial and taking only the linear terms, we get

$$\boldsymbol{f}^{\mathrm{e}} = \boldsymbol{f}^{\mathrm{e\,tr}} - \sum_{\circ=(\alpha,\beta)} \Delta\zeta^\circ \psi^\circ \boldsymbol{m}^\circ \otimes \boldsymbol{N}^\circ \cdot \boldsymbol{F}_n^{\mathrm{e\,tr}-1}, \tag{8.122}$$

where

$$\Delta\zeta^\circ = \zeta^\circ - \zeta_n^\circ \| \boldsymbol{f}^{\mathrm{e\,tr}} \cdot \boldsymbol{m}_n^\circ \| , \qquad \circ = (\alpha, \beta) , \tag{8.123}$$

and $\boldsymbol{f}^{\mathrm{e\,tr}} = \boldsymbol{F}^{\mathrm{e\,tr}} \cdot \boldsymbol{F}_n^{\mathrm{e\,tr}-1}$. The linearized equation for $\boldsymbol{f}^{\mathrm{e}}$ has a predictor-corrector format that resembles the incremental form developed earlier for a crystal deforming in single slip, see (8.85). To determine the co-rotational incremental slips, we impose the two consistency conditions

$$R_\circ = \psi^\circ[(\boldsymbol{m}^\circ \otimes \boldsymbol{n}^\circ) : \boldsymbol{\tau}] - \Big(\tau_{Yn} + \sum_{\bullet=(\alpha,\beta)} \Delta\zeta^\bullet h_{\mathcal{D}}\Big) \to 0 , \qquad \circ = (\alpha, \beta) . \tag{8.124}$$

Box 8.2 summarizes the algorithm. For a duplex system, the problem reduces to the solution of two equations in two unknowns, which is a much smaller system to solve than the eleven equations required to obtain the exact solution. To extend the algorithm to multislip systems, it is only necessary to change the range of α and β to reflect the total number of unknown incremental slips. The example below demonstrates the performance of this algorithm through some stress-point simulations.

Step 1. Initialize $\boldsymbol{f}^{\mathrm{e\,tr}} = \mathbf{1} + \partial\Delta\boldsymbol{u}_c/\partial\boldsymbol{x}_{cn}$, $\zeta^\circ = \zeta_n^\circ$ for $\circ = (\alpha, \beta)$
Step 2. Rotate $\boldsymbol{m}^\circ = \boldsymbol{f}^{\mathrm{e\,tr}} \cdot \boldsymbol{m}_n^\circ / \| \boldsymbol{f}^{\mathrm{e\,tr}} \cdot \boldsymbol{m}_n^\circ \|$ for $\circ = (\alpha, \beta)$
Step 3. Solve $\boldsymbol{f}^{\mathrm{e}} = \boldsymbol{f}^{\mathrm{e\,tr}} - \sum_{\circ=(\alpha,\beta)} \Delta\zeta^\circ \psi^\circ \boldsymbol{m}^\circ \otimes \boldsymbol{N}^\circ \cdot \boldsymbol{F}_n^{\mathrm{e\,tr}-1}$
Step 4. Compute $\boldsymbol{f}^{\mathrm{e}} \to \boldsymbol{b}^{\mathrm{e}} \to \boldsymbol{\tau}$
Step 5. Iterate $R_{(\alpha)} \to 0$ and $R_{(\beta)} \to 0$ for $\Delta\zeta^{(\alpha)}$ and $\Delta\zeta^{(\beta)}$, and exit.

Box 8.2 Linearized algorithm for calculating the incremental co-rotational slips $\Delta\zeta^{(\alpha)}$ and $\Delta\zeta^{(\beta)}$ in the finite deformation range

As an illustration, we revisit the example presented in Section 8.5 for an f.c.c. crystal with Euler angles $\theta = 0$ and $\phi = 18°$ in the reference configuration. The crystal is subjected to two deformation histories: a volume-preserving extension where the conforming field varies according to the relations

$$\widetilde{u}_1 = kX_1 , \qquad \widetilde{u}_2 = -0.5kX_2 , \qquad \widetilde{u}_3 = -0.5kX_3 ,$$

and simple shearing where the conforming displacement field varies according to the relations

$$\widetilde{u}_1 = kX_2 , \qquad \widetilde{u}_2 = 0 , \qquad \widetilde{u}_3 = 0 .$$

All the other crystal properties are the same as in the example of Section 8.5.

Figure 8.9 shows the stress-strain curves for $h_{\mathcal{D}} = 0$ (perfect plasticity) and $h_{\mathcal{D}} > 0$ (Taylor hardening). In these simulations, we only consider the first two critical slip systems (i.e., we suppress the other systems) to illustrate obvious patterns: For the isochoric tension test, the effect of finite deformation on the overall crystal response is to introduce geometric softening, wherein the crystal rotation facilitates easier slip on the active system(s). This is obvious from the $h_{\mathcal{D}} = 0$ simulation where the relevant component of the second Piola-Kirchhoff stress $\boldsymbol{S}$ (see Section 5.4) softens to a value equal to zero after applying the relevant component of the Green-Lagrange strain $\boldsymbol{E} = (\boldsymbol{C}-\mathbf{1})/2$ (see Section 5.2) of 25%, signifying complete geometric softening. In contrast, simple shearing introduces geometric hardening, wherein the crystal rotation makes shearing on the relevant slip system(s) more difficult. These geometrical effects are not accommodated by the infinitesimal-deformation simulations reported in Fig. 8.6.

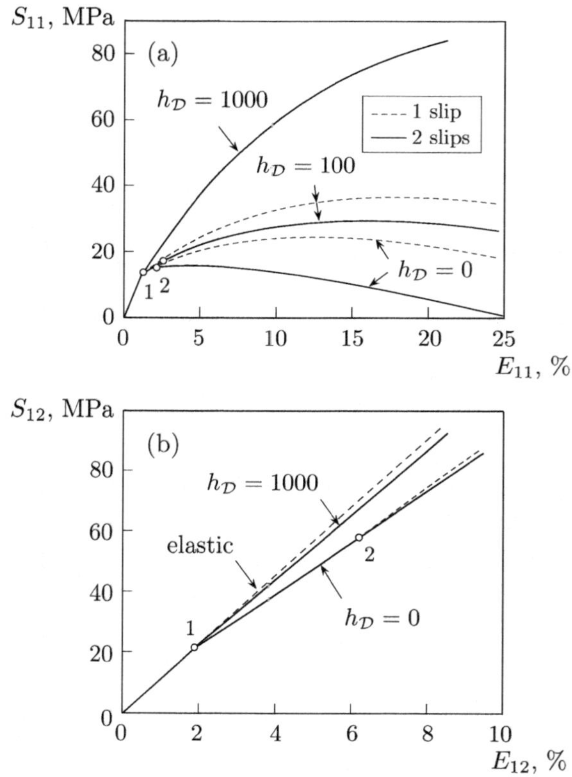

Fig. 8.9 Stress-point simulations on a f.c.c. crystal with Euler angles $\theta = 0$ and $\phi = 18°$ in the reference configuration: (a) isochoric extension test, and (b) isochoric simple shear test. Smaller-font numbers denote independent slip systems detected.

8.11 Twisting and Stretching of a Hollow Cylinder

We consider a hollow cylinder, shown in Fig. 8.10, with a height of 4 m, outer diameter of 2 m, and a thickness of 0.1 m. The finite element mesh has 5,148 nodes and 2,560 eight-node hexahedral elements, all integrated with the B-bar option. The cylinder is clamped at both its top and bottom ends while the inner and outer vertical faces are assumed to be traction-free. The top end is axially stretched by 0.25 cm and torsionally twisted by 0.7° while holding the bottom end fixed. Similar simulations, but without stretching the cylinder, are presented by Borja and Rahmani (2012). Crystal orientation is defined by Euler angles $\theta = \phi = 0$, with Young's modulus $E = 15$ GPa, Poisson's ratio $\nu = 0.37$, initial yield strength $\tau_{Y0} = 10$ MPa, and hardening parameter $h = -1$ MPa (softening). An infinitesimal formulation is used in the simulation.

Figure 8.10 shows the resulting final plastic strain contour. In the first simulation where the crystal orientation is uniform, four vertical deformation bands emerge from the imposed deformation, as shown in Fig. 8.10a. These bands did not form randomly, but rather, they are determined by the relative crystal orientation with respect to the direction of stretching and twisting. In the second simulation, a small imperfection is embedded in the cylinder

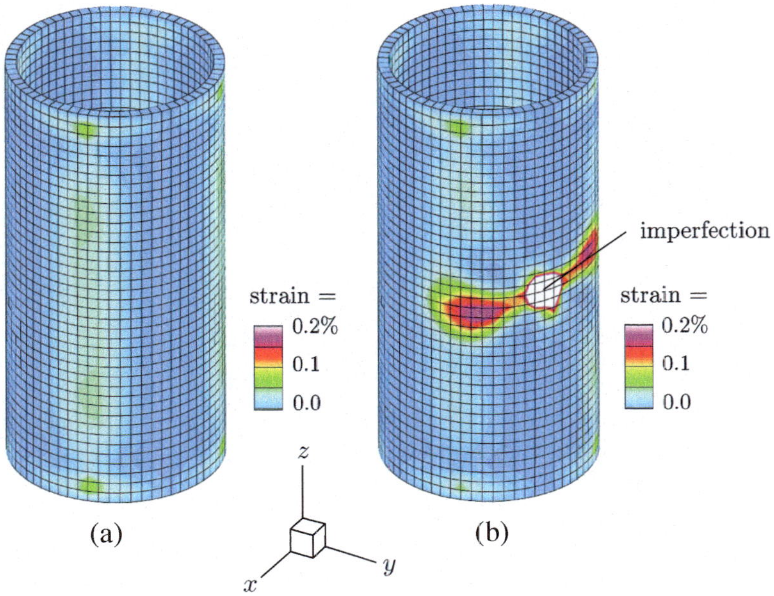

Fig. 8.10 Deformation bands forming in a hollow cylinder subjected to torsional twisting: (a) uniform crystal orientation with $\theta = \phi = 0$; (b) non-uniform crystal orientation with an imperfection of $\theta = 20°$ and $\phi = 0$ in four adjacent hexahedral elements (see white window). Color bar is second invariant of deviatoric plastic strain in percent.

by rotating the same crystal in four adjacent finite elements (see the white window in Fig. 8.10b) to an orientation defined by Euler angles $\theta = 20°$ and $\phi = 0$. This imperfection generates more intense localized deformation and complementary deformation bands propagating away from the imperfection.

Table 8.1 shows the convergence profile of Newton iterations at various stages of loading. Note that the convergence of the iterations is faster for the case where the crystal is uniformly oriented, and slower when there is an imperfection. The latter convergence profile may be attributed to increased difficulty in identifying the active slip systems in the neighborhood of the imperfection.

Table 8.1 Combined twisting and stretching of a hollow cylinder: convergence profile of Newton iterations. Tabulated errors at different crystal orientations are based on the relative norm of residual force vector, $\|\boldsymbol{r}^k\|/\|\boldsymbol{r}^0\|$

Percent Load	Iteration	Uniform	Non-Uniform
50%	1	1.00e+00	1.00e+00
	2	9.69e−16	9.48e−16
75%	1	1.00e+00	1.00e+00
	2	9.48e−16	1.69e−09
	3	—	9.39e−11
	4		1.31e−16
100%	1	1.00e+00	1.00e+00
	2	7.39e−08	1.06e−07
	3	8.01e−09	8.44e−08
	4	6.09e−10	5.99e−09
	5	4.15e−11	6.65e−10
	6	1.22e−16	9.39e−11
	7	—	2.38e−12
	8		1.93e−12
	9		2.44e−15

8.12 Summary

Crystal plasticity theory describes the inelastic deformation behavior of materials with crystalline microstructures, such as metals and igneous rocks. The inelastic response of these materials depends on the crystal orientation relative to the loading directions. Ductility of a material depends on both the atomic packing and the structure of the slip systems. Crystal plasticity theory is described by multiple yield constraints with significant redundancy, necessitating a robust filtering algorithm to remove the redundant constraints. An 'ultimate algorithm' for filtering the redundant constraints may be used for this purpose. An important effect of geometric nonlinearity is to introduce either geometric softening or geometric hardening into the overall crystal response. Both effects play a key role in the prediction of localized deformation and instability (discussed in the next chapter) of materials with crystalline microstructures.

9 Bifurcation

9.1 Introduction

Localized deformation in solids occurs in many applications and in different forms. Examples of localized deformation include Lüders bands in metals (Nádai 1931), cracking in concrete (Bažant and Planas 1997), and flow localization in bulk metallic glasses (Flores and Dauskardt 2001). Figure 9.1 shows images of shear bands in steel and bulk metallic glass at different scales. Shear bands have received considerable attention over the years because of their distinctive and intriguing style characterized by significant shear offset within a very narrow zone. However, compaction bands, characterized by a significant volume decrease with little or no shear offset, are now also recognized as a complementary deformation style particularly in high-porosity rocks (Issen and Rudnicki 2000). Both shear bands and compaction bands are understood as examples of a larger class of so-called deformation bands (Aydin et al. 2006, Borja and Aydin 2004, Borja 2004).

Deformation bands, typically tabular in form, are generally regarded as a result of material instability, but material instability does not always result in a deformation band. Pore collapse in high-porosity rocks could be pervasive enough to generate cataclastic flow, where diffuse compaction occurs throughout the specimen (Wong et al. 1997). An opposite mechanism is void growth in metal specimens subjected to a high ratio of isotropic extension-to-deviatoric shear stress, called triaxiality (Agarwal et al. 2003). From a larger scale, the deformation mode in the specimen may appear diffuse and exhibit no localized deformation patterns. From a smaller scale (e.g., micron-scale and smaller), the same material could exhibit more complex patterns of localized deformation.

Bifurcation theory of continuum mechanics is often used to detect the onset of instability in the material response. The theory identifies a stress state at which the solution could lose uniqueness. The tangential stress-strain tensor plays a key role in the theory of bifurcation, since it is used to determine whether or not a material would continue to exhibit a 'stable' response. Theory of plasticity provides a mathematical description of the tangential stress-strain tensor through the continuum elastoplastic tangent operator $\boldsymbol{c}^{\mathrm{ep}}$. This chapter thus uses the unique mathematical structure of the tensor $\boldsymbol{c}^{\mathrm{ep}}$ to

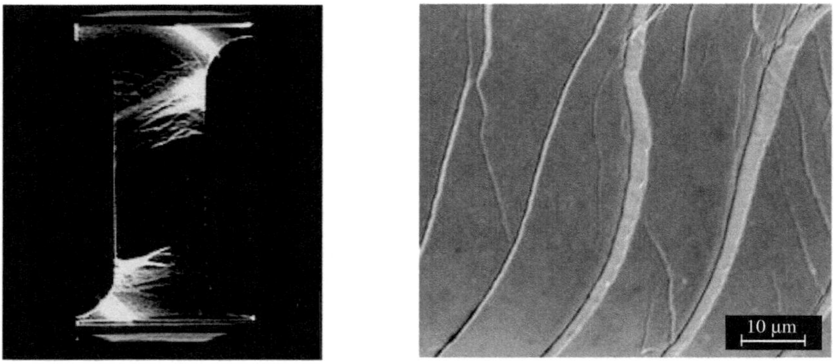

(a) (b)

Fig. 9.1 Strain localization images: (a) Lüders bands in 1018 annealed steel polished to a metallographic finish (image courtesy of Mike Meier and Aaron Broumas); (b) SEM image of shear steps formed by the propagation of shear bands during rolling of a bulk metallic glass (image courtesy of Katharine Flores)

predict bifurcation and identify many common instability modes. 'Flutter instability,' a form of instability corresponding to a vibrational motion of increasing amplitude (Bigoni 2000), is not covered in this chapter.

9.2 Stability of Incrementally Linear Solids

If the tangential response of a material is the same irrespective of the direction of the incremental load, then the response is said to be incrementally linear. A nonlinear elastic solid is an example of a material exhibiting an incrementally linear response. An elastoplastic solid in a state of yield is an example of a material exhibiting an incrementally nonlinear response: the tangential moduli are different for plastic loading than for elastic unloading. If there is only one yield constraint for the solid, then the tangent operator can have two branches: one for loading and another for unloading. When there are two or more yield criteria to be satisfied, such as in crystal plasticity where the slip systems constitute two or more yield constraints, then the tangent operator could have more than two branches. In this section, we establish conditions for the stability of an incrementally linear solid.

Consider a solid with domain $\mathcal{B}$ and surface boundary $\partial\mathcal{B}$. Assume that $\partial\mathcal{B}$ admits the usual decomposition $\partial\mathcal{B} = \overline{\partial\mathcal{B}_t \cup \partial\mathcal{B}_u}$ and $\varnothing = \partial\mathcal{B}_t \cap \partial\mathcal{B}_u$, where $\partial\mathcal{B}_t$ and $\partial\mathcal{B}_u$ are, respectively, portions of the boundary where surface tractions and displacements are prescribed. For quasistatic loading the boundary-value problem states that

$$\left.\begin{array}{rl}\nabla\cdot\dot{\boldsymbol{\sigma}}+\dot{\boldsymbol{b}}=\mathbf{0} & \text{in } \mathcal{B}\\ \dot{\boldsymbol{u}}=\dot{\boldsymbol{u}}_0 & \text{on } \partial\mathcal{B}_u\\ \dot{\boldsymbol{\sigma}}\cdot\boldsymbol{n}=\dot{\boldsymbol{t}}_0 & \text{on } \partial\mathcal{B}_t\end{array}\right\}, \tag{9.1}$$

where $\boldsymbol{b}$ is the weight density vector, $\boldsymbol{n}$ is the outward unit normal vector to $\partial\mathcal{B}$, $\boldsymbol{u}_0$ is the prescribed boundary displacement, $\boldsymbol{t}_0$ is the prescribed surface traction, and the superimposed dot denotes a time derivative. Note that the boundary-value problem is stated in incremental, or rate, form, rather than in total form. Stability analysis deals with very small changes in the stress state, necessitating that the problem be stated in rate form. We want to find a point in the solution where more than one local response could emerge. This means that, apart from the local solution $(\dot{\boldsymbol{\sigma}}, \dot{\boldsymbol{u}})$ satisfying (9.87), we might also find a different solution $(\dot{\boldsymbol{\sigma}}^*, \dot{\boldsymbol{u}}^*)$ satisfying the same governing equations. This corresponds to a local bifurcation point, and the alternative solution $(\dot{\boldsymbol{\sigma}}^*, \dot{\boldsymbol{u}}^*)$ is the associated local bifurcated response.

Equation (2.64) of Chapter 2 generalizes to 3D as follows

$$\int_{\mathcal{B}} \nabla\cdot[(\dot{\boldsymbol{u}}^*-\dot{\boldsymbol{u}})\cdot(\dot{\boldsymbol{\sigma}}^*-\dot{\boldsymbol{\sigma}})]\,\mathrm{d}V = \int_{\mathcal{B}} (\dot{\boldsymbol{\epsilon}}^*-\dot{\boldsymbol{\epsilon}}):(\dot{\boldsymbol{\sigma}}^*-\dot{\boldsymbol{\sigma}})\,\mathrm{d}V + \int_{\mathcal{B}} (\dot{\boldsymbol{u}}^*-\dot{\boldsymbol{u}})\cdot[\nabla\cdot(\dot{\boldsymbol{\sigma}}^*-\dot{\boldsymbol{\sigma}})]\,\mathrm{d}V\,. \tag{9.2}$$

The second integral on the right-hand side vanishes since the two stresses satisfy the governing partial differential equation. The term on the left-hand side can be written as a surface integral, and also vanishes. Thus, any pair of possible solutions must satisfy the condition

$$\int_{\mathcal{B}} (\dot{\boldsymbol{\epsilon}}^*-\dot{\boldsymbol{\epsilon}}):(\dot{\boldsymbol{\sigma}}^*-\dot{\boldsymbol{\sigma}})\,\mathrm{d}V = 0\,. \tag{9.3}$$

Uniqueness is guaranteed for every point and for every pair of stresses and strains linked by the constitutive equation if

$$(\dot{\boldsymbol{\epsilon}}^*-\dot{\boldsymbol{\epsilon}}):(\dot{\boldsymbol{\sigma}}^*-\dot{\boldsymbol{\sigma}}) > 0 \tag{9.4}$$

for all $\boldsymbol{x}\in\mathcal{B}$. This last condition is sometimes called the second-order work for stability (Darve et al. 2004).

A more familiar result is obtained if the material response is incrementally linear in the sense

$$\dot{\boldsymbol{\sigma}} = \boldsymbol{c}:\dot{\boldsymbol{\epsilon}}\,, \qquad \dot{\boldsymbol{\sigma}}^* = \boldsymbol{c}:\dot{\boldsymbol{\epsilon}}^*\,, \tag{9.5}$$

i.e., the tangential stress-strain tensor $\boldsymbol{c}$ is the same irrespective of the incremental load path. In this case, the local uniqueness condition (9.4) coincides with the local stability condition

$$\dot{\boldsymbol{\phi}}:\boldsymbol{c}:\dot{\boldsymbol{\phi}} > 0 \tag{9.6}$$

for any symmetric second-order tensor $\dot{\boldsymbol{\phi}}\neq\mathbf{0}$ (in the present discussion, $\dot{\boldsymbol{\phi}}$ may be considered as the 'jump' in the strain rate, i.e., $\dot{\boldsymbol{\phi}} = \dot{\boldsymbol{\epsilon}}^* - \dot{\boldsymbol{\epsilon}}$). Note that this definition of stability is not related to the stability theory of Lyapunov for dynamical systems (Lyapunov 1966).

The tangent stress-strain tensor $\boldsymbol{c}$ can be expressed as the sum of a symmetric part and a skew-symmetric part,

$$\boldsymbol{c} = \mathrm{sym}(\boldsymbol{c}) + \mathrm{skw}(\boldsymbol{c})\,, \tag{9.7}$$

where $\mathrm{sym}(c_{ijkl}) = (c_{ijkl} + c_{klij})/2$ (major symmetry), and $\mathrm{skw}(c_{ijkl}) = (c_{ijkl} - c_{klij})/2$. The skew-symmetric component has no effect on the local stability condition since $\dot{\boldsymbol{\phi}} : \mathrm{skw}(\boldsymbol{c}) : \dot{\boldsymbol{\phi}} = 0$. Hence, the local stability condition reduces to $\dot{\boldsymbol{\phi}} : \mathrm{sym}(\boldsymbol{c}) : \dot{\boldsymbol{\phi}} > 0$, and the condition for the loss of uniqueness is

$$\det[\mathrm{sym}(\boldsymbol{c})] = 0\,. \tag{9.8}$$

A physical meaning of the loss of uniqueness may be appreciated from an experimental program employing an arbitrary loading (Chambon 2005, Nova 1994). For example, consider a cylindrical specimen of rock compressed axially while maintaining the radial stress constant. Condition (9.8) then signals the loss of ability to control the various components of the test, such as the application of an axial compression.

A stronger condition for the loss of stability is given by the stationary-stress criterion, $\boldsymbol{c} : \dot{\boldsymbol{\phi}} = \boldsymbol{0}$. For non-trivial solutions to exist we must have

$$\det(\boldsymbol{c}) = 0\,. \tag{9.9}$$

The determinant condition shown above is identical to the loss of uniqueness criterion if $\boldsymbol{c}$ has the major symmetry. However, if $\boldsymbol{c} \neq \mathrm{sym}(\boldsymbol{c})$, then the loss of uniqueness condition is not equivalent to a stationary stress, as can be inferred from the relation

$$\boldsymbol{c} : \dot{\boldsymbol{\phi}} = [\mathrm{sym}(\boldsymbol{c}) + \mathrm{skw}(\boldsymbol{c})] : \dot{\boldsymbol{\phi}} = \mathrm{skw}(\boldsymbol{c}) : \dot{\boldsymbol{\phi}} \neq \boldsymbol{0}\,, \tag{9.10}$$

since $\mathrm{sym}(\boldsymbol{c}) : \dot{\boldsymbol{\phi}} = \boldsymbol{0}$ is implied in (9.8). If $\boldsymbol{c}$ has no major symmetry, then condition (9.8) generally precedes condition (9.9).

Suppose that

$$\dot{\boldsymbol{\phi}} = \dot{\varphi}\,\mathrm{sym}(\boldsymbol{m} \otimes \boldsymbol{n})\,, \tag{9.11}$$

where $\boldsymbol{m}$ and $\boldsymbol{n}$ are unit vectors, and $\dot{\varphi} \in \mathbb{R}^1$; then we have a planar wave (Achenbach 1987). Substituting into (9.6) and noting that $\dot{\varphi}^2 > 0$ results in the strong ellipticity condition

$$\boldsymbol{m} \cdot \boldsymbol{A} \cdot \boldsymbol{m} > 0\,, \tag{9.12}$$

where $\boldsymbol{A} = \boldsymbol{A}(\boldsymbol{n})$ is called the acoustic tensor, with components $A_{ik} = n_k c_{ijkl} n_l$. Appropriately enough, the name of $\boldsymbol{A}$ was derived from its relation to the propagation of planar waves. The strong ellipticity condition is attained from one of many forms of the tensor $\dot{\boldsymbol{\phi}}$, so it follows that a stable response is strongly elliptic but not vice versa. Furthermore, since $\boldsymbol{m} \cdot \boldsymbol{A} \cdot \boldsymbol{m} = \boldsymbol{m} \cdot \mathrm{sym}(\boldsymbol{A}) \cdot \boldsymbol{m}$, the loss of strong ellipticity occurs when

$$\min|_{\boldsymbol{n}} \det\big(\mathrm{sym}(\boldsymbol{A})\big) = 0\,. \tag{9.13}$$

Here, the minimization of the set of determinants is concerned with a condition that is initially strongly elliptic, and the critical point in the solution is sought at which equation (9.13) is satisfied for the first time. Finding the critical state involves testing for all possible values of $\boldsymbol{n}$ and choosing the one that yields the smallest value of the determinant. If the minimum value of determinant becomes equal to zero, then we say that strong ellipticity is lost.

If $\boldsymbol{A} \cdot \boldsymbol{m} = \boldsymbol{0}$, then $\boldsymbol{m} \cdot \boldsymbol{A} \cdot \boldsymbol{m} = 0$. The former condition represents the loss of ellipticity, inasmuch as $\det(\boldsymbol{A}) \neq 0$ defines an elliptic system. Loss of ellipticity occurs when

$$\mathcal{F} = \min|_{\boldsymbol{n}} \det(\boldsymbol{A}) = 0\,. \tag{9.14}$$

A physical interpretation of this last equation may be seen by expanding

$$\boldsymbol{A} \cdot \boldsymbol{m} = [\boldsymbol{c} : \mathrm{sym}(\boldsymbol{m} \otimes \boldsymbol{n})] \cdot \boldsymbol{n} = \frac{1}{\dot{\varphi}}(\dot{\boldsymbol{\sigma}}^* - \dot{\boldsymbol{\sigma}}) \cdot \boldsymbol{n} = \boldsymbol{0}\,. \tag{9.15}$$

Hence, loss of ellipticity signals a stationary traction vector on a plane with unit normal $\boldsymbol{n}$. Later, we shall associate this plane with a deformation band, and also associate $\mathcal{F}$ with a so-called localization function. Loss of strong ellipticity and loss of ellipticity occur at the same time if $\boldsymbol{A}$ is symmetric. If $\boldsymbol{A}$ is not symmetric, then loss of strong ellipticity would come first.

9.3 Stability of Elastoplastic Solids

Consider an elastoplastic solid with one yield constraint. The tangent stress-strain tensor $\boldsymbol{c}$ for this solid can have two forms, one for elastic unloading and another for plastic loading,

$$\boldsymbol{c} = \begin{cases} \boldsymbol{c}^{\mathrm{e}} & \text{for elastic unloading}\,, \\ \boldsymbol{c}^{\mathrm{ep}} & \text{for plastic loading}\,. \end{cases} \tag{9.16}$$

Thus, the mechanical response is incrementally nonlinear in the sense that two possible solutions, $(\dot{\boldsymbol{\sigma}}, \dot{\boldsymbol{u}})$ and $(\dot{\boldsymbol{\sigma}}^*, \dot{\boldsymbol{u}}^*)$, could be linked by two different tangent constitutive tensors. In this case, equation (9.6) is no longer an appropriate local stability condition. An appropriate local stability condition should account different combinations of loading and unloading occurring on the two alternative solutions, and, in general, cannot be established by a simple expression such as the one given by equation (9.6) for an incrementally linear solid. However, we can always establish stability criteria using bounds imposed by so-called comparison solids (Hill 1958, Raniecki and Bruhns 1981). A comparison solid is a fictitious solid with a local stability condition given by a known analytical expression. It is defined in such a way that when the comparison solid is stable, then the real material is also stable.

In the following discussion we assume that $\boldsymbol{c}^{\mathrm{e}}$ is positive-definite with major and minor symmetries. The following abbreviated notations are used throughout the remainder of this chapter:

$$\boldsymbol{f} := \frac{\partial f}{\partial \boldsymbol{\sigma}} , \qquad \boldsymbol{g} := \frac{\partial g}{\partial \boldsymbol{\sigma}} , \tag{9.17}$$

where f and g are the yield and plastic potential functions, respectively. Furthermore, $\boldsymbol{c}^{\mathrm{ep}}$ takes the standard form:

$$\boldsymbol{c}^{\mathrm{ep}} = \boldsymbol{c}^{\mathrm{e}} - \boldsymbol{c}^{\mathrm{p}} , \quad \boldsymbol{c}^{\mathrm{p}} = \frac{1}{\chi}\boldsymbol{c}^{\mathrm{e}} : \boldsymbol{g} \otimes \boldsymbol{f} : \boldsymbol{c}^{\mathrm{e}} , \quad \chi = \boldsymbol{f} : \boldsymbol{c}^{\mathrm{e}} : \boldsymbol{g} + H > 0 , \tag{9.18}$$

where H is the generalized plastic modulus. Conditions for loading and unloading will play an important role in the following stability analysis, and here they are determined according to the criteria

$$\operatorname{sign}(\boldsymbol{f} : \boldsymbol{c}^{\mathrm{e}} : \dot{\boldsymbol{\epsilon}}) = \begin{cases} (-) & \text{for elastic unloading,} \\ (+) & \text{for plastic loading.} \end{cases} \tag{9.19}$$

We identify the expression $\boldsymbol{c}^{\mathrm{e}} : \dot{\boldsymbol{\epsilon}}$ as the elastic stress predictor; its inner product with $\boldsymbol{f}$ determines whether the angle between them is acute (plastic loading) or obtuse (elastic unloading). This definition of loading/unloading condition is equivalent to the Kuhn-Tucker conditions.

Associative Plasticity

The Hill (1958) comparison solid for associative plasticity is an incrementally linear solid possessing only the loading branch of the elastoplastic response. For this reason, it is also called the 'in-loading comparison solid.' The stress and strain increments are linked uniquely by the tangent stress-strain tensor $\boldsymbol{c}^{\mathrm{ep}}$, so the local stability condition is given by (9.6) with the tangent constitutive tensor replaced by $\boldsymbol{c}^{\mathrm{ep}}$:

$$\mathcal{L}_1 = \dot{\boldsymbol{\phi}} : \boldsymbol{c}^{\mathrm{ep}} : \dot{\boldsymbol{\phi}} = \dot{\boldsymbol{\phi}} : (\boldsymbol{c}^{\mathrm{e}} - \boldsymbol{c}^{\mathrm{p}}) : \dot{\boldsymbol{\phi}} > 0 . \tag{9.20}$$

The comparison solid can only trigger the loading branch, so the consistency condition must be satisfied by both alternative solutions. For an incremental deformation $\dot{\boldsymbol{\epsilon}}$, we have

$$\boldsymbol{f} : \boldsymbol{c}^{\mathrm{e}} : (\dot{\boldsymbol{\epsilon}} - \dot{\lambda}\boldsymbol{f}) - \dot{\lambda}H = 0 . \tag{9.21}$$

Solving for $\dot{\lambda}$ gives

$$\dot{\lambda} = \frac{1}{\chi} \boldsymbol{f} : \boldsymbol{c}^{\mathrm{e}} : \dot{\boldsymbol{\epsilon}} \geq 0 \,, \tag{9.22}$$

where $\chi = \boldsymbol{f} : \boldsymbol{c}^{\mathrm{e}} : \boldsymbol{f} + H > 0$. Similarly, for loading direction $\dot{\boldsymbol{\epsilon}}^* = \dot{\boldsymbol{\epsilon}} + \dot{\boldsymbol{\phi}}$ we have

$$\boldsymbol{f} : \boldsymbol{c}^{\mathrm{e}} : (\dot{\boldsymbol{\epsilon}} + \dot{\boldsymbol{\phi}} - \dot{\bar{\lambda}} \boldsymbol{f}) - \dot{\bar{\lambda}} H = 0 \,, \tag{9.23}$$

from which we obtain

$$\dot{\bar{\lambda}} = \dot{\lambda} + \dot{\bar{\bar{\lambda}}} \geq 0 \,, \qquad \dot{\bar{\bar{\lambda}}} = \frac{1}{\chi} \boldsymbol{f} : \boldsymbol{c}^{\mathrm{e}} : \dot{\boldsymbol{\phi}} \geq 0 \,. \tag{9.24}$$

Note that $\dot{\boldsymbol{\epsilon}}$ and $\dot{\boldsymbol{\phi}}$ are two independent kinematical variables, so the constraint $\dot{\bar{\lambda}} \geq 0$ can hold if and only if both $\dot{\lambda}$ and $\dot{\bar{\bar{\lambda}}}$ are nonnegative.

We now turn to the real material and show that if the comparison solid is stable, then the real solid is also stable. There are four possible combinations of the real solid response representing two possible tangent stress-strain tensors for each of the two alternative solutions. The first combination is when both $\dot{\boldsymbol{\epsilon}}$ and $\dot{\boldsymbol{\epsilon}}^*$ trigger the loading branch (loading/loading combination). This corresponds to the same 'in-loading comparison solid' discussed previously for which the local stability condition is given by (9.20). Of interest is when $\dot{\boldsymbol{\epsilon}}$ triggers the loading branch and $\dot{\boldsymbol{\epsilon}}^*$ triggers the unloading branch (loading/unloading combination). The relevant mathematical conditions are $\boldsymbol{f} : \boldsymbol{c}^{\mathrm{e}} : \dot{\boldsymbol{\epsilon}} > 0$ and $\boldsymbol{f} : \boldsymbol{c}^{\mathrm{e}} : \dot{\boldsymbol{\epsilon}}^* < 0$. Note that these two conditions also imply that $\boldsymbol{f} : \boldsymbol{c}^{\mathrm{e}} : \dot{\boldsymbol{\phi}} < 0$. In this case the alternative solutions are linked by the constitutive equations $\dot{\boldsymbol{\sigma}} = \boldsymbol{c}^{\mathrm{ep}} : \dot{\boldsymbol{\epsilon}}$ and $\dot{\boldsymbol{\sigma}}^* = \boldsymbol{c}^{\mathrm{e}} : \dot{\boldsymbol{\epsilon}}^*$, and the uniqueness condition (9.4) becomes

$$\mathcal{P} = \dot{\boldsymbol{\phi}} : (\dot{\boldsymbol{\sigma}}^* - \dot{\boldsymbol{\sigma}}) = \dot{\boldsymbol{\phi}} : \boldsymbol{c}^{\mathrm{e}} : \dot{\boldsymbol{\phi}} + \dot{\boldsymbol{\phi}} : \boldsymbol{c}^{\mathrm{p}} : \dot{\boldsymbol{\epsilon}} > 0 \,. \tag{9.25}$$

Subtracting (9.20) from the above equation gives

$$\begin{aligned} \mathcal{P} - \mathcal{L}_1 &= \dot{\boldsymbol{\phi}} : \boldsymbol{c}^{\mathrm{p}} : \dot{\boldsymbol{\phi}} + \dot{\boldsymbol{\phi}} : \boldsymbol{c}^{\mathrm{p}} : \dot{\boldsymbol{\epsilon}} \\ &= \frac{1}{\chi} \underbrace{(\dot{\boldsymbol{\phi}} : \boldsymbol{c}^{\mathrm{e}} : \boldsymbol{f})}_{<0} \underbrace{(\boldsymbol{f} : \boldsymbol{c}^{\mathrm{e}} : \dot{\boldsymbol{\epsilon}}^*)}_{<0} > 0 \,. \end{aligned} \tag{9.26}$$

This gives $\mathcal{P} > \mathcal{L}_1$, which means that the loading/unloading combination cannot precede the loading/loading combination. Similarly, we can show that the unloading/loading combination cannot precede the loading/loading combination. Finally, the unloading/unloading combination will not lead to any bifurcation since $\boldsymbol{c}^{\mathrm{e}}$ is positive-definite.

The tensor $\boldsymbol{c}^{\mathrm{ep}}$ is singular when $H = 0$ (perfect plasticity). This can be verified from equation (9.20) by replacing $\dot{\boldsymbol{\phi}}$ with $\boldsymbol{f}$, yielding the following identity

$$\mathcal{L}_1 = \boldsymbol{f} : \left(\boldsymbol{c}^{\mathrm{e}} - \frac{\boldsymbol{c}^{\mathrm{e}} : \boldsymbol{f} \otimes \boldsymbol{f} : \boldsymbol{c}^{\mathrm{e}}}{\boldsymbol{f} : \boldsymbol{c}^{\mathrm{e}} : \boldsymbol{f}} \right) : \boldsymbol{f} \equiv 0 \,. \tag{9.27}$$

Therefore, for associative plasticity the local stability condition is $H > 0$. When $H = 0$ and the strain rate is applied in the direction of the plastic flow, i.e., when $\dot{\boldsymbol{\epsilon}} \propto \boldsymbol{f}$, then

$$\dot{\boldsymbol{\sigma}} = \boldsymbol{c}^{\mathrm{ep}} : \dot{\boldsymbol{\epsilon}} = \left(\boldsymbol{c}^{\mathrm{e}} - \frac{\boldsymbol{c}^{\mathrm{e}} : \boldsymbol{f} \otimes \boldsymbol{f} : \boldsymbol{c}^{\mathrm{e}}}{\boldsymbol{f} : \boldsymbol{c}^{\mathrm{e}} : \boldsymbol{f}}\right) : k\boldsymbol{f} \equiv \boldsymbol{0} . \tag{9.28}$$

This means that the incremental stress predictor is offset by the incremental plastic corrector when the incremental strain is applied in the direction of the plastic flow.

Non-associative Plasticity

For non-associative plasticity, the Raniecki-Bruhns (1981) comparison solid provides the local stability condition

$$\mathcal{L}_2 = \dot{\boldsymbol{\phi}} : \widetilde{\boldsymbol{c}}^{\mathrm{ep}} : \dot{\boldsymbol{\phi}} = \dot{\boldsymbol{\phi}} : (\boldsymbol{c}^{\mathrm{e}} - \widetilde{\boldsymbol{c}}^{\mathrm{p}}) : \dot{\boldsymbol{\phi}} > 0 , \tag{9.29}$$

where

$$\widetilde{\boldsymbol{c}}^{\mathrm{p}} = \frac{1}{4r\chi}\boldsymbol{c}^{\mathrm{e}} : (\boldsymbol{g} + r\boldsymbol{f}) \otimes (\boldsymbol{g} + r\boldsymbol{f}) : \boldsymbol{c}^{\mathrm{e}} , \qquad \forall r > 0 , \tag{9.30}$$

and $\chi = \boldsymbol{g} : \boldsymbol{c}^{\mathrm{e}} : \boldsymbol{f} + H > 0$. We recover the local stability condition developed in the previous section when $\boldsymbol{g} = \boldsymbol{f}$ and $r = 1$, see equation (9.20). For non-associative plasticity, the comparison solid is an incrementally linear solid having a fictitious constitutive tangent tensor given by $\widetilde{\boldsymbol{c}}^{\mathrm{ep}}$.

We now test the $\mathcal{L}_2$-stability criterion for different possible loading combinations. First, consider the 'in-loading comparison solid' for which the two alternative solutions follow the loading branch. The local stability condition imposed by the loading/loading combination is given by

$$\mathcal{U} = \dot{\boldsymbol{\phi}} : \boldsymbol{c}^{\mathrm{ep}} : \dot{\boldsymbol{\phi}} = \dot{\boldsymbol{\phi}} : \left(\boldsymbol{c}^{\mathrm{e}} - \frac{1}{\chi}\boldsymbol{c}^{\mathrm{e}} : \boldsymbol{g} \otimes \boldsymbol{f} : \boldsymbol{c}^{\mathrm{e}}\right) : \dot{\boldsymbol{\phi}} > 0 . \tag{9.31}$$

Note that this equation is not the same as equation (9.29). In fact, subtracting $\mathcal{L}_2$ from $\mathcal{U}$ gives

$$\mathcal{U} - \mathcal{L}_2 = \frac{[\dot{\boldsymbol{\phi}} : \boldsymbol{c}^{\mathrm{e}} : (\boldsymbol{g} - r\boldsymbol{f})]^2}{4r\chi} \geq 0 . \tag{9.32}$$

Therefore, $\mathcal{U} \geq \mathcal{L}_2$, which means that $\mathcal{L}_2$ serves as a lower bound for $\mathcal{U}$.

To show that the $\mathcal{L}_2$-condition also delimits the stability under combined loading and unloading, we take $b_1 \equiv \boldsymbol{f} : \boldsymbol{c}^{\mathrm{e}} : \dot{\boldsymbol{\epsilon}} > 0$ and $b_2 \equiv \boldsymbol{f} : \boldsymbol{c}^{\mathrm{e}} : \dot{\boldsymbol{\epsilon}}^* < 0$. Then $\dot{\boldsymbol{\sigma}} = \boldsymbol{c}^{\mathrm{ep}} : \dot{\boldsymbol{\epsilon}}$ and $\dot{\boldsymbol{\sigma}}^* = \boldsymbol{c}^{\mathrm{e}} : \dot{\boldsymbol{\epsilon}}^*$, from which the local stability condition for loading/unloading becomes

$$\mathcal{P} = \dot{\boldsymbol{\phi}} : \boldsymbol{c}^{\mathrm{e}} : \dot{\boldsymbol{\phi}} + \frac{1}{\chi}\dot{\boldsymbol{\phi}} : \boldsymbol{c}^{\mathrm{e}} : \boldsymbol{g} \otimes \boldsymbol{f} : \boldsymbol{c}^{\mathrm{e}} : \dot{\boldsymbol{\epsilon}} > 0 . \tag{9.33}$$

For brevity in notations, we set $a = \dot{\boldsymbol{\phi}} : \boldsymbol{c}^{\mathrm{e}} : \boldsymbol{g}$. Subtracting $\mathcal{L}_2$ from $\mathcal{P}$ and multiplying the result by $4r\chi$ gives

$$\begin{aligned} 4r\chi(\mathcal{P} - \mathcal{L}_2) &= 4rab_1 + [a + r(b_2 - b_1)]^2 \\ &= [a + r(b_1 + b_2)]^2 - 4r^2 b_1 b_2 \geq 0 \,. \end{aligned} \tag{9.34}$$

Therefore, $\mathcal{L}_2 \leq \mathcal{P}$, which means that $\mathcal{L}_2$ also serves as a lower bound for $\mathcal{P}$.

One may ask why we have not used the function $\mathcal{U}$ as a lower bound delimiter for the stability region. The reason is because when the plastic flow is non-associative, the condition $\boldsymbol{f} : \boldsymbol{c} : \dot{\boldsymbol{\epsilon}} > 0$ does not imply $\boldsymbol{g} : \boldsymbol{c} : \dot{\boldsymbol{\epsilon}} > 0$, so it is not possible to show that the local stability condition for the in-loading comparison solid is an acceptable lower bound. Mathematically, it is possible that the loading/unloading combination could in fact precede the loading/loading combination depending on the degree of non-associativity of the plastic flow. Probing the loading/unloading combination, however, is not straightforward because the $\mathcal{P}$-stability condition must be checked for all possible values of $\dot{\boldsymbol{\phi}}$ and $\dot{\boldsymbol{\epsilon}}$. Thus, local stability based on the $\mathcal{L}_2$ criterion provides a 'safe' bound for delineating the stable region for non-associative plastic flow.

There is one as-yet undefined variable in the expression for $\mathcal{L}_2$, which is the free parameter $r > 0$. In general, r can be any positive real number; however, there is an 'optimal' value for this parameter that maximizes the range of the stability region. To figure this optimal value, we rewrite (9.29) as

$$\dot{\boldsymbol{\phi}} : \boldsymbol{c}^{\mathrm{e}} : \dot{\boldsymbol{\phi}} - \frac{1}{4r\chi}[(\dot{\boldsymbol{\phi}} : \boldsymbol{c}^{\mathrm{e}} : (\boldsymbol{g} + r\boldsymbol{f})]^2 > 0 \,. \tag{9.35}$$

By the Cauchy-Schwartz inequality, we have

$$[(\dot{\boldsymbol{\phi}} : \boldsymbol{c}^{\mathrm{e}} : (\boldsymbol{g} + r\boldsymbol{f})]^2 \leq (\dot{\boldsymbol{\phi}} : \boldsymbol{c}^{\mathrm{e}} : \dot{\boldsymbol{\phi}})[(\boldsymbol{g} + r\boldsymbol{f}) : \boldsymbol{c}^{\mathrm{e}} : (\boldsymbol{g} + r\boldsymbol{f})] \,. \tag{9.36}$$

Thus, (9.35) is satisfied by the stronger condition

$$(\dot{\boldsymbol{\phi}} : \boldsymbol{c}^{\mathrm{e}} : \dot{\boldsymbol{\phi}})\Big[1 - \frac{1}{4r\chi}(\boldsymbol{g} + r\boldsymbol{f}) : \boldsymbol{c}^{\mathrm{e}} : (\boldsymbol{g} + r\boldsymbol{f})\Big] \geq 0 \,. \tag{9.37}$$

Since $\boldsymbol{c}^{\mathrm{e}}$ is positive definite, we have

$$\chi \geq \frac{1}{4r}(\boldsymbol{g} + r\boldsymbol{f}) : \boldsymbol{c}^{\mathrm{e}} : (\boldsymbol{g} + r\boldsymbol{f}) \,, \tag{9.38}$$

from which the optimal value of r is obtained as

$$r_{\mathrm{opt}} = \left(\frac{\boldsymbol{g} : \boldsymbol{c}^{\mathrm{e}} : \boldsymbol{g}}{\boldsymbol{f} : \boldsymbol{c}^{\mathrm{e}} : \boldsymbol{f}}\right)^{1/2} . \tag{9.39}$$

Together with equations (9.29) and (9.30), this optimal value of r can be used to maximize the range of the local stability region for the comparison solid.

It must be pointed out that the $\mathcal{L}_2$-stability condition simply ensures that when the stress state is inside the stability region the real solid is locally stable, but it does not imply, particularly when the flow rule is non-associative, that the real solid is locally unstable when the stress state falls outside this region. Furthermore, it must also be pointed out that the unique structure of $\boldsymbol{c}^{\mathrm{ep}}$ in the non-associative case also produces a singularity when $H = 0$, as can be seen from (9.31) by replacing $\dot{\boldsymbol{\phi}}$ with $\boldsymbol{g}$:

$$\mathcal{U} = \boldsymbol{g} : \left(\boldsymbol{c}^{\mathrm{e}} - \frac{\boldsymbol{c}^{\mathrm{e}} : \boldsymbol{g} \otimes \boldsymbol{f} : \boldsymbol{c}^{\mathrm{e}}}{\boldsymbol{f} : \boldsymbol{c}^{\mathrm{e}} : \boldsymbol{g}} \right) : \boldsymbol{g} \equiv 0 \,. \tag{9.40}$$

However, because $\mathcal{U}$ is not the stability delimiter for non-associative plasticity, it is also possible for a hardening response ($H > 0$) to be locally unstable when the stress state falls outside the local stability region.

Exercise 9.1. Consider an elastoplastic solid with non-associative plastic flow. Assume two alternative solutions represented by $\dot{\boldsymbol{\epsilon}}$ and $\dot{\boldsymbol{\epsilon}}^* = \dot{\boldsymbol{\epsilon}} + \dot{\boldsymbol{\phi}}$. Take $b_1 \equiv \boldsymbol{f} : \boldsymbol{c}^{\mathrm{e}} : \dot{\boldsymbol{\epsilon}} < 0$ and $b_2 \equiv \boldsymbol{f} : \boldsymbol{c}^{\mathrm{e}} : \dot{\boldsymbol{\epsilon}}^* > 0$ (unloading/loading combination). Show that the local stability condition for unloading/loading is bounded from below by the $\mathcal{L}_2$ stability criterion.

9.4 Deformation Bands

In Section 9.2 we noted that the loss of ellipticity may be linked to an emerging deformation band. We now pursue this idea further and consider different possible loading combinations leading to the development of a deformation band. Consider a rectangular block shown in Fig. 9.2 and recall from Chapter 7 the following kinematics of a deformation band

$$\boldsymbol{v}(\boldsymbol{x}) = \begin{cases} \overline{\boldsymbol{v}} & \text{if } c \leq 0\,; \\ \overline{\boldsymbol{v}} + \widetilde{\boldsymbol{v}}/h & \text{if } 0 \leq c \leq h\,; \\ \overline{\boldsymbol{v}} + \widetilde{\boldsymbol{v}} & \text{if } c \geq h\,, \end{cases} \tag{9.41}$$

where $\overline{\boldsymbol{v}}$ is the continuous component of velocity field and $\widetilde{\boldsymbol{v}}$ is the velocity jump (assumed constant) over the thickness h of the band. The strain rate is given by

$$\dot{\boldsymbol{\epsilon}}(\boldsymbol{x}) = \begin{cases} \dot{\overline{\boldsymbol{\epsilon}}} & \text{outside the band}\,; \\ \dot{\overline{\boldsymbol{\epsilon}}} + \operatorname{sym}(\widetilde{\boldsymbol{v}} \otimes \boldsymbol{n})/h & \text{inside the band}\,, \end{cases} \tag{9.42}$$

where $\dot{\overline{\boldsymbol{\epsilon}}} = \operatorname{sym}(\boldsymbol{\nabla}\overline{\boldsymbol{v}})$ and $\boldsymbol{n}$ is the unit normal vector perpendicular to the band. The break in the slope of the velocity distribution produces 'weak' discontinuities at $c = 0$ and $c = h$, see Chapter 7. We note at the outset that the definition of 'inside' and 'outside' the band is arbitrary since the only aspect of interest is the development of a discontinuity in the velocity distribution.

For purposes of stability analysis, it suffices to consider only one of the two band faces, and denote the tangent stress-strain tensors as $\boldsymbol{c}$ and $\boldsymbol{c}^*$ for

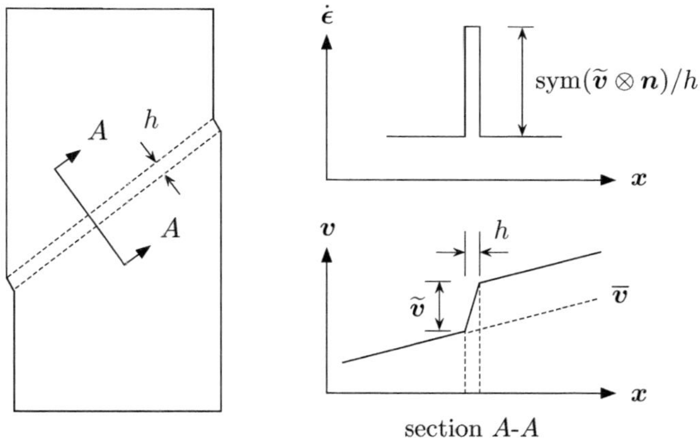

Fig. 9.2 Deformation band with thickness h: weak discontinuity is $h > 0$ resulting in a jump in the strain rate $\dot{\boldsymbol{\epsilon}}$; strong discontinuity is $h \to 0$ resulting in a jump in the velocity field $\boldsymbol{v}$

points just outside and just inside this face. The corresponding Cauchy stress rate tensors are

$$\dot{\boldsymbol{\sigma}} = \boldsymbol{c} : \dot{\overline{\boldsymbol{\epsilon}}}\,, \qquad \dot{\boldsymbol{\sigma}}^* = \boldsymbol{c}^* : \left(\dot{\overline{\boldsymbol{\epsilon}}} + \dot{\varphi}\,\mathrm{sym}(\boldsymbol{m} \otimes \boldsymbol{n})\right), \tag{9.43}$$

where $\dot{\varphi} = |\widetilde{\boldsymbol{v}}|/h$ and $\boldsymbol{m}$ is the unit vector in the direction of $\widetilde{\boldsymbol{v}}$. A necessary condition for an emerging deformation band is

$$\boldsymbol{n} \cdot (\dot{\boldsymbol{\sigma}}^* - \dot{\boldsymbol{\sigma}}) = \boldsymbol{0}\,, \tag{9.44}$$

i.e., the traction rates on each side of the band face must be continuous. Because the tangent operator in an elastoplastic material can have two branches, the bifurcation condition must consider different possible loading combinations on each side of the band.

Loading/Loading Bifurcation

Consider a loading/loading bifurcation mode in which plastic loading occurs on both sides of an emerging band. Rudnicki and Rice (1975) explored this mode of bifurcation in the context of pressure-sensitive dilatant materials. Setting $\boldsymbol{c} = \boldsymbol{c}^* = \boldsymbol{c}^{\mathrm{ep}}$, the bifurcation condition (9.44) becomes

$$\dot{\varphi}\boldsymbol{A}^{\mathrm{ep}} \cdot \boldsymbol{m} = \boldsymbol{0}\,, \tag{9.45}$$

where $\boldsymbol{A}^{\mathrm{ep}} = \boldsymbol{n} \cdot \boldsymbol{c}^{\mathrm{ep}} \cdot \boldsymbol{n}$ (with components $A^{\mathrm{ep}}_{ik} = n_j c^{\mathrm{ep}}_{ijkl} n_l$) is the elastoplastic acoustic tensor. A trivial solution, not of interest in the present discussion, occurs when $\varphi = 0$ (i.e., no deformation band). However, a non-trivial solution of interest in this section occurs when

$$\mathcal{F} = \min|_{\boldsymbol{n}} \det(\boldsymbol{A}^{\text{ep}}) = 0\,. \tag{9.46}$$

The condition coincides with the loss of ellipticity as defined in Section 9.2, cf. equation (9.14). The solution to the bifurcation problem requires that we search for the critical orientation $\boldsymbol{n}$ at which the localization function $\mathcal{F}$ is minimized. We then have bifurcation if this function vanishes, and the value of $\boldsymbol{n}$ at the bifurcation point determines the orientation of an emerging deformation band.

The bifurcation point for an emerging deformation band is usually described by a so-called critical plastic modulus H_{LL}, where the subscripts 'LL' denote loading/loading. Let

$$\boldsymbol{A}^{\text{e}} = \boldsymbol{n} \cdot \boldsymbol{c}^{\text{e}} \cdot \boldsymbol{n}\,, \qquad \boldsymbol{\alpha} = \boldsymbol{n} \cdot \boldsymbol{c}^{\text{e}} : \boldsymbol{g}\,, \qquad \boldsymbol{\beta} = \boldsymbol{f} : \boldsymbol{c}^{\text{e}} \cdot \boldsymbol{n}\,. \tag{9.47}$$

Then we can write the elastoplastic acoustic tensor as

$$\boldsymbol{A}^{\text{ep}} = \boldsymbol{A}^{\text{e}} - \frac{1}{\chi_{\text{LL}}} \boldsymbol{\alpha} \otimes \boldsymbol{\beta}\,, \tag{9.48}$$

where $\chi_{\text{LL}} = \boldsymbol{f} : \boldsymbol{c}^{\text{e}} : \boldsymbol{g} + H_{\text{LL}}$. The bifurcation condition is

$$\left(\boldsymbol{A}^{\text{e}} - \frac{1}{\chi_{\text{LL}}} \boldsymbol{\alpha} \otimes \boldsymbol{\beta}\right) \cdot \boldsymbol{m} = \boldsymbol{0}\,. \tag{9.49}$$

We see that $\boldsymbol{A}^{\text{e}} \cdot \boldsymbol{m} = k\boldsymbol{\alpha}$, or $\boldsymbol{m} = k{\boldsymbol{A}^{\text{e}}}^{-1} \cdot \boldsymbol{\alpha}$, where k is a nonzero scalar multiplier. Substituting into the bifurcation condition yields

$$k\left(\boldsymbol{A}^{\text{e}} - \frac{1}{\chi_{\text{LL}}} \boldsymbol{\alpha} \otimes \boldsymbol{\beta}\right) \cdot {\boldsymbol{A}^{\text{e}}}^{-1} \cdot \boldsymbol{\alpha} = \boldsymbol{0}\,. \tag{9.50}$$

Provided that $k\boldsymbol{\alpha} \neq \boldsymbol{0}$, this gives

$$\chi_{\text{LL}} = \boldsymbol{\alpha} \cdot {\boldsymbol{A}^{\text{e}}}^{-1} \cdot \boldsymbol{\beta}\,, \tag{9.51}$$

which yields the following expression for the critical plastic modulus for loading/loading bifurcation:

$$H_{\text{LL}} = \chi_{\text{LL}} - \boldsymbol{f} : \boldsymbol{c}^{\text{e}} : \boldsymbol{g}\,. \tag{9.52}$$

Closed-form expressions are available for the critical plastic modulus H_{LL} for some simple cohesive-frictional material models available in the literature, such as the von Mises, Mohr-Coulomb, and Drucker-Prager plasticity models (Runesson et al. 1991).

Unloading/Loading Bifurcation

Next, consider a bifurcation mode with 'unloading outside' and 'loading inside' the band. Rice and Rudnicki (1980) demonstrated that this mode is less critical than the loading/loading bifurcation mode described previously. Instead of presenting the same proof as the one given by Rice and Rudnicki, we demonstrate the same result in a slightly different way as follows.

Consider a bifurcation mode with $\boldsymbol{f} : \boldsymbol{c}^{\mathrm{e}} : \dot{\boldsymbol{\epsilon}} < 0$ (unloading outside) and $\boldsymbol{f} : \boldsymbol{c}^{\mathrm{e}} : \dot{\boldsymbol{\epsilon}}^* > 0$ (loading inside), where $\dot{\boldsymbol{\epsilon}}^* = \dot{\boldsymbol{\epsilon}} + \dot{\varphi}\mathrm{sym}(\boldsymbol{m} \otimes \boldsymbol{n})$. The relevant stress rates are $\dot{\boldsymbol{\sigma}} = \boldsymbol{c}^{\mathrm{e}} : \dot{\boldsymbol{\epsilon}}$ and $\dot{\boldsymbol{\sigma}}^* = \boldsymbol{c}^{\mathrm{ep}} : \dot{\boldsymbol{\epsilon}}^*$. The bifurcation condition (9.44) becomes

$$\dot{\varphi}\Big(\boldsymbol{A}^{\mathrm{e}} - \frac{1}{\chi_{\mathrm{UL}}}\boldsymbol{\alpha} \otimes \boldsymbol{\beta}\Big) \cdot \boldsymbol{m} = \frac{1}{\chi_{\mathrm{UL}}}(\boldsymbol{f} : \boldsymbol{c}^{\mathrm{e}} : \dot{\boldsymbol{\epsilon}})\boldsymbol{\alpha}\,, \tag{9.53}$$

where $\chi_{\mathrm{UL}} = \boldsymbol{f} : \boldsymbol{c}^{\mathrm{e}} : \boldsymbol{g} + H_{\mathrm{UL}}$, and subscript "UL" indicates an unloading/loading mode. The coefficient tensor has an inverse provided by the Sherman-Morrison formula

$$\Big(\boldsymbol{A}^{\mathrm{e}} - \frac{1}{\chi_{\mathrm{UL}}}\boldsymbol{\alpha} \otimes \boldsymbol{\beta}\Big)^{-1} = \boldsymbol{A}^{\mathrm{e}-1} + \frac{\boldsymbol{A}^{\mathrm{e}-1} \cdot \boldsymbol{\alpha} \otimes \boldsymbol{\beta} \cdot \boldsymbol{A}^{\mathrm{e}-1}}{\chi_{\mathrm{UL}} - \chi_{\mathrm{LL}}}\,. \tag{9.54}$$

This yields the following analytical expression for the velocity jump,

$$\dot{\varphi}\boldsymbol{m} = \Big(\frac{\boldsymbol{f} : \boldsymbol{c}^{\mathrm{e}} : \dot{\boldsymbol{\epsilon}}}{\chi_{\mathrm{UL}} - \chi_{\mathrm{LL}}}\Big)\boldsymbol{A}^{\mathrm{e}-1} \cdot \boldsymbol{\alpha}\,. \tag{9.55}$$

The inequalities for unloading outside and loading inside the band yield the following constraint:

$$\boldsymbol{f} : \boldsymbol{c}^{\mathrm{e}} : \big[\dot{\boldsymbol{\epsilon}} + \dot{\varphi}\mathrm{sym}(\boldsymbol{m} \otimes \boldsymbol{n})\big] = \underbrace{\boldsymbol{f} : \boldsymbol{c}^{\mathrm{e}} : \dot{\boldsymbol{\epsilon}}}_{<0}\Big(1 + \frac{\chi_{\mathrm{LL}}}{\chi_{\mathrm{UL}} - \chi_{\mathrm{LL}}}\Big) > 0\,. \tag{9.56}$$

Hence, we must have

$$1 + \frac{\chi_{\mathrm{LL}}}{\chi_{\mathrm{UL}} - \chi_{\mathrm{LL}}} \equiv \frac{\chi_{\mathrm{UL}}}{\chi_{\mathrm{UL}} - \chi_{\mathrm{LL}}} < 0\,. \tag{9.57}$$

However, the plastic divisors χ_{LL} and χ_{UL} must always be positive, so it follows that $\chi_{\mathrm{UL}} < \chi_{\mathrm{LL}}$, or $H_{\mathrm{UL}} < H_{\mathrm{LL}}$. This means that an unloading/loading bifurcation mode requires a smaller plastic modulus than the loading/loading mode; therefore, it cannot precede the loading/loading bifurcation mode, thus completing the proof. On a related note, Borre and Maier (1989) demonstrated that the singularity of the elastoplastic acoustic tensor is a sufficient condition for bifurcation.

Figure 9.3 shows a schematic representation of the local stability regions for an elastoplastic solid. The innermost region is the local stability region provided by either the Hill (for associative flow rule) or Raniecki-Bruhns (non-associative flow rule) comparison solid. The next outermost stability region is defined by the in-loading comparison solid, which coincides with the local stability region if the flow rule is associative. Since a stable region is always elliptic but not vice versa, the local stability region is contained within the ellipticity limit. Furthermore, the stable region for the in-loading comparison solid is contained in the ellipticity limit since the planar-wave mode of deformation for the latter type of bifurcation is only a special case of a more general deformation mode considered in the former bifurcation mode.

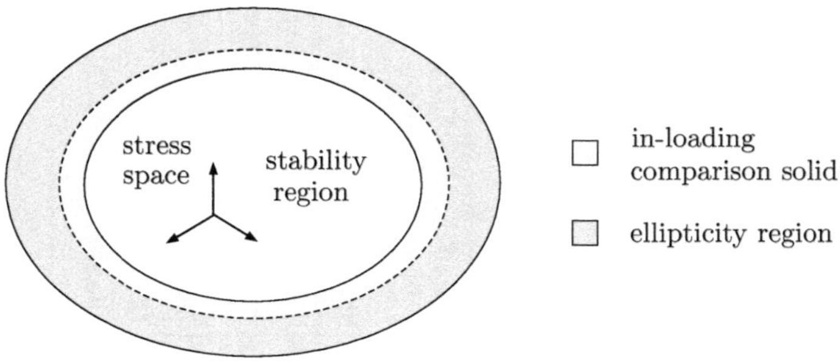

Fig. 9.3 Schematic representation of the local stability regions for elastoplastic solids

Exercise 9.2. Consider an elastoplastic solid with non-associative plastic flow. Assume a deformation band-type bifurcation with strain rates $\dot{\boldsymbol{\epsilon}}$ outside and $\dot{\boldsymbol{\epsilon}}^* = \dot{\boldsymbol{\epsilon}} + \dot{\varphi}\mathrm{sym}(\boldsymbol{m} \otimes \boldsymbol{n})$ inside the band. Take $\boldsymbol{f} : \boldsymbol{c}^{\mathrm{e}} : \dot{\boldsymbol{\epsilon}} > 0$ and $\boldsymbol{f} : \boldsymbol{c}^{\mathrm{e}} : \dot{\boldsymbol{\epsilon}}^* < 0$ (loading outside and unloading inside). Show that $H_{\mathrm{LU}} < H_{\mathrm{LL}}$, which means that the loading/unloading bifurcation mode cannot precede the loading/loading bifurcation mode.

9.5 Spectral Representation

Under the assumption of isotropy, the bifurcation condition (9.14) simplifies to a simpler form that facilitates a more efficient search of the critical bifurcation point. We recall the spectral form of the tangent stress-strain matrix for isotropic materials

$$\begin{aligned}\boldsymbol{c} &= \sum_{A=1}^{3}\sum_{B=1}^{3} a_{AB}\boldsymbol{m}^{(A)} \otimes \boldsymbol{m}^{(B)} \\ &+ \frac{1}{2}\sum_{A=1}^{3}\sum_{B\neq A} \gamma_{AB}(\boldsymbol{m}^{(AB)} \otimes \boldsymbol{m}^{(AB)} + \boldsymbol{m}^{(AB)} \otimes \boldsymbol{m}^{(BA)}), \qquad (9.58)\end{aligned}$$

where a_{AB}'s are the tangential moduli in principal axes, $\boldsymbol{m}^{(A)} = \boldsymbol{n}^{(A)} \otimes \boldsymbol{n}^{(A)}$, $\boldsymbol{m}^{(AB)} = \boldsymbol{n}^{(A)} \otimes \boldsymbol{n}^{(B)}$, and $\gamma_{AB} = (\sigma_B - \sigma_A)/(\epsilon^{\mathrm{e}}_B - \epsilon^{\mathrm{e}}_A)$. The first double summations represent the material moduli in principal axes, whereas the second represent the spin of the principal axes. We recall that $\gamma_{AB} > 0$ for linear isotropic elasticity.

Consider the spectral decomposition of any second-order symmetric tensor $\dot{\boldsymbol{\phi}} \neq \mathbf{0}$:

$$\dot{\boldsymbol{\phi}} = \sum_{A=1}^{3} \dot{\phi}_A \boldsymbol{n}^{(A)} \otimes \boldsymbol{n}^{(A)} + \sum_{A=1}^{3} \sum_{B \neq A} \omega_{AB}(\phi_B - \phi_A) \boldsymbol{n}^{(A)} \otimes \boldsymbol{n}^{(B)} . \tag{9.59}$$

The local stability condition (9.6) for an incrementally linear response is

$$\dot{\boldsymbol{\phi}} : \boldsymbol{c} : \dot{\boldsymbol{\phi}} = \sum_{A=1}^{3} \sum_{B=1}^{3} \dot{\phi}_A a_{AB} \dot{\phi}_B + \sum_{A=1}^{3} \sum_{B \neq A} \gamma_{AB} \omega_{AB}^2 (\phi_B - \phi_A)^2 > 0 . \tag{9.60}$$

The second double summations represent the energy generated by the spins of the principal axes, and is nonnegative. The critical condition occurs when these spins are zero, in which case, the local stability condition simplifies to the form

$$\sum_{A=1}^{3} \sum_{B=1}^{3} \dot{\phi}_A a_{AB} \dot{\phi}_B > 0 . \tag{9.61}$$

Similar stability criteria may be derived for elastoplastic solids using the spectral form of the tangent operator for the appropriate comparison solids.

Next, we consider an emerging deformation band with orientation defined by the unit normal vector $\boldsymbol{n}$. Since the unit eigenvectors $\boldsymbol{n}^{(A)}$ form linearly independent bases, we can write $\boldsymbol{n}$ as a linear combination of these eigenvectors (Andrade and Borja 2006):

$$\boldsymbol{n} = \sum_{A=1}^{3} \psi_A \boldsymbol{n}^{(A)} , \tag{9.62}$$

where ψ_A, $A = 1, 2, 3$, are some scalar coefficients satisfying the condition $\sum_{A=1}^{3} \psi_A^2 = 1$, i.e., $\psi_A = \boldsymbol{n}^{(A)} \cdot \boldsymbol{n}$. In terms of ψ_A, we can express the elastoplastic acoustic tensor $\boldsymbol{A}^{\mathrm{ep}}$ in spectral form as

$$\boldsymbol{A}^{\mathrm{ep}} = \sum_{A=1}^{3} \sum_{B=1}^{3} \mathcal{A}_{AB} \boldsymbol{n}^{(A)} \otimes \boldsymbol{n}^{(B)} , \tag{9.63}$$

where

$$\mathcal{A}_{AB} = \begin{cases} \psi_A^2 a_{AA}^{\mathrm{ep}} + \frac{1}{2} \sum_{C \neq A} \psi_C^2 \gamma_{AC} & \text{if } A = B , \\ \psi_A (a_{AB}^{\mathrm{ep}} + \frac{1}{2} \gamma_{AB}) \psi_B & \text{if } A \neq B . \end{cases} \tag{9.64}$$

The determinant of a second-order tensor is invariant with respect to rotation, so the determinant of $\boldsymbol{A}^{\mathrm{ep}}$ and the determinant of the 3×3 matrix $\boldsymbol{\mathcal{A}} = [\mathcal{A}_{AB}]$ must be the same.

Consider the spherical angles θ and ϕ shown in Fig. 9.4, and note that

$$\psi_1 = \sin\theta \sin\phi\,, \qquad \psi_2 = \cos\phi\,, \qquad \psi_3 = \cos\theta\sin\phi\,. \tag{9.65}$$

Let $\boldsymbol{x} = \{\theta, \phi\}^{\mathsf{T}}$ and $\boldsymbol{\psi} = \{\psi_1, \psi_2, \psi_3\}^{\mathsf{T}}$; we can then write the determinant function as

$$\overline{\mathcal{F}} = \overline{\mathcal{F}}(\boldsymbol{x}) = \det(\boldsymbol{\mathcal{A}})\,. \tag{9.66}$$

In other words, the determinant of the acoustic tensor is now viewed as a function of the spherical angles (Oliver et al. 2010). Therefore, the stationary point of the localization function can be obtained by differentiating with respect to these angles.

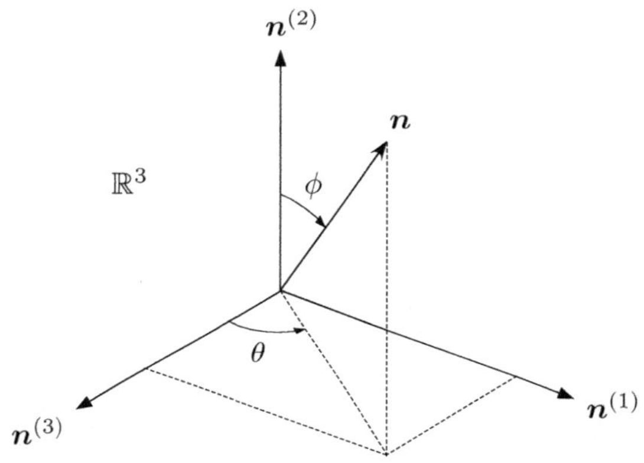

Fig. 9.4 Spherical angles θ and ϕ defining orientation of unit normal vector $\boldsymbol{n}$ perpendicular to the band face reckoned with respect to principal directions $\boldsymbol{n}^{(1)}, \boldsymbol{n}^{(2)}, \boldsymbol{n}^{(3)}$

The critical values of the spherical angles at the stationary point may be obtained from the optimality conditions

$$\boldsymbol{r}(\boldsymbol{x}) = \frac{\partial \overline{\mathcal{F}}}{\partial \boldsymbol{x}} = \frac{\partial \overline{\mathcal{F}}}{\partial \boldsymbol{\psi}} \cdot \frac{\partial \boldsymbol{\psi}}{\partial \boldsymbol{x}} = \boldsymbol{0}\,, \tag{9.67}$$

where $\boldsymbol{r} = \{\partial\overline{\mathcal{F}}/\partial\theta, \partial\overline{\mathcal{F}}/\partial\phi\}^{\mathsf{T}}$. We use the identity $\partial\overline{\mathcal{F}}/\partial\boldsymbol{\mathcal{A}} = \overline{\mathcal{F}}\boldsymbol{\mathcal{A}}^{-\mathsf{T}}$ for the terms on the right-hand side, and write

$$\boldsymbol{r}(\boldsymbol{x}) = \overline{\mathcal{F}}\boldsymbol{\mathcal{A}}^{-\mathsf{T}} : \frac{\partial \boldsymbol{\mathcal{A}}}{\partial \boldsymbol{\psi}} \cdot \frac{\partial \boldsymbol{\psi}}{\partial \boldsymbol{x}} = \boldsymbol{0}\,, \tag{9.68}$$

where $\partial\boldsymbol{\mathcal{A}}/\partial\boldsymbol{\psi}$ is a third-order tensor with components (no sum)

$$\frac{\partial \mathcal{A}_{AB}}{\partial \psi_C} = \begin{cases} 2\psi_A \delta_{AC} a^{\mathrm{ep}}_{AA} + \underbrace{\psi_C \gamma_{CA}}_{C \neq A} & \text{if } A = B\,, \\ \delta_{AC}(a^{\mathrm{ep}}_{AB} + \frac{1}{2}\gamma_{AB})\psi_B + \psi_A(a^{\mathrm{ep}}_{AB} + \frac{1}{2}\gamma_{AB})\delta_{BC} & \text{if } A \neq B\,. \end{cases} \tag{9.69}$$

The relevant derivatives are

$$\frac{\partial \boldsymbol{\psi}}{\partial \boldsymbol{x}} = \begin{bmatrix} \cos\theta\sin\phi & \sin\theta\cos\phi \\ 0 & -\sin\phi \\ -\sin\theta\sin\phi & \cos\theta\cos\phi \end{bmatrix} . \tag{9.70}$$

A first approximation of the roots is facilitated by a coarse sweep over half a unit ball, $[0,\pi]\times[0,\pi]$, to determine the point where $\|\boldsymbol{r}\|$ is closest to zero or where $\overline{\mathcal{F}}$ attains its global minimum. If the global minimum of $\overline{\mathcal{F}}$ is nowhere close to zero, then bifurcation is unlikely to occur and there is no reason to pursue a finer search. However, if the global minimum is sufficiently close to zero, or if $\overline{\mathcal{F}}$ reverses in sign, then bifurcation is likely to occur and the spherical angles must be calculated with greater accuracy. This can be done by solving the residual equation $\boldsymbol{r}(\boldsymbol{x}) = \boldsymbol{0}$ exactly. Because of the highly nonlinear nature of the optimality conditions, we resort to Newton iteration to determine the critical values of the spherical angles at the bifurcation point.

To perform the Newton iteration, we need to construct the following 2×2 Jacobian matrix

$$\boldsymbol{r}'(\boldsymbol{x}) = \frac{\partial^2 \overline{\mathcal{F}}}{\partial \boldsymbol{x}\partial \boldsymbol{x}} . \tag{9.71}$$

Bifurcation analysis is carried out at a fixed state of stress, typically at the converged stress configuration of each load step, so the function $\overline{\mathcal{F}}$ depends solely on the spherical angles. Using the chain rule, the Jacobian matrix is obtained as

$$\frac{\partial^2 \overline{\mathcal{F}}}{\partial \boldsymbol{x}\partial \boldsymbol{x}} = \left(\frac{\partial \boldsymbol{\psi}}{\partial \boldsymbol{x}}\right)^{\mathsf{T}} \cdot \boldsymbol{\mathcal{B}} \cdot \frac{\partial \boldsymbol{\psi}}{\partial \boldsymbol{x}} + \frac{\partial \overline{\mathcal{F}}}{\partial \boldsymbol{\psi}} \cdot \frac{\partial^2 \boldsymbol{\psi}}{\partial \boldsymbol{x}\partial \boldsymbol{x}} , \tag{9.72}$$

where

$$\begin{aligned} \boldsymbol{\mathcal{B}} := \frac{\partial^2 \overline{\mathcal{F}}}{\partial \boldsymbol{\psi}\partial \boldsymbol{\psi}} &= \overline{\mathcal{F}}\Big(\boldsymbol{\mathcal{A}}^{-\mathsf{T}} : \frac{\partial \boldsymbol{\mathcal{A}}}{\partial \boldsymbol{\psi}}\Big) \otimes \Big(\boldsymbol{\mathcal{A}}^{-\mathsf{T}} : \frac{\partial \boldsymbol{\mathcal{A}}}{\partial \boldsymbol{\psi}}\Big) \\ &\quad + \overline{\mathcal{F}}\Big(\boldsymbol{\mathcal{A}}^{-1} \cdot \frac{\partial \boldsymbol{\mathcal{A}}}{\partial \boldsymbol{\psi}} \cdot \boldsymbol{\mathcal{A}}^{-\mathsf{T}}\Big) : \frac{\partial \boldsymbol{\mathcal{A}}}{\partial \boldsymbol{\psi}} \\ &\quad + \overline{\mathcal{F}} \boldsymbol{\mathcal{A}}^{-\mathsf{T}} : \frac{\partial^2 \boldsymbol{\mathcal{A}}}{\partial \boldsymbol{\psi}\partial \boldsymbol{\psi}} . \end{aligned} \tag{9.73}$$

Relevant expressions for the two tensors with second derivatives, $\partial^2\boldsymbol{\psi}/\partial\boldsymbol{x}\partial\boldsymbol{x}$ and $\partial^2\boldsymbol{\mathcal{A}}/\partial\boldsymbol{\psi}\partial\boldsymbol{\psi}$, are needed. We leave the derivation of an expression for the former tensor as an exercise. As for the latter tensor, we write (no sum):

$$\frac{\partial^2 \mathcal{A}_{AB}}{\partial \psi_C \partial \psi_D} = \begin{cases} 2\delta_{AD}\delta_{AC} a^{\mathrm{ep}}_{AA} + \underbrace{\delta_{CD}\gamma_{CA}}_{C\neq A} & \text{if } A = B , \\ \delta_{AC}(a^{\mathrm{ep}}_{AB} + \frac{1}{2}\gamma_{AB})\delta_{BD} + \delta_{AD}(a^{\mathrm{ep}}_{AB} + \frac{1}{2}\gamma_{AB})\delta_{BC} & \text{if } A \neq B . \end{cases} \tag{9.74}$$

The fourth-order tensor defined above is very sparse, allowing a straightforward evaluation of the final Jacobian matrix. Furthermore, for a wide class of elastoplastic models at least one of the elements of the vector $\boldsymbol{\psi}$ is zero (Rudnicki and Rice 1975). For example, it can be shown that the von Mises, Drucker-Prager, and Mohr-Coulomb models always localize within one of three principal stress planes when used in combination with an isotropic elasticity tensor. These important features are not used in other search algorithms such as the ones proposed by Ortiz et al. (1987) and Mosler (2005).

Note that whereas the present calculations aim to identify the singular point for the acoustic tensor $\boldsymbol{\mathcal{A}}$, the algorithm described above would not work when this acoustic tensor is in fact singular. In reality, one may expect the localization function to be either greater than zero or less than zero, but never exactly equal to zero because of the nature of the numerical calculations.

Exercise 9.3. Under the assumption of isotropy, construct the spectral form of the tangential stress-strain tensor $\widetilde{\boldsymbol{c}}^{\mathrm{ep}}$ defined in (9.29), and write an expression for the elastoplastic material moduli $\widetilde{a}^{\mathrm{ep}}_{AB}$ in principal axes for the comparison solid.

Exercise 9.4. Derive the elements of the third-order tensor $\partial^2\boldsymbol{\psi}/\partial\boldsymbol{x}\partial\boldsymbol{x}$ from equation (9.70). This tensor should have 12 elements.

9.6 E-modes

Eigenmodes (or e-modes) define the instantaneous kinematics of deformation at the point of bifurcation, as well as describe the specific mode of bifurcation of a given stressed element. As an illustration, consider the moment when the elastoplastic stress-strain tensor becomes singular, i.e., when $H = 0$. The e-mode is defined by the instantaneous plastic flow direction, as can be seen from

$$\dot{\boldsymbol{\sigma}} = \boldsymbol{c}^{\mathrm{ep}} : \boldsymbol{g} = \left(\boldsymbol{c}^{\mathrm{e}} - \frac{\boldsymbol{c}^{\mathrm{e}} : \boldsymbol{g} \otimes \boldsymbol{f} : \boldsymbol{c}^{\mathrm{e}}}{\boldsymbol{f} : \boldsymbol{c}^{\mathrm{e}} : \boldsymbol{g}}\right) : \boldsymbol{g} \equiv \boldsymbol{0}\,. \tag{9.75}$$

In J_2 plasticity the e-mode is an isochoric deformation with $\mathrm{tr}(\boldsymbol{g}) = 0$. In crystal plasticity the plastic flow is of the form $\boldsymbol{g} = \mathrm{sym}(\boldsymbol{m} \otimes \boldsymbol{n})$, and so, in the limit of perfect plasticity,

$$\boldsymbol{n} \cdot \dot{\boldsymbol{\sigma}} = \boldsymbol{n} \cdot \boldsymbol{c}^{\mathrm{ep}} : \mathrm{sym}(\boldsymbol{m} \otimes \boldsymbol{n}) \equiv \boldsymbol{A}^{\mathrm{ep}} \cdot \boldsymbol{m} \equiv \boldsymbol{0}\,, \tag{9.76}$$

which implies that $\boldsymbol{A}^{\mathrm{ep}}$ is singular whenever $\boldsymbol{c}^{\mathrm{ep}}$ is singular.

A more systematic approach for finding the e-mode of a singular stress-strain tensor is as follows. Consider the homogeneous equation $\boldsymbol{c} : \dot{\boldsymbol{\epsilon}} = \boldsymbol{0}$, where $\boldsymbol{c}$ has the spectral form shown in (9.58). The singularities of $\boldsymbol{c}$ are of two types according to the following equations, see Ogden (1984):

$$\text{(a)} \qquad \det(a_{AB}) = 0\,, \tag{9.77}$$

$$\text{(b)} \qquad \sigma_A = \sigma_B\,, \quad A \neq B\,. \tag{9.78}$$

Type (a) corresponds to the singularity of the 3×3 matrix in principal axes, whereas type (b) singularity gives rise to indeterminate spins. Each type of singularity generates a specific e-mode.

Type (a) singularity freezes the spins of the principal axes. The associated e-mode is of the form

$$\dot{\boldsymbol{\epsilon}} = \sum_{A=1}^{3} \dot{\epsilon}_A \boldsymbol{n}^{(A)} \otimes \boldsymbol{n}^{(A)} , \tag{9.79}$$

where $\{\dot{\epsilon}_A\}$ is the e-mode of the singular matrix $[a_{AB}]$. In this case, bifurcation occurs at fixed principal axes. If $[a_{AB}]$ is not singular, then the homogeneous equation $\boldsymbol{c} : \dot{\boldsymbol{\epsilon}} = \boldsymbol{0}$ can only admit the trivial solution $\dot{\epsilon}_A = 0$ for $A = 1, 2, 3$. The corresponding e-mode is

$$\dot{\boldsymbol{\epsilon}} = \omega_{AB}(\epsilon_B - \epsilon_A)\boldsymbol{n}^{(A)} \otimes \boldsymbol{n}^{(B)} \tag{9.80}$$

for all $A \neq B$ where $\sigma_A = \sigma_B$. This is a pure shearing e-mode at fixed principal values and produces rotation of the principal axes. If the singularities of $\boldsymbol{c}$ are of both types (a) and (b), then the e-mode is given by the full tensor

$$\dot{\boldsymbol{\epsilon}} = \sum_{A=1}^{3} \dot{\epsilon}_A \boldsymbol{n}^{(A)} \otimes \boldsymbol{n}^{(A)} + \sum_{A=1}^{3} \sum_{B \neq A} \omega_{AB}(\epsilon_B - \epsilon_A)\boldsymbol{n}^{(A)} \otimes \boldsymbol{n}^{(B)} . \tag{9.81}$$

Note that, irrespective of the type of singularity, the e-mode has no unique norm.

We now consider the general e-mode of a singular acoustic tensor. The e-mode defines a deformation band of the form $\mathrm{sym}(\boldsymbol{m} \otimes \boldsymbol{n})$, where $\boldsymbol{n}$ is the unit vector identifying the zero minimum determinant of $\boldsymbol{A}$, and

$$\boldsymbol{m} = \pm \frac{\boldsymbol{A}^{\mathrm{e}-1} \cdot \boldsymbol{\alpha}}{\|\boldsymbol{A}^{\mathrm{e}-1} \cdot \boldsymbol{\alpha}\|} \tag{9.82}$$

is the eigenvector of $\boldsymbol{A}$, see Sec. 9.4. The correct sign of $\boldsymbol{m}$ is determined by imposing the loading/loading condition (9.24),

$$\boldsymbol{f} : \boldsymbol{c}^{\mathrm{e}} : \mathrm{sym}(\boldsymbol{m} \otimes \boldsymbol{n}) \geq 0 . \tag{9.83}$$

The trace of $\mathrm{sym}(\boldsymbol{m} \otimes \boldsymbol{n})$, equal to $\boldsymbol{m} \cdot \boldsymbol{n}$, determines the type of deformation band formed at the bifurcation point. Various possibilities include (Aydin et al. 2006, Borja and Aydin 2004):

$$\left.\begin{array}{rl} \boldsymbol{m} \cdot \boldsymbol{n} = 1 : & \text{dilation band} \\ 0 < \boldsymbol{m} \cdot \boldsymbol{n} < 1 : & \text{dilatant shear band} \\ \boldsymbol{m} \cdot \boldsymbol{n} = 0 : & \text{shear band} \\ -1 < \boldsymbol{m} \cdot \boldsymbol{n} < 0 : & \text{compactive shear band} \\ \boldsymbol{m} \cdot \boldsymbol{n} = -1 : & \text{compaction band} \end{array}\right\} \tag{9.84}$$

In a shear band the instantaneous relative movement is tangent to the band. In dilation and compaction bands the instantaneous relative movement is

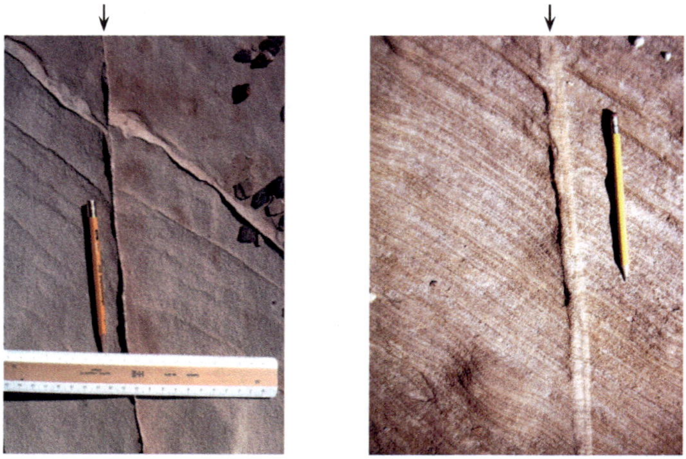

(a) (b)

Fig. 9.5 Deformation bands (marked by arrow): (a) a shear band in the Entrada Sandstone, San Rafael Desert, Utah; and (b) a compaction band in the Aztec Sandstone at Valley of Fire, Nevada. After Borja and Aydin (2004).

normal to the band. Mixed-mode bands are the most common type, where shearing combines with either compaction or dilation. We should note that the mode of deformation pertains only to the moment of bifurcation and may or may not persist at post-bifurcation. For example, a compactive shear band could eventually become a pure shear band when there are no more pores inside the band to compact. Figure 9.5 shows examples of shear and compaction bands in two sandstones. Because some significant volume of pores needs to collapse to form a compaction band, this type of band is usually thicker than a shear band.

Compaction band is an intriguing deformation style amenable to simple mathematical interpretation for the case of an isotropic solid. Consider the condition $\boldsymbol{m} = -\boldsymbol{n} = -\sum_{A=1}^{3} \psi_A \boldsymbol{n}^{(A)}$. The bifurcation condition simplifies to the form

$$\boldsymbol{A}^{\text{ep}} \cdot \boldsymbol{n} = \sum_{A=1}^{3} \sum_{B=1}^{3} \mathcal{A}_{AB} \psi_B \boldsymbol{n}^{(A)} = \boldsymbol{0}\,, \tag{9.85}$$

where the $\mathcal{A}_{AB}$'s are given in (9.64). This is satisfied when $\boldsymbol{n}$ is parallel to one of three principal eigenvectors. For example, assume that $\boldsymbol{n}$ is parallel to $\boldsymbol{n}^{(A)}$ so that $\psi_A = 1$ and $\psi_B = \psi_C = 0$. We see that the spin terms in (9.64) drop out, and we are left with the condition

$$a^{\text{ep}}_{AA} = 0 \qquad (\text{no sum})\,. \tag{9.86}$$

In other words, a compaction (or dilation) band would form when one of the diagonal elements of the matrix $[a^{\text{e}}_{AB}]$ vanishes; the direction of the compaction (or dilation) is determined by the index of the specific diagonal element of the matrix $[a^{\mathrm{e}}_{AB}]$ that vanishes.

9.7 Finite Deformation Bifurcation

Extending the local stability criterion to the finite deformation range is fairly straightforward provided that we use appropriate measures of stress and deformation. Consider a solid body defined by domain $\mathcal{B}$ and bounded by surface $\partial\mathcal{B}$ in the reference configuration, and assume the usual decomposition $\partial\mathcal{B} = \overline{\partial\mathcal{B}_{\boldsymbol{t}} \cup \partial\mathcal{B}_{\boldsymbol{u}}}$ and $\varnothing = \partial\mathcal{B}_{\boldsymbol{t}} \cap \partial\mathcal{B}_{\boldsymbol{u}}$. The governing equations are

$$\left.\begin{aligned} \mathrm{DIV}(\dot{\boldsymbol{P}}) + \dot{\boldsymbol{B}} = \boldsymbol{0} \quad & \text{in } \mathcal{B} \\ \dot{\boldsymbol{u}} = \dot{\boldsymbol{u}}_0 \quad & \text{on } \partial\mathcal{B}_{\boldsymbol{u}} \\ \dot{\boldsymbol{P}}\cdot\boldsymbol{N} = \dot{\boldsymbol{t}}_0 \quad & \text{on } \partial\mathcal{B}_{\boldsymbol{t}} \end{aligned}\right\}, \tag{9.87}$$

where $\boldsymbol{B}$ is the body force vector per unit reference volume, $\boldsymbol{N}$ is the outward unit normal vector to $\partial\mathcal{B}$ in the reference configuration, $\dot{\boldsymbol{u}}_0$ is the prescribed boundary velocity, $\boldsymbol{t}_0$ is the prescribed nominal surface traction (force per unit area in the reference configuration), and the superimposed dot implies a material time derivative. The local solution is given by $(\dot{\boldsymbol{P}}, \dot{\boldsymbol{u}})$, but we are also interested in an emerging alternative solution $(\dot{\boldsymbol{P}}^*, \dot{\boldsymbol{u}}^*)$.

Equation (9.2) now takes the form

$$\begin{aligned} \int_{\mathcal{B}} \mathrm{DIV}[(\dot{\boldsymbol{u}}^* - \dot{\boldsymbol{u}})\cdot(\dot{\boldsymbol{P}}^* - \dot{\boldsymbol{P}})]\,\mathrm{d}V = \int_{\mathcal{B}} (\dot{\boldsymbol{F}}^* - \dot{\boldsymbol{F}}) : (\dot{\boldsymbol{P}}^* - \dot{\boldsymbol{P}})\,\mathrm{d}V \\ + \int_{\mathcal{B}} (\dot{\boldsymbol{u}}^* - \dot{\boldsymbol{u}})\cdot \mathrm{DIV}(\dot{\boldsymbol{P}}^* - \dot{\boldsymbol{P}})\,\mathrm{d}V\,. \end{aligned} \tag{9.88}$$

The second integral on the right-hand side vanishes since the two stresses satisfy the governing partial differential equation, and so does the term on the left-hand side. Thus, any pair of possible solutions must satisfy the condition

$$(\dot{\boldsymbol{F}}^* - \dot{\boldsymbol{F}}) : (\dot{\boldsymbol{P}}^* - \dot{\boldsymbol{P}}) > 0\,. \tag{9.89}$$

For incrementally linear materials the stress rate tensor $\dot{\boldsymbol{P}}$ is related to the rate of deformation gradient $\dot{\boldsymbol{F}}$ via the same two-point tangential tensor A defined in Chapter 5. The local stability condition then reduces to the form

$$\dot{\boldsymbol{\phi}} : \mathsf{A} : \dot{\boldsymbol{\phi}} > 0\,. \tag{9.90}$$

If $\dot{\boldsymbol{\phi}}$ is given by

$$\dot{\boldsymbol{\phi}} = \dot{\varphi}\boldsymbol{m}\otimes\boldsymbol{N}\,, \tag{9.91}$$

then the local stability condition reduces to the strong ellipticity condition

$$\boldsymbol{m}\cdot\boldsymbol{A}\cdot\boldsymbol{m} > 0\,, \tag{9.92}$$

where $\boldsymbol{A} = \boldsymbol{A}(\boldsymbol{N})$ is the acoustic tensor with components $A_{ij} = N_A \mathsf{A}_{iAjB} N_B$.

Throughout the remainder of this section, we shall limit the discussions to bifurcation into a deformation band. In the process, we shall see a very striking similarity with the kinematics of crystal slip in the finite deformation range discussed in Chapter 8. There is a major difference between the two problems, however: in a crystal the slips are evolving, whereas in bifurcation the slip is just nucleating. Therefore, in the latter case the deformation is understood to be smooth and continuous up until the inception of a deformation band. To make this point more lucid, we shall review some of the basic kinematics of a deformation band in the finite deformation range with the aid of Fig. 9.6.

Consider a smooth surface $\mathcal{S}_0 \in \mathcal{B}$ (the master side) and denote any point on $\mathcal{S}_0$ by $\boldsymbol{Y}$ parameterized in terms of the local coordinates ξ^1 and ξ^2, i.e., $\boldsymbol{Y} = \boldsymbol{Y}(\xi^1, \xi^2)$. The unit normal to $\mathcal{S}_0$ in the direction of $\widehat{\mathcal{S}}_0$ (the slave side) is

$$\boldsymbol{N} = \boldsymbol{N}(\xi^1, \xi^2) = \frac{\boldsymbol{Y}_{,1} \times \boldsymbol{Y}_{,2}}{\|\boldsymbol{Y}_{,1} \times \boldsymbol{Y}_{,2}\|} \,. \tag{9.93}$$

We consider any material point with position vector $\boldsymbol{X}$ in $\mathcal{B}$. Inside the deformation band, the mapping takes the form

$$\boldsymbol{X}(\xi^1, \xi^2, c) = \boldsymbol{Y}(\xi^1, \xi^2) + c\boldsymbol{N}(\xi^1, \xi^2) \,, \tag{9.94}$$

where $0 \leq c \leq h_0$. Thus, $\boldsymbol{X}(\xi^1, \xi^2, 0) = \boldsymbol{Y}(\xi^1, \xi^2)$ defines the master band face $\mathcal{S}_0$, and $\boldsymbol{X}(\xi^1, \xi^2, h_0)$ defines the slave band face $\widehat{\mathcal{S}}_0$. Here, h_0 is the thickness of an emerging band in the reference configuration.

Next we define the velocity field of any material point with position vector $\boldsymbol{X}$ by a ramp-like relation

$$\boldsymbol{V} = \begin{cases} \overline{\boldsymbol{V}} & \text{if } c \leq 0 \,, \\ \overline{\boldsymbol{V}} + c\widetilde{\boldsymbol{V}}/h_0 & \text{if } 0 \leq c \leq h_0 \,, \\ \overline{\boldsymbol{V}} + \widetilde{\boldsymbol{V}} & \text{if } c \geq 0 \,, \end{cases} \tag{9.95}$$

where $\overline{\boldsymbol{V}}$ is the continuous part and $\widetilde{\boldsymbol{V}}$ is the relative velocity of the opposite band faces. The corresponding rates of deformation gradient are

$$\dot{\boldsymbol{F}}(\boldsymbol{X}) = \begin{cases} \dot{\overline{\boldsymbol{F}}} & \text{outside the band,} \\ \dot{\overline{\boldsymbol{F}}} + (\widetilde{\boldsymbol{V}} \otimes \boldsymbol{N})/h_0 & \text{inside the band.} \end{cases} \tag{9.96}$$

Since the deformation is smooth and continuous up until bifurcation, the identity $\boldsymbol{l} = \dot{\boldsymbol{F}} \cdot \boldsymbol{F}^{-1}$ continues to hold, with $\boldsymbol{F} = \overline{\boldsymbol{F}}$ up until bifurcation. Therefore, we have

$$\boldsymbol{l} = \begin{cases} \bar{\boldsymbol{l}} & \text{outside the band,} \\ \bar{\boldsymbol{l}} + (\widetilde{\boldsymbol{V}} \otimes \boldsymbol{N} \cdot \boldsymbol{F}^{-1})/h_0 & \text{inside the band,} \end{cases} \tag{9.97}$$

where $\bar{\boldsymbol{l}} = \dot{\overline{\boldsymbol{F}}} \cdot \boldsymbol{F}^{-1} = \dot{\overline{\boldsymbol{F}}} \cdot \overline{\boldsymbol{F}}^{-1}$ is the spatial velocity gradient.

We can also write the spatial velocity gradient inside the band by a similar re-parameterization of $\mathcal{S} = \phi(\mathcal{S}_0)$ in the current configuration. Denoting any

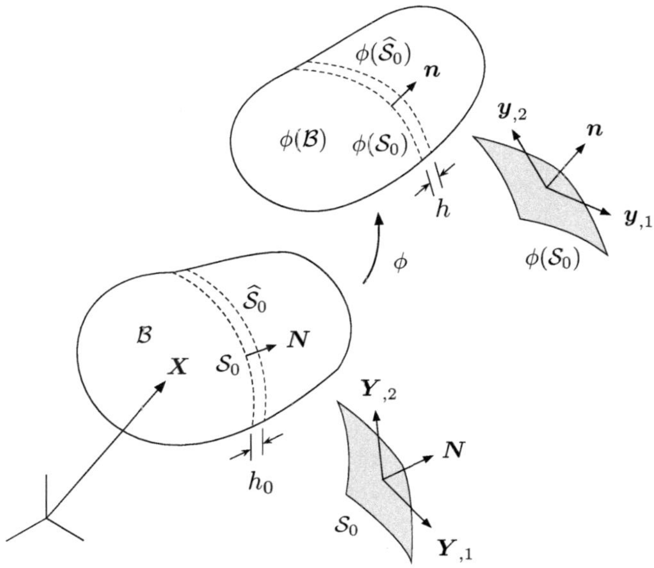

Fig. 9.6 Normal parameterization of deformation band geometry with respect to reference and current configurations

point on $\mathcal{S}$ by $\boldsymbol{y} = \boldsymbol{y}(\zeta^1, \zeta^2)$, where ζ^1 and ζ^2 are the local coordinates on $\mathcal{S}$, then the unit normal to $\mathcal{S}$ is

$$\boldsymbol{n} = \boldsymbol{n}(\zeta^1, \zeta^2) = \frac{\boldsymbol{y}_{,1} \times \boldsymbol{y}_{,2}}{\|\boldsymbol{y}_{,1} \times \boldsymbol{y}_{,2}\|} . \tag{9.98}$$

Any point $\boldsymbol{x}$ inside the deformation band is then defined by the mapping

$$\boldsymbol{x}(\zeta^1, \zeta^2, c) = \boldsymbol{y}(\zeta^1, \zeta^2) + c\boldsymbol{n}(\zeta^1, \zeta^2) , \tag{9.99}$$

where $0 \le c \le h$, and h is the thickness of the band in the current configuration. The velocity gradient is then given by

$$\boldsymbol{l} = \begin{cases} \bar{\boldsymbol{l}} & \text{outside the band,} \\ \bar{\boldsymbol{l}} + (\widetilde{\boldsymbol{V}} \otimes \boldsymbol{n})/h & \text{inside the band.} \end{cases} \tag{9.100}$$

Comparing the two expressions for the velocity gradient yields the relations

$$\frac{\boldsymbol{n}}{h} = \frac{\boldsymbol{N} \cdot \boldsymbol{F}^{-1}}{h_0} \equiv \frac{\boldsymbol{F}^{-\mathsf{T}} \cdot \boldsymbol{N}}{h_0} . \tag{9.101}$$

This identity has an interesting geometrical meaning if we assume that h and h_0 are both small. Consider an infinitesimal reference volume $\mathrm{d}V = h_0 \mathrm{d}A$ defined by the band thickness in the reference configuration, and the associated push-forward volume $\mathrm{d}v = h\mathrm{d}a$ in the current configuration, then

$$h\mathrm{d}a = Jh_0\mathrm{d}A\,, \tag{9.102}$$

where $J = \mathrm{d}v/\mathrm{d}V$ is the usual Jacobian. Substituting into (9.101) yields

$$J\boldsymbol{F}^{-\mathsf{T}} \cdot \boldsymbol{N}\mathrm{d}A = \boldsymbol{n}\mathrm{d}a\,, \tag{9.103}$$

and we recover Nanson's formula.

With these preliminaries in hand, we can now develop a bifurcation condition leading to a deformation band. For the nominal traction rate to be continuous on $\widehat{\mathcal{S}}$, we must have

$$(\dot{\boldsymbol{P}}^* - \dot{\boldsymbol{P}}) \cdot \boldsymbol{N} = \boldsymbol{0}\,, \tag{9.104}$$

where $\boldsymbol{P}^*$ and $\boldsymbol{P}$ are the first Piola-Kirchhoff stress tensors on each side of $\widetilde{\mathcal{S}}_0$. If the two stress rates are related to the corresponding rates of deformation gradient via the same tangent operator A (i.e., if the solid response is incrementally linear), then the condition for a continuous nominal traction rate is

$$\frac{1}{h_0}\boldsymbol{A} \cdot \widetilde{\boldsymbol{V}} = \boldsymbol{0}\,, \qquad A_{ij} = N_A \mathsf{A}_{iAjB} N_B\,. \tag{9.105}$$

In this case, nontrivial solutions can exist if

$$\det(\boldsymbol{A}) = 0\,. \tag{9.106}$$

Because $\boldsymbol{N}$ determines the critical orientation of the band, $\boldsymbol{A} = \boldsymbol{A}(\boldsymbol{N})$ is called the Lagrangian acoustic tensor.

Next, we recall from Chapter 5 the push-forward transformation

$$\mathsf{a}_{ikjl} = F_{kA}F_{lB}\mathsf{A}_{iAjB}\,. \tag{9.107}$$

Substituting into the expression for the Lagrangian acoustic tensor $\boldsymbol{A}$ and using (9.101) gives

$$A_{ij} = N_A F^{-1}_{Ak}\mathsf{a}_{ikjl}N_B F^{-1}_{Bl} = \left(\frac{h_0}{h}\right)^2 n_k \mathsf{a}_{ikjl} n_l\,. \tag{9.108}$$

The jump condition (9.105) can be written in the alternative form

$$\left(\frac{h_0}{h^2}\right)\boldsymbol{a} \cdot [\![\boldsymbol{v}]\!] = \boldsymbol{0}\,, \qquad a_{ij} = n_k \mathsf{a}_{ikjl} n_l\,. \tag{9.109}$$

Nontrivial solutions can exist if

$$\det(\boldsymbol{a}) = 0\,. \tag{9.110}$$

Because $\boldsymbol{n}$ determines the critical orientation of the band, $\boldsymbol{a} = \boldsymbol{a}(\boldsymbol{n})$ is called the Eulerian acoustic tensor.

9.8 Persistent Shear Band

We revisit the example of plane strain compression of a rectangular sand specimen presented in Section 6.11. This time, we are interested in delineating the evolution of the localization function as a function of vertical compression of the specimen. The localization function is calculated from the formula $\mathcal{F} = \det(\boldsymbol{a})$, where $\boldsymbol{a}$ is the Eulerian acoustic tensor defined in equation (9.109). Because the plane of the shear band is well defined, the search for the stationary point of $\mathcal{F}$ can be made by a sweep over half a unit circle (as opposed to a sweep over half a unit ball for three-dimensional problems). We caution that the following simulations are valid only up to the bifurcation point and slightly beyond it; the analysis cannot be extended too far beyond this point or the solution will exhibit mesh sensitivity. For the simulation of responses well past the bifurcation point, it is essential that the mesh-sensitivity issue be addressed through the finite element enhancement techniques discussed in Chapter 7, for example.

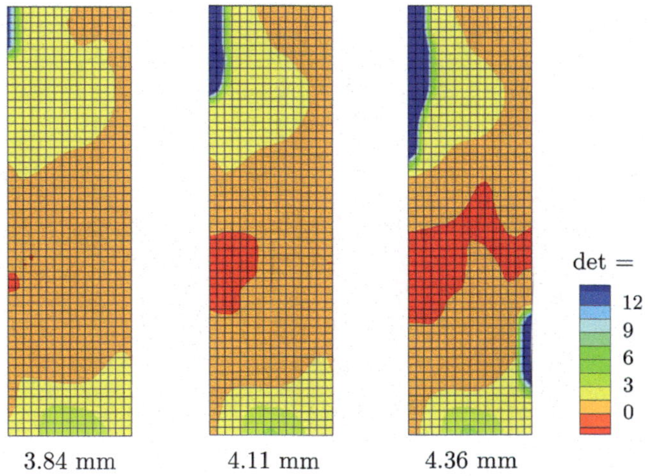

Fig. 9.7 Evolution of localization function for a strongly heterogeneous rectangular sand specimen in plane strain vertical compression. Initial specific volume varies from 1.3 to 2.0; numbers in mm denote vertical compression. Color bar is normalized localization function.

The imposed heterogeneity in density is the same as in the example of Section 6.11, i.e., the sand is strongly heterogeneous with specific volume varying from 1.3 to 2.0. The localization function is then expected to exhibit erratic variation where shear bands would appear to nucleate at different locations only to die out as more critical zones of localized deformation emerge.

Ultimately, however, the critical points of localized deformation are expected to converge to a certain dominant pattern called persistent shear band. This is illustrated in Fig. 9.7 where a shear band (red zone) appears to nucleate at different points on the left vertical face of the specimen. Because of the high density contrast in the loose layer, the shear band is not well defined in the beginning, and in fact, a conjugate shear band appears to emerge from the right vertical face, in competition with the more dominant shear band emerging from the left side. This pattern of localized deformation is remarkably similar to the one reported by Borja et al. (2013a) from a similar finite element simulation but using constant strain triangular elements.

To demonstrate the impact of density contrast on the localized deformation pattern, we consider the same density distribution reported in Section 6.11 but proportionately adjust the specific volume to fall within the range 1.4 to 1.8, thus delineating a slightly less heterogeneous specimen (essentially, this entails adjusting the color bar shown in Fig. 6.10 to reflect this narrower range of values of the specific volume). Figure 9.8 shows a more predictable evolution of the localization function, in which a single persistent shear band nucleates from the looser layer on the left side and propagates upward to the right. We remark that apart from the heterogeneity in density, it has been demonstrated lately that the persistent shear band can also be influenced by fluid flow and the spatial variation of degree of saturation within an unsaturated porous material (Borja et al. 2013b).

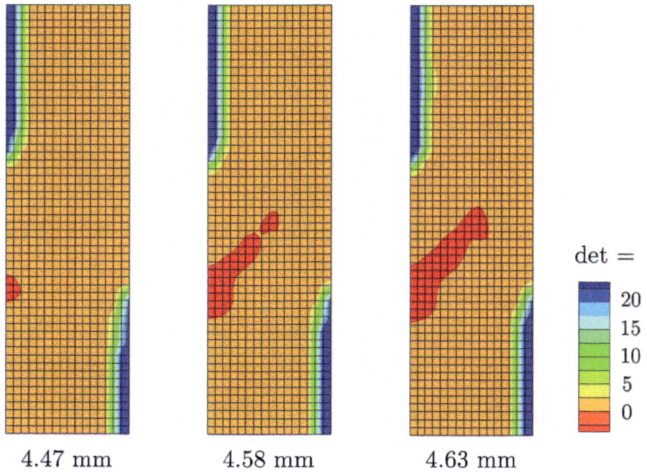

Fig. 9.8 Evolution of localization function for a mildly heterogeneous rectangular sand specimen in plane strain vertical compression. Initial specific volume varies from 1.4 to 1.8; numbers in mm denote vertical compression. Color bar is normalized localization function.

9.9 Summary

Elastoplastic solids are an incrementally nonlinear material for which there exist more than one tangent constitutive tensor for a given state of stress. Comparison solids may be used to delineate bounds of the local stability region for elastoplastic materials. Loss of ellipticity is a type of instability associated with an emerging tabular deformation band. Specific examples of a deformation band include shear band, compaction band, and mixed-mode band. Closed-form solutions for the inception of a deformation band are available for simple plasticity models in the infinitesimal deformation range; however, more robust search algorithms are needed to accommodate more elaborate constitutive models and/or finite deformation effects.

References

Achenbach, J.D.: Wave Propagation in Elastic Solids. North-Holland, New York (1987)

Agarwal, H., Gokhale, A.M., Graham, S., Horstemeyer, M.F.: Void growth in 6061-aluminum alloy under triaxial stress state. Materials Science and Engineering A 341, 35–42 (2003)

Anand, L., Kothari, M.: A computational procedure for rate-independent crystal plasticity. Journal of the Mechanics and Physics of Solids 44, 525–558 (1996)

Andrade, J.E., Borja, R.I.: Capturing strain localization in dense sands with random density. International Journal for Numerical Methods in Engineering 67, 1531–1564 (2006)

Argyris, J.H., Faust, G., Szimmat, J., Warnke, E.P., Willam, K.J.: Recent developments in the finite element analysis of prestressed concrete reactor vessels. Nuclear Engineering and Design 28, 42–75 (1974)

Armero, F., Garikipati, K.: Recent advances in the analysis and numerical simulation of strain localization in inelastic solids. In: Owen, D.R.J., Oñate, E., Hinton, E. (eds.) Proceedings of Computational Plasticity IV, pp. 547–561. CIMNE, Barcelona (1995)

Asaro, R.J.: Crystal plasticity. Journal of Applied Mechanics 50, 921–934 (1983)

Asaro, R.J., Rice, J.R.: Strain localization in ductile single crystals. Journal of the Mechanics and Physics of Solids 25, 309–338 (1977)

Aydin, A., Borja, R.I., Eichhubl, P.: Geological and mathematical framework for failure modes in granular rock. Journal of Structural Geology 28, 83–98 (2006)

Barenblatt, G.I.: The mathematical theory of equilibrium cracks in brittle fracture. Advances in Applied Mechanics 7, 55–129 (1962)

Baud, P., Klein, E., Wong, T.-F.: Compaction localization in porous sandstones: spatial evolution of damage and acoustic emission activity. Journal of Structural Geology 26, 603–604 (2004)

Bauschinger, J.: Ueber die Quercontraction und -Dilatation bei der Längenausdehnung und -Zusammendrückung prismatischer Körper. Leipzig (1879)

Bažant, Z.P., Planas, J.: Fracture and Size Effect in Concrete and Other Quasibrittle Materials. CRC Press, New York (1997)

Belytschko, T., Black, T.: Elastic crack growth in finite elements with minimal remeshing. International Journal for Numerical Methods in Engineering 45, 601–620 (1999)

Bigoni, D.: Bifurcation and instability of non-associative elastoplastic solids. In: Petryk, H. (ed.) Material Instabilities in Elastic and Plastic Solids. CISM Courses and Lectures, vol. 414, pp. 1–52. Springer, Wien New York (2000)

Bilby, B.A., Cottrell, A.H., Swindon, K.H.: The spread of plastic yield from a notch. Proceedings of the Royal Society of London, Ser. A 272, 304–314 (1963)

Bochev, P.B., Dohrmann, C.R., Gunzburger, M.D.: Stabilization of low-order mixed finite elements for the Stokes equations. SIAM Journal on Numerical Analysis 44, 82–101 (2006)

Bochev, P.B., Dohrmann, C.R.: A computational study of stabilized, low-order C0 finite element approximations of Darcy equations. Computational Mechanics 38, 323–333 (2006)

Borja, R.I., Song, X., Wu, W.: Critical state plasticity, Part VII: Triggering a shear band variably saturated porous media. Computer Methods in Applied Mechanics and Engineering (2013a), `http://dx.doi.org/10.1016/j.cma.2013.03.008`

Borja, R.I., Song, X., Rechenmacher, A.L., Abedi, S., Wu, W.: Shear band in sand with spatially varying density. Journal of the Mechanics and Physics of Solids 61, 219–234 (2013b)

Borja, R.I., Rahmani, H.: Computational aspects of elasto-plastic deformation in polycrystalline solids. Journal of Applied Mechanics 79, 031024 (9 pages) (2012), doi:10.1115.4005898

Borja, R.I.: Assumed enhanced strain and the extended finite element methods: A unification of concepts. Computer Methods in Applied Mechanics and Engineering 197, 2789–2803 (2008)

Borja, R.I., Foster, C.D.: Continuum mathematical modeling of slip weakening in geological systems. Journal of Geophysical Research 112 (2007), doi:10.1029/2005JB004056

Borja, R.I.: Conditions for instabilities in collapsible solids including volume implosion and compaction banding. Acta Geotechnica 1(2006), 107–122 (2006)

Borja, R.I., Andrade, J.E.: Critical state plasticity. Part VI: Meso-scale finite element simulation of strain localization in discrete granular materials. Computer Methods in Applied Mechanics and Engineering 195, 5115–5140 (2006)

Borja, R.I., Aydin, A.: Computational modeling of deformation bands in granular media, I: Geological and mathematical framework. Computer Methods in Applied Mechanics and Engineering 193, 2667–2698 (2004)

Borja, R.I.: Computational modeling of deformation bands in granular media, II: Numerical simulations. Computer Methods in Applied Mechanics and Engineering 193, 2699–2718 (2004)

Borja, R.I., Sama, K.M., Sanz, P.F.: On the numerical integration of three-invariant elastoplastic constitutive model. Computer Methods in Applied Mechanics and Engineering 192, 1227–1258 (2003)

Borja, R.I.: Bifurcation of elastoplastic solids to shear band mode at finite strain. Computer Methods in Applied Mechanics and Engineering 191, 5287–5314 (2002a)

Borja, R.I.: Finite element simulation of strain localization with large deformation: Capturing strong discontinuity using a Petrov-Galerkin multiscale formulation. Computer Methods in Applied Mechanics and Engineering 191, 2949–2978 (2002b)

Borja, R.I., Regueiro, R.A.: Strain localization in frictional materials exhibiting displacement jumps. Computer Methods in Applied Mechanics and Engineering 190, 2555–2580 (2001)

Borja, R.I.: A finite element model for strain localization analysis of strongly discontinuous fields based on standard Galekin approximation. Computer Methods in Applied Mechanics and Engineering 190, 1529–1549 (2000)

Borja, R.I., Tamagnini, C., Amorosi, A.: Coupling plasticity and energy-conserving elasticity models for clays. Journal of Geotechnical and Geoenvironmental Engineering 123, 948–957 (1997)

Borja, R.I., Tamagnini, C.: Cam-Clay plasticity, Part III: Extension of the infinitesimal model to include finite strains. Computer Methods in Applied Mechanics and Engineering 155, 73–95 (1998)

Borja, R.I., Wren, J.R.: Discrete micromechanics of elastoplastic crystals. International Journal for Numerical Methods in Engineering 36, 3815–3840 (1993)

Borja, R.I., Lee, S.R.: Cam-Clay plasticity, Part I: Implicit integration of elasto-plastic constitutive relations. Computer Methods in Applied Mechanics and Engineering 78, 49–72 (1990)

Borre, G., Maier, G.: On linear versus nonlinear flow rules in strain localization analysis. Meccanica 24, 36–41 (1989)

Boswell, L.F., Chen, Z.: A general criterion for plain concrete. International Journal of Solids and Structures 23, 621–630 (1987)

Brace, W.F.: Some new measurements of linear compressibility of rocks. Journal of Geophysical Research 70, 391–398 (1965)

Brezzi, F.: A discourse on the stability conditions for mixed finite element formulations. Computer Methods in Applied Mechanics and Engineering 82, 27–57 (1990)

Broek, D.: Elementary Engineering Fracture Mechanics. Kluwer Academic Publishers, Massachusetts (1986)

Budiansky, B., Wu, T.T.: Theoretical prediction of plastic strains of polycrystals. In: Rosenberg, R.M. (ed.) Proceedings of the Fourth U.S. National Congress on Applied Mechanics, pp. 1175–1185. ASME, New York (1962)

Butterfield, R.: A natural compression law for soils. Géotechnique 29, 469–480 (1979)

Chambon, R.: Some theoretical results about second order work, uniqueness, existence and controllability independent of the constitutive equation. Journal of Engineering Mathematics 52, 53–61 (2005)

Chen, W.F., Han, D.J.: Plasticity for Structural Engineers. Springer, Berlin (1988)

Chen, W.F.: Plasticity in Reinforced Concrete. McGraw-Hill, New York (1982)

Chester, F.M., Higgs, N.G.: Multi-mechanism frictions constitutive model for ultra-fine quartz gouge at hypocentral conditions. Journal of Geophysical Research 97, 1859–1870 (1992)

Chester, F.M., Evans, J.P., Beigel, R.L.: Internal structure and weakening mechanisms of the San Andreas fault. Journal of Geophysical Research 98, 771–786 (1993)

Coleman, B.D., Gurtin, M.E.: Thermodynamics with internal state variables. The Journal of Chemical Physics 47, 597–613 (1967)

Coulomb, C.A.: Essai sur une application des règles des maximis et minimis à quelques problèmes de statique relatifs à l'architecture. Mémoires de Mathématique et de Physique. Académie des Sciences, Paris 7, 343–382 (1773)

Dafalias, Y.F.: Plastic spin: necessity or redundancy? International Journal of Plasticity 14, 909–931 (1998)

Darve, F., Servant, G., Laouafa, F., Khoa, H.D.V.: Failure in geomaterials: continuous and discrete analyses. Computer Methods in Applied Mechanics and Engineering 193, 3057–3085 (2004)

de Saint-Venant, B.: Mémoire sur l'établissement des équations différentielles des mouvement interieurs opérés dans les corps solides ductiles au dela des limites où l'élasticité pourrait les ramener à leur premier état. Comptes Rendus de l'Académie des Sciences, Paris 70, 473–480 (1870)
Di Toro, G., Goldsby, D.L., Tullis, T.E.: Friction falls towards zero in quartz rock as slip velocity approaches seismic rates. Nature 427, 436–439 (2004)
Dieterich, J.H.: Time dependent friction and the mechanics of stick slip. Pure and Applied Geophysics 116, 790–806 (1978)
Dieterich, J.H.: Modeling of rock friction 1. Experimental results and constitutive equations. Journal of Geophysical Research 84, 2161–2168 (1979)
Drucker, D.C., Prager, W.: Soil mechanics and plastic analysis or limit design. Quarterly of Applied Mathematics 10, 157–165 (1952)
DiMaggio, F.L., Sandler, I.S.: Material model for granular soils. Journal of the Engineering Mechanics Division ASCE 97, 935–350 (1971)
Dugdale, D.S.: Yielding of steel sheets containing slits. Journal of the Mechanics and Physics of Solids 8, 100–104 (1960)
Flores, K.M., Dauskardt, R.H.: Mean stress effects on flow localization and failure in bulk metallic glass. Acta Materialia 49, 2527–2537 (2001)
Flores, K.M., Dauskardt, R.H.: Local heating associated with crack tip plasticity in Zr-Ti-Ni-Cu-Be bulk amorphous metals. Journal of Materials Research 14, 638–643 (1999)
Foster, C.D., Borja, R.I., Regueiro, R.A.: Embedded strong discontinuity finite elements for fractured geomaterials with variable friction. International Journal for Numerical Methods in Engineering 72, 549–581 (2007)
Gudehus, G.: Elastoplastische Stoffgleichungen für trockenen Sand. Ingenieur-Archiv. 42, 151–169 (1973)
Gurtin, M.E.: An Introduction to Continuum Mechanics. Academic Press, Orlando (1981)
Hanks, T.C.: Earthquake stress drops, ambient tectonic stresses ad stresses that drive plate motions. Pure and Applied Geophysics 115, 441–458 (1977)
Hansen, B.: Line ruptures regarded as narrow rupture zones. Basic equations based on kinematic considerations. In: Brussels Conference 58 on Earth Pressure Problems, Brussels, vol. 1, pp. 39–48 (September1958)
Hardin, B.O., Richart Jr., F.E.: Elastic wave velocities in granular soils. Journal of the Soil Mechanics and Foundations Division, ASCE 89, 33–65 (1963)
Hashiguchi, K., Ueno, M.: Elasto-plastic constitutive laws of granular materials. In: Murayama, S., Schofield, A.N. (eds.) Constitutive Equations of Soils, Proceedings of the Ninth International Conference on Soil Mechanics and Foundation Engineering, Specialty Session 9, Tokyo, pp. 73–82 (1977)
Hencky, H.: Zur Theorie plastischer Deformationen und der hierdurch im Material hervorgerufenen Nachspannungen. Zeitschrift für Angewandte Mathematik und Mechanik 4, 323–334 (1924)
Hill, R., Rice, J.R.: Constitutive analysis of elastic-plastic crystals at arbitrary strain. Journal of the Mechanics and Physics of Solids 20, 401–413 (1972)
Hill, R.: A general theory of uniqueness and stability in elastic-plastic solids. Journal of the Mechanics and Physics of Solids 6, 236–249 (1958)
Hill, R.: The Mathematical Theory of Plasticity. Oxford University Press, Oxford (1950)
Houlsby, G.T.: The use of a variable shear modulus in elastic-plastic models for clays. Computers and Geotechnics 1, 3–13 (1985)

Hughes, T.J.R.: The Finite Element Method. Prentice-Hall, Inc., Englewood-Cliffs (1987)

Hughes, T.J.R.: Numerical implementation of constitutive models: Rate independent deviatoric plasticity. In: Nemat-Nasser, S., Asaro, R., Hegemier, G. (eds.) Theoretical Foundations for Large Scale Computations of Nonlinear Material Behavior, pp. 29–57. Martinus Nijhoff Publishers, Dordrecht (1984)

Hughes, T.J.R.: Generalization of selective integration procedures to anisotropic and nonlinear media. International Journal for Numerical Methods in Engineering 15, 1413–1418 (1980)

Ida, Y.: Cohesive force across the tip of a longitudinal shear crack and Griffiths specific surface energy. Journal of Geophysical Research 77, 3796–3805 (1972)

Issen, K.A., Rudnicki, J.W.: Conditions for compaction bands in porous rock. Journal of Geophysical Research 105, 21529–21536 (2000)

Jaeger, J.C., Cook, N.G.W.: Fundamentals of Rock Mechanics. Chapman and Hall Ltd., London (1976)

Jefferies, M.G.: Nor-Sand: a simple critical state model for sand. Géotechnique 43, 91–103 (1993)

Jiang, J., Pietruszczak, S.: Convexity of yield loci for pressure sensitive materials. Computers and Geotechnics 5, 51–63 (1988)

Kang, H.D., Willam, K.J.: Localization characteristics of triaxial concrete model. Journal of Engineering Mechanics 125, 941–950 (1999)

Karush, W.: Minima of Functions of Several Variables with Inequalities as Side Constraints. M.Sc. Dissertation, Department of Mathematics, University of Chicago, Chicago, Illinois (1939)

Koiter, W.T.: General theorems for elastic plastic solids. Progress in Solid Mechanics 1, 182–221 (1960)

Koiter, W.T.: Stress-strain relations, uniqueness and variational theorems for elastic-plastic materials with a singular yield surface. Quarterly of Applied Mathematics 11, 350–354 (1953)

Kondner, R.L.: Hyperbolic stress-strain response: cohesive soils. Journal of Soil Mechanics and Foundations Division, ASCE 89, 115–144 (1963)

Krieg, R.D., Key, S.W.: Implementation of a time dependent plasticity theory into structural computer programs. In: Stricklin, J.A., Saczalski, K.J. (eds.) Constitutive Equations in Viscoplasticity: Computational and Engineering Aspects, AMD-20. ASME, New York (1976)

Kuhn, H.W., Tucker, A.W.: Nonlinear programming. In: Proceedings of the Second Berkeley Symposium on Mathematical Statistics and Probability, pp. 481–492. University of California Press (1951)

Labuz, J.F., Riedel, J.J., Dai, S.T.: Shear fracture in sandstone under plane-strain compression. Engineering Fracture Mechanics 73, 820–828 (2006)

Lade, P.V., Kim, M.K.: Single hardening constitutive model for frictional materials II. Yield criterion and plastic work contours. Computers and Geotechnics 6, 13–29 (1988)

Lade, P.V., Duncan, J.M.: Elastoplastic stress-strain theory for cohesionless soil. Journal of Geotechnical Engineering Division, ASCE 101, 1037–1053 (1975)

Lambe, T.W., Whitman, R.V.: Soil Mechanics. Wiley, New York (1969)

Lapusta, N., Rice, J.R., Ben-Zion, Y., Zhen, G.: Elastodynamic analysis for slow tectonic loading with spontaneous rupture episodes on faults with rate- and state-dependent friction. Journal of Geophysical Research 105, 23765–23789 (2000)

Launay, P., Gachon, H.: Strain and ultimate strength of concrete under triaxial stress. Special Publication SP-34(1), 269–282 (1972)

Laursen, T.A.: Computational Contact and Impact Mechanics. Springer, Heidelberg (2003)

Lee, E.H.: Elastic-plastic deformations at finite strains. Journal of Applied Mechanics 36, 1–6 (1969)

Lévy, M.: Extrait du Mémoire sur les équations générales des mouvements intérieurs des corps solides ductiles au delà des limites où l'élasticité pourrait les ramener à leur premier état. Journal de Mathématiques Pures et Appliquées II sér 16, 369–372 (1871)

Linder, C., Armero, F.: Finite elements with embedded strong discontinuities for the modeling of failure in solids. International Journal for Numerical Methods in Engineering 72, 1391–1433 (2007)

Liu, F., Borja, R.I.: A contact algorithm for frictional crack propagation with the extended finite element method. International Journal for Numerical Methods in Engineering 76, 1489–1512 (2008)

Liu, F., Borja, R.I.: An extended finite element framework for slow-rate frictional faulting with bulk plasticity and variable friction. International Journal for Numerical and Analytical Methods in Geomechanics 33, 1535–1560 (2009)

Liu, F., Borja, R.I.: Stabilized low-order finite elements for frictional contact with the extended finite element method. Computer Methods in Applied Mechanics and Engineering 199, 2456–2471 (2010)

Lode, W.: Versuche über den Einfluß der mittleren Hauptspannung auf das Fließen der Metalle Eisen, Kupfer und Nickel. Zeitschrift für Physik 36, 913–939 (1926)

Lubliner, J.: Plasticity Theory. Dover Publications, Inc., New York (2008)

Lyapunov, A.M.: Stability of Motion. Academic Press, New York (1966)

Malvern, L.E.: Introduction to the Mechanics of a Continuous Medium. Prentice-Hall, Englewood Cliffs (1969)

Marone, C., Kilgore, B.: Scaling of the critical slip distance for seismic faulting with shear strain in fault zones. Nature 362, 618–621 (1993)

Marsden, J.E., Hughes, T.J.R.: Mathematical Foundations of Elasticity. Prentice-Hall, Englewood Cliffs (1983)

Matsuoka, H., Nakai, T.: Stress-deformation and strength characteristics of soil under three different principal stresses. Proceedings of the Japanese Society of Civil Engineers 232, 59–70 (1974)

Menétrey, P., Willam, K.J.: Triaxial failure criterion for concrete and its generalization. ACI Structural Journal 92, 311–318 (1995)

Meschke: Consideration of aging of shotcrete in the context of a 3-D viscoplastic material model. International Journal for Numerical Methods in Engineering 39, 3123–3143 (1996)

Miehe, C., Schröder, J.: A comparative study of stress update algorithms for rate-independent and rate-dependent crystal plasticity. International Journal for Numerical Methods in Engineering 50, 273–298 (2001)

Mills, L.L., Zimmerman, R.M.: Compressive strength of plain concrete under multiaxial loading conditions. ACI Journal 67, 802–807 (1970)

Mohr, O.: Welche Umstände bedingen die Elastizitätsgrenze und den Bruch eines Materiales? Zeitschrift des Vereines Deutscher Ingenieure 44, 1524–1530, 1572–1577 (1900)

Moreau, J.J.: Evolution problem associated with a moving convex set in a Hilbert space. Journal of Differential Equations 26, 347–374 (1977)

Mosler, J.: Numerical analyses of discontinuous material bifurcation: strong and weak discontinuities. Computer Methods in Applied Mechanics and Engineering 194, 979–1000 (2005)

Nádai, A.: Plasticity. McGraw-Hill, New York (1931)

Nayak, G.C., Zienkiewicz, O.C.: Elasto-plastic stress analysis. A generalization for various constitutive relations including strain softening. International Journal for Numerical Methods in Engineering 5, 113–135 (1972)

Nova, R.: Controllability of the incremental response of soil specimens subjected to arbitrary loading programmes. Journal of the Mechanical Behavior of Materials 5, 221–243 (1994)

Nowinski, J.L.: Applications of Functional Analysis in Engineering. Plenum Press, New York (1981)

Ogden, R.W.: Non-linear Elastic Deformations. Ellis Horwood Ltd., Chichester (1984)

Okubo, P.G., Dieterich, J.H.: Effects of physical fault properties on frictional instabilities produced on simulated faults. Journal of Geophysical Research 89, 5817–5827 (1984)

Oliver, J., Huespe, A.E., Cante, J.C., Díaz, G.: On the numerical resolution of the discontinuous material bifurcation problem. International Journal for Numerical Methods in Engineering 83, 786–804 (2010)

Ortiz, M., Leroy, Y., Needleman, A.: A finite element method for localized failure analysis. Computer Methods in Applied Mechanics and Engineering 61, 189–214 (1987)

Owen, D.R.J., Hinton, E.: Finite Elements in Plasticity: Theory and Practice. Pineridge Press Limited, Swansea (1980)

Palmer, A.C., Rice, J.R.: The growth of slip surfaces in the progressive failure of overconsolidated clay. Proceedings of the Royal Society of London, Ser. A 332, 527–548 (1973)

Peirce, D., Asaro, R.J., Needleman, A.: An analysis of nonuniform and localized deformation in ductile single crystals. Acta Metallurgica 30, 1087–1119 (1982)

Perić, D., Ayari, M.A.: On the analytical solutions for the three-invariant Cam clay model. International Journal of Plasticity 18, 1061–1082 (2002)

Prager, W.: The theory of plasticity: A survey of recent achievement. Proceedings of the Institute of Mechanical Engineers, London 169, 41–57 (1955)

Prandtl, L.: Spannungsverteilung in plastischen Körpern. In: Biezeno, C.B., Burgers, J.M. (eds.) Proceeding of the 1st International Congress in Applied Mechanics, Delft, pp. 43–54 (1924)

Raniecki, B., Bruhns, O.T.: Bounds to bifurcation stresses in solids with non-associated plastic flow law at finite strain. Journal of the Mechanics and Physics of Solids 29, 153–172 (1981)

Ramberg, W., Osgood, W.R.: Description of stress-strain curves by three parameters. National Advisory Committee for Aeronautics, Technical Note No. 902 (1945)

Regueiro, R.A., Borja, R.I.: Plane strain finite element analysis of pressure-sensitive plasticity with strong discontinuity. International Journal of Solids and Structures 38, 3647–3672 (2001)

Reuss, A.: Berücksichtigung der elastischen Formänderung in der Plastizitätstheorie. Zeitschrift für Angewandte Mathematik und Mechanik 10, 266–274 (1930)

Rice, J.R., Cocco, M.: Seismic fault rheology and earthquake dynamics. In: Handy, M.R., Hirth, G., Hovius, N. (eds.) Tectonic Faults: Agents of Change on a Dynamic Earth, Dahlem Workshop 95, pp. 99–137. The MIT Press, Cambridge (2007)

Rice, J.R., Rudnicki, J.W.: A note on some features of the theory of localization of deformation. International Journal of Solids and Structures 16, 597–605 (1980)

Rice, J.R.: Inelastic constitutive relations for solids: An internal-variable theory and its application to metal plasticity. Journal of the Mechanics and Physics of Solids 19, 433–455 (1971)

Reynolds, O.: On the dilatancy of media composed of rigid particles in contact. Philosophical Magazine 20, 469–481 (1885)

Roscoe, K.H., Burland, J.H.: On the generalized stress-strain behaviour of 'wet' clay. In: Heyman, J., Leckie, F.A. (eds.) Engineering Plasticity, pp. 535–609. Cambridge University Press, Cambridge (1968)

Rowe, P.W.: Theoretical meaning and observed values of deformation parameters for soil. In: Proceedings of the Roscoe Memorial Symposium on Stress-Strain Behaviour of Soils, Cambridge, Foulis, Henley-on-Thames, pp. 143–194 (1972)

Rudnicki, J.W., Rice, J.R.: Conditions for the localization of deformation in pressure-sensitive dilatant materials. Journal of the Mechanics and Physics of Solids 23, 371–394 (1975)

Ruina, A.L.: Slip instability and state variable friction laws. Journal of Geophysical Research 88, 10359–10370 (1983)

Rummel, F., Alheid, H.J., Frohn, C.: Dilatancy and fracture- induced velocity changes in rock and their relation to frictional sliding. Pure and Applied Geophysics 116, 743–764 (1978)

Runesson, K., Ottosen, N.S., Perić, D.: Discontinuous bifurcations of elastic-plastic solutions at plane stress and plane strain. International Journal of Plasticity 7, 99–121 (1991)

Sanz, P.F., Borja, R.I., Pollard, D.D.: Mechanical aspects of thrust faulting driven by far-field compression and their implications for fold geometry. Acta Geotechnica 2, 17–31 (2007)

Schmid, E., Boas, W.: Kristallplastizität mit besonderer Berücksichtigung der Metalle. Springer, Berlin (1935)

Schofield, A., Wroth, P.: Critical State Soil Mechanics. McGraw-Hill, New York (1968)

Scholz, C.H.: The Mechanics of Earthquakes and Faulting. Cambridge University Press, New York (1990)

Simmons, G., Brace, W.F.: Comparison of static and dynamic measurements of compressibility of rocks. Journal of Geophysical Research 70, 5649–5656 (1965)

Simo, J.C., Hughes, T.J.R.: Computational Inelasticity. Springer, New York (1998)

Simo, J.C.: Numerical analysis and simulation of plasticity. In: Ciarlet, P.G., Lions, J.L. (eds.) Handbook of Numerical Analysis, vol. VI, pp. 183–499. Elsevier Science B.V. (1998)

Simo, J.C., Oliver, J.: A new approach to the analysis and simulation of strain softening in solids. In: Bažant, Z.P., Bittnar, Z., Jirásek, M., Mazars, J. (eds.) Fracture and Damage in Quasibrittle Structures, E&FN Spon, London (1994)

Simo, J.C.: Algorithms for static and dynamics multiplicative plasticity that preserve the classical return mapping schemes of the infinitesimal theory. Computer Methods in Applied Mechanics and Engineering 99, 61–112 (1992)

Simo, J.C., Laursen, T.A.: An augmented Lagrangian treatment of contact problems involving friction. Computers & Structures 42, 97–116 (1992)

Simo, J.C., Taylor, R.L.: Consistent tangent operators for rate-independent elastoplasticity. Computer Methods in Applied Mechanics and Engineering 48, 101–118 (1985)

Spaepen, F.: Homogeneous flow of metallic glasses: A free volume perspective. Scripta Materialia 54, 363–367 (2006)

Stakgold, I.: Green's Functions and Boundary Value Problems, 2nd edn. John Wiley & Sons, New York (1998)

Stroud, M.A.: The behaviour of sands at low stress levels in the simple-shear apparatus, Ph.D. Thesis, Cambridge University (1971)

Tamagnini, C., Castellanza, R., Nova, R.: A generalized backward Euler algorithm for the numerical integration of an isotropic hardening elastoplastic model for mechanical and chemical degradation of bonded geomaterials. International Journal for Numerical and Analytical Methods in Geomechanics 26, 963–1004 (2002)

Tatsuoka, F., Ishihara, K.: Undrained deformation and liquefaction of sand under cyclic stresses. Soils and Foundations 15, 29–44 (1975)

Taylor, G.I.: Plastic strain in metals. Journal Institute of Metals 62, 307–324 (1938)

Taylor, G.R., Quinney, H.: The plastic distortion of metals. Philosophical Transactions of the Royal Society A 230, 323–362 (1931)

Tresca, H.: Mémoire sur l'écoulement des corps solides soumis à de fortes pressions. Comptes Rendus hebdomadaires des séances de l'Académie des Sciences, Paris 59, 754–758 (1864)

Vermeer, P.A., de Borst, R.: Non-associated plasticity for soils, concrete and rock. Heron 29, 1–64 (1984)

von Mises, R.: Mechanik der festen Körper im plastisch-deformablen Zustand, Nachrichten von der Gesellschaft der Wissenschaften zu Göttingen. Mathematisch-Physikalische Klasse 1, 582–592 (1913)

White, J.A., Borja, R.I.: Block-preconditioned Newton-Krylov solvers for fully-coupled flow and geomechanics. Computational Geosciences 15, 647–659 (2011)

Wiley, C.R.: Advanced Engineering Mathematics, 4th edn. McGraw-Hill Book Company, New York (1975)

Wilkins, M.L.: Calculation of elastic-plastic flow. In: Alder, B., et al. (eds.) Methods of Computational Physics 3. Academic Press, New York (1964)

Willam, K.J., Warnke, E.P.: Constitutive model for triaxial behavior of concrete. Concrete Structures Subjected to Triaxial Stresses. In: International Association for Bridges and Structural Engineering, Bergamo, Italy (1974)

Willam, K.J.: Constitutive models for engineering materials. Encyclopedia of Physical Science and Technology, 3rd edn., vol. 3, pp. 603–633. Academic Press (2002)

Wong, T.-F., David, C., Zhu, W.: The transition from brittle faulting to cataclastic flow in porous sandstones: Mechanical deformation. Journal of Geophysical Research 102, 3009–3025 (1997)

Wong, T.-F.: Shear fracture energy of westerly granite from post- failure behavior. Journal of Geophysical Research 87, 990–1000 (1982)

Wriggers, P.: Computational Contact Mechanics, 2nd edn. Springer, New York (2006)

Wright, W.J., Saha, R., Nix, W.D.: Deformation mechanisms of the $Zr_{40}Ti_{14}Ni_{10}Cu_{12}Be_{24}$ bulk metallic glass. Materials Transactions 42, 642–649 (2001)

Zytynski, M., Randolph, M.F., Nova, R., Wroth, C.P.: On modelling the unloading-reloading behaviour of soils. International Journal for Numerical and Analytical Methods in Geomechanics 2, 87–93 (1978)

Index